Sustainable Development Goals Series

The **Sustainable Development Goals Series** is Springer Nature's inaugural cross-imprint book series that addresses and supports the United Nations' seventeen Sustainable Development Goals. The series fosters comprehensive research focused on these global targets and endeavours to address some of society's greatest grand challenges. The SDGs are inherently multidisciplinary, and they bring people working across different fields together and working towards a common goal. In this spirit, the Sustainable Development Goals series is the first at Springer Nature to publish books under both the Springer and Palgrave Macmillan imprints, bringing the strengths of our imprints together.

The Sustainable Development Goals Series is organized into eighteen subseries: one subseries based around each of the seventeen respective Sustainable Development Goals, and an eighteenth subseries, "Connecting the Goals", which serves as a home for volumes addressing multiple goals or studying the SDGs as a whole. Each subseries is guided by an expert Subseries Advisor with years or decades of experience studying and addressing core components of their respective Goal.

The SDG Series has a remit as broad as the SDGs themselves, and contributions are welcome from scientists, academics, policymakers, and researchers working in fields related to any of the seventeen goals. If you are interested in contributing a monograph or curated volume to the series, please contact the Publishers: Zachary Romano [Springer; zachary.romano@springer.com] and Rachael Ballard [Palgrave Macmillan; rachael.ballard@palgrave.com].

Adnan Trakic · Ridoan Karim ·
Pervaiz K. Ahmed
Editors

Energy Justice

Affordable, Reliable, Sustainable and Modern Energy for All

Springer

Editors
Adnan Trakic
Department of Business Law
and Taxation
Monash University Malaysia
Subang Jaya, Selangor, Malaysia

Ridoan Karim
Department of Business Law
and Taxation
Monash University Malaysia
Subang Jaya, Selangor, Malaysia

Pervaiz K. Ahmed
Institute of Global Strategy
and Competitiveness
Sunway University
Subang Jaya, Selangor, Malaysia

ISSN 2523-3084 ISSN 2523-3092 (electronic)
Sustainable Development Goals Series
ISBN 978-981-97-6058-9 ISBN 978-981-97-6059-6 (eBook)
https://doi.org/10.1007/978-981-97-6059-6

This Springer imprint is published by the registered company Springer Nature Singapore Pte Ltd.
The registered company address is: 152 Beach Road, #21-01/04 Gateway East, Singapore 189721, Singapore

Foreword

Climate change is growing every year. Today people are now normalised in terms of hearing of the impacts of climate change. Across the world the majority of climate change results from the energy sector through carbon dioxide emissions. That is why this book is welcome in that it focuses directly on the energy sector and trying to ensure improved *just outcomes* from this sector. Further, the primary focus is on Asia where there is less literature to date, and this adds more originality to the book.

Energy justice literature has been growing now for over a decade. More recently, in several countries and institutions, it is now crossing from theory into practice—in the United States, Colombia, South Africa, the United States and the United Nations. Further literature is needed and more solutions to different energy injustices need research. This book commendably provides such research contributions. It covers a range of the key issues on energy justice and also contributes further literature in key areas such as in ASEAN countries, energy taxation and in financial and Islamic approaches to energy justice.

Every chapter is good contribution but for me, the structure of the book makes it an important contribution to the field. It begins with thc human rights-based approach to energy justice and also explores the relationship between energy and environmental justice. Later there is a very interesting chapter on access to electricity but which explores the issues in a more socio-economic perspective with an additional focus to look at education. This is also a much-needed analytical perspective needed on education and energy justice.

There are further interesting chapters that utilise a range of the energy justice principles to explore important new topics such as hydrogen and how investment is emerging in the sector (and the challenges this brings). Further chapters explore areas where much-needed energy justice research is needed which includes on litigation, nuclear energy and the indigenous community of the Orang Asli.

Overall, this text should be welcomed into the energy justice literature and energy literature more broadly. It is opening the door to much-needed research on energy justice in ASEAN primarily and with comparative research and insight from a range of other countries. The editorial team can be commended for bringing a representative range of high-quality scholars together. Finally, I look forward to following all their careers into the future after this exciting publication.

Biarritz, France
May 2024

Raphael J. Heffron

Raphael J. Heffron, BA, MA, M.Litt., M.Phil., M.Sc., Ph.D., Barrister-at-Law Professor of Energy Justice and the Social Contract, Universite de Pau et des Pays de l'Adour, E2S UPPA, CNRS, TREE, Pau, France.

EU Jean Monnet Professor in the Just Transition to a Low-Carbon Economy (2019–2022)—Senior Counsel @ HYPERLINK "https://www.janson.be/en/lawyers/raphael-heffron" Janson, Brussels, Belgium.

Senior Fellow of the UK Higher Education Academy—Fellow of the Royal Society of Arts—Fellow of the Royal Society of Edinburgh's Young Academy of Scotland.

Visiting Professor Queen Mary University of London, Eduardo Mondlane University (Mozambique), University of Brawijaja (Indonesia), University of Western Australia and Associate Researcher, Energy Policy Research Group, University of Cambridge (UK) & Oxford Institute for Energy Studies (UK).

Preface

The terms "equality," "equity" and "justice" are being used more frequently, especially in discussions about socio-economic developments. This increased usage reflects a growing awareness and concern for issues related to discrimination, inequality and human rights. Some people criticise the overuse of these terms, suggesting that they are sometimes applied imprecisely. Despite this, the broad and flexible nature of these terms allows them to be combined with other words to provide a context of fairness and well-being. For example, the words "energy" are used on the one hand, and "energy equality," "energy equity" and "energy justice" are used on the other. The flexibility of these terms enables nuanced discussions in different fields, including energy. Using these terms in conjunction with specific sectors like energy can significantly enhance the understanding and approach to related challenges.

Among the three terms "equality," "equity" and "justice," the term "justice" stands out in that it incorporates the first two. Justice integrates the principles of equality and equity by ensuring fair access to energy. Like equality, justice aims to ensure that everyone has access to resources and opportunities. However, it goes further by addressing the barriers that prevent fair access, which aligns with the principles of equity. On the other hand, like equity, justice recognises that different individuals and communities have different needs and circumstances. It seeks to distribute resources and opportunities in a way that accounts for these differences, ensuring fair and inclusive outcomes. Justice also involves addressing historical and systemic issues that have led to disparities. It focuses on creating systemic changes that promote long-term fairness and equity. Lastly, justice ensures that diverse voices are heard and considered in decision-making processes, contributing to fair and inclusive governance. Hence, justice is a more holistic concept that incorporates the principles of both equality and equity, aiming to create a fair and inclusive society by addressing both immediate needs and underlying systemic issues. It is thus unsurprising that the term "energy justice" is used more frequently than "energy equality" or "energy equity."

We live in a world grappling with planetary health and sustainability issues. How, then, does energy justice relate to these challenges? In our opinion, the two concepts are deeply interconnected, as both energy justice and sustainability aim to create a fair, equitable and inclusive society. Energy justice seeks to ensure fair access to energy resources, while sustainability focuses on meeting the needs of the present without compromising the ability of future generations to meet their own needs. Both concepts prioritise long-term well-being and the responsible use of resources. Energy justice cannot be fully achieved without considering the environmental impact of energy production and consumption. Sustainability emphasises the importance of protecting the environment and promoting the use of renewable energy sources, and sustainable practices help mitigate environmental degradation, which disproportionately affects marginalised communities, a key concern of energy justice.

Both energy justice and sustainability require a long-term perspective. Addressing energy justice involves considering future generations and ensuring that current actions do not create new injustices. Sustainability inherently involves planning for the long-term health of the planet and its inhabitants, aligning with the goals of energy justice. However, "energy justice" and "sustainability" in conjunction with each other have received little attention or exploration, especially from a multidisciplinary perspective. This provoked us, as editors, to explore and understand energy justice through a multidisciplinary lens, with a particular focus on the link between energy justice and sustainability.

In the initial stage of embarking on this project, we struggled to find the correct lens of "sustainability." This struggle led us all the way back to the UN's Sustainable Development Goal 7 (SDG 7). SDG 7 focuses on ensuring access to affordable, reliable, sustainable and modern energy for all, and it aligns closely with the principles of both energy justice and sustainability. Using SDG 7 as a framing lens allowed us to integrate the principles of sustainability. This led us to reflect on the interconnections between affordable, reliable, sustainable and modern energy, and the broader goals of social equity and environmental stewardship. Following this, we invited authors who could contribute perspectives from their home disciplines, including the sciences, law, economics and other social sciences.

We, the editors, believe very strongly that energy justice transcends the mere provision of clean energy, and this cannot be done without addressing persistent challenges related to environmental, social and governance issues within the energy sector. And that is why we are pleased to present a book that considers the intersection of energy justice, sustainable development and human rights. In doing so, the book recognises the inextricable link between fair access to clean energy and the upholding of fundamental human rights. This recognition is a positive step towards developing energy solutions that prioritise both the planet and its people. We hope that the readers of the book will gain insights into the enormous challenges that humanity faces and the need to do this by charting a future in which justice is inclusively provided to all.

Subang Jaya, Malaysia

Adnan Trakic
Ridoan Karim
Pervaiz K. Ahmed

Contents

Editors and Contributors

About the Editors

Adnan Trakic is Associate Professor and Head of Business Law and Taxation Department at the School of Business, Monash University Malaysia. He graduated with an LLB (Hons), a Master of Comparative Laws (MCL), a Doctor of Philosophy (Ph.D.) from the International Islamic University Malaysia (IIUM) and Graduate Certificate in Higher Education (GCHE) from Monash University. He has taught and researched in the fields of conventional and Islamic finance law, business law and dispute resolution. A co-author of several books, his published work extends to many articles, book chapters and other contributions. e-mail: adnan.trakic@monash.edu

Ridoan Karim is Lecturer at the School of Business, Monash University Malaysia, where he teaches and conducts research in the fields of business and cyber law. He is Passionate Academic and Researcher who has published widely in peer-reviewed journals and presented papers at numerous national and international conferences. His areas of interest include "energy justice," "science, technology and law," "privacy and data protection law" and "Asian and comparative law," among others. Dr. Karim has served as Consultant and Fellow on projects funded by the Ministry of Education Malaysia, University of Malaya and the Monash Data Futures Institute (MDFI). He earned his Doctor of Philosophy (Ph.D.) from the University of Malaya, Malaysia, as well as a Master of Business Administration (MBA) from the University of Chichester, UK, a Master of Comparative Laws (MCL) from International Islamic University Malaysia (IIUM) and a Bachelor of Laws (LLB) from BRAC University, Bangladesh. e-mail: ridoan.karim@monash.edus

Pervaiz K. Ahmed is Distinguished Professor and Director of the Sunway Institute for Global Strategy and Competitiveness, Sunway University. Pervaiz is Management Scholar exploring innovation and strategic dynamics of entrepreneurial change. His outlook on management is coloured by moral responsibility, particularly ethics of care and sustainability. Pervaiz has held several senior positions in the UK before coming to Malaysia. He has published extensively in international journals and is Regular Keynote Speaker with a number of academic awards for his research. In addition, he has served as Editor and sits on the editorial boards of several international journals. He has extensive experience working with and advising blue-chip companies and public sector organisations, such as Unilever, Ford, AT&T, NCR, British Telecommunications and the NHS in Europe. He has also been involved with corporate clients in Asia, as well as government agencies, such as the Singapore National Productivity Council. In recent times Pervaiz has been involved with the Government of Dubai's Public Sector Innovation and Improvement Initiative and also served on projects commissioned by Economic Planning Unit (EPU), Ministry of International Trade (MITI) and APEC Secretariat. e-mail: pervaiz@sunway.edu.my

Contributors

Muhamad Hassan Ahmad International Islamic University Malaysia, Kuala Lumpur, Malaysia

Pervaiz K. Ahmed Monash University Malaysia, Kuala Lumpur, Malaysia

Francesco Cannas Assistant Professor, Law and Taxation, EDHEC Business School, Promenade des Anglais, France;
Assistant Professor, Tax Law, University of Turin, Turin, Italy

Lilian M. L. Chan Monash University Malaysia, Kuala Lumpur, Malaysia

Julio C. Colón Killeen, TX, USA

Farahdilah Ghazali School of Business and Economics, Universiti Putra Malaysia, Serdang, Selangor, Malaysia

Evgeny Guglyuvatyy Department of Business Law and Taxation, School of Business, Monash University, Kuala Lumpur, Malaysia

Sabina Hodžić Faculty of Tourism and Hospitality Management, University of Rijeka, Opatija, Croatia

Mohammad Naqib Eishan Jan International Islamic University Malaysia, Kuala Lumpur, Malaysia

Ridoan Karim Monash University Malaysia, Kuala Lumpur, Malaysia;
Department of Business Law and Taxation, School of Business, Monash University Malaysia, Subang Jaya, Selangor, Malaysia

Pauline Khoo Monash University Malaysia, Kuala Lumpur, Malaysia

Diane Kraal Monash University, Melbourne, VIC, Australia

Grace H. Y. Lee Monash University Malaysia, Kuala Lumpur, Malaysia

Mohamed Hanipa Maidin International Islamic University Malaysia (IIUM), Kuala Lumpur, Malaysia

Ashgar Ali Ali Mohamed International Islamic University Malaysia, Kuala Lumpur, Malaysia

Maizatun Mustafa International Islamic University Malaysia, Kuala Lumpur, Malaysia

Jasmin Omercic Assistant Professor of Business Studies, City University (CU) - Ulster University, Lusail, Qatar

Rajesh Ramachandran Monash University Malaysia, Kuala Lumpur, Malaysia

Manjeevan Seera School of Business, Monash University Malaysia, Kuala Lumpur, Malaysia

Mirzet Šeho School of Business, Monash University Malaysia, Bandar Sunway, Malaysia

Audrey K. L. Siah Monash University Malaysia, Kuala Lumpur, Malaysia

Richard Taylor Monash University, Melbourne, VIC, Australia

Adnan Trakic Monash University Malaysia, Kuala Lumpur, Malaysia

Part I

Fundamentals of Energy Justice

Energy Justice: What is It and Why It Matters?

1

Adnan Trakic, Ridoan Karim, and Pervaiz K. Ahmed

Abstract

This chapter explores the concept of energy justice and its importance in today's energy landscape. It highlights the global climate crisis and the urgent need for sustainable energy solutions, referencing the United Nations' call to achieve Sustainable Development Goal (SDG) 7 for affordable, reliable, sustainable, and modern energy for all. Energy justice is built on three principles: distributive justice (equitable distribution of energy resources), procedural justice (inclusive participation in energy decisions), and recognition justice (addressing the needs of vulnerable groups). By establishing a foundation for the book, this chapter sets the stage for further discussions on the multifaceted aspects of energy justice, guiding readers toward a more inclusive and sustainable energy future.

Keywords

Energy justice · Sustainable energy · Climate crisis · SDG 7 · Inclusive energy · Sustainable future

1.1 Introduction

The UN Secretary-General António Guterres, on 30 November 2023, said on the stage of the United Nations Climate Change Conference (COP28) summit that 'we are living through climate collapse in real-time—and the impact is devastating' (UN News 2023). For us, it adds to nothing new. While world leaders debate climate change policies at COP28, the United Nations (UN) has always been persistent with their Sustainable Development Goal (SDG) 7, which calls for a realisation of just energy transition. SDG 7 is about ensuring access to clean and affordable energy, which is key to helping impoverished and deprived segments of society and developing agriculture, business, communications, education, healthcare, and transportation (United Nations 2022).

Let us take a deeper look into SDG 7. This goal has five extended targets to be achieved by 2030 (United Nations 2022; Küfeoğlu 2022):

- ensure universal access to affordable, reliable, and modern energy,
- increase substantially the share of renewable energy in the global energy mix,

A. Trakic (✉) · R. Karim · P. K. Ahmed
Monash University Malaysia, Kuala Lumpur, Malaysia
e-mail: adnan.trakic@monash.edu

R. Karim
e-mail: ridoan.karim@monash.edu

P. K. Ahmed
e-mail: pervaiz@sunway.edu.my

A. Trakic et al. (eds.), *Energy Justice*, Sustainable Development Goals Series,
https://doi.org/10.1007/978-981-97-6059-6_1

- double the global rate of improvement in energy efficiency,
- enhance international cooperation to facilitate access to clean energy research and technology, and
- expand infrastructure and upgrade technology for modern and sustainable energy services for all in developing countries, particularly least developed countries, small island developing States, and land-locked developing countries.

This goal, along with others, was adopted by world leaders in September 2015 at a historic UN Summit (Küfeoğlu 2022). Almost a decade has passed since then, and world leaders have been repeating the same speech with new stories and dreams every year. Although these stories are beautiful to hear, in reality, very little progress has been made so far. At least, the data shows that countries' policies have largely failed to achieve the year-to-year goals in this regard.

Energy is the dominant contributor to climate change, accounting for around 60% of global greenhouse gas emissions (Oreggioni et al. 2021). Since 1990, global emissions of CO_2 have increased by more than 46% (Oreggioni et al. 2021). About 789 million people still live without electricity, and the estimation is that 650 million people could lack electricity access in 2030 (IISD-SDG Knowledge Hub 2019). Although Latin America, the Caribbean, Eastern Asia and South-Eastern Asia are progressively approaching universal access to electricity, Sub-Saharan Africa lags behind, accounting for 70% of the global deficit (The World Bank 2021). About 3 billion people still rely on wood, coal, charcoal or animal waste for cooking and heating (Tucho and Kumsa 2020). Over the 2010–2018 period, progress has remained largely stagnant, with the rate of increase in access to clean cooking even decelerating since 2012 in some countries, falling behind population growth (Tucho and Kumsa 2020). Under current and planned policies, 2.3 billion people will still be deprived of access to clean cooking fuels and technologies in 2030 (Tucho and Kumsa 2020).

Among all this bad news, do we have something to cheer about? Perhaps we do. Commendable success has been made in the contribution of renewables to the energy mix, increasing energy efficiency and international financial flows to the energy market. The share of renewables in the global energy mix reached 17.3% of final energy consumption in 2017, up from 17.2% in 2016 to 16.3% in 2010 (The World Bank 2020a). Renewable consumption is growing faster than global energy consumption, continuing a trend in evidence since 2011 (The World Bank 2020a). Most of the growth in renewable energy has occurred in the wind and solar power sectors (World Resources Institute 2021). Global primary energy intensity and energy efficiency improved by 1.7% in 2017 (IEA 2019). It has also been found that the international financial flows to developing countries supporting clean and renewable energy have increased (Qamruzzaman and Jianguo 2020). Although total flows have doubled since 2010, reaching USD$21.4 billion in 2017, only 12% reached the least-developed countries, which are the furthest from achieving the various SDG 7 targets (The World Bank 2020b).

That being said, the COVID-19 crisis and the Russia-Ukraine war slowed down the developments relating to renewable energy for electricity generation and improvements in energy efficiency (The World Bank 2020b). The global efforts remain insufficient to reach the key targets of SDG 7 by 2030, especially if the war continues to disrupt electrification efforts. To meet the SDGs targets by 2030, countries must safeguard the gains attained before the COVID-19 outbreak and the war. World leaders must prioritise leaving no one behind, given that a large proportion of the population still lives without electricity. This will require stronger political commitment, long-term energy planning, increased public and private financing, and adequate policy and fiscal incentives to spur faster deployment of new technologies.

This is exactly where the concept of energy justice is vital in highlighting the inequalities in energy access and the impact of energy policies on various communities. This chapter lays the foundation for the book by discussing the principles of energy justice and explaining why it is crucial in addressing real-world energy challenges. It emphasises the significant role of energy in climate change and the persistent disparities in energy access globally. Additionally, it sheds light on the complex barriers hindering sustainable energy development. This chapter analyses the diverse facets of energy justice and serves as a guide for the forthcoming chapters of the book, steering discussions toward a more inclusive and sustainable energy future.

1.2 What is 'Energy Justice' and Why It Matters?

During a recent conversation with family and friends in Europe, the term 'energy justice' was brought up. They shared their experience about the winter following the Russia-Ukraine war, which was unlike any other. In many parts of Europe, people lacked access to sufficient energy to heat their homes (Minion and Banerjee 2024). The main concern was not the drastic increase in utility bills, but rather the shortage of coal, gas, and firewood stockpiles, as the supply was unable to meet the demand. The problem was exacerbated by panic buying. However, even without it, countries that prohibit the export of a particular commodity due to significant shortages and high domestic demand, and where alternative supply sources are not viable, face a serious challenge.

It is not wise to tell those who are affected by extreme winter temperatures that we ought to reduce the production of coal, oil, and gas. Their immediate concern is survival, and without proper heating, they may face serious illnesses or even death (Prontera 2024). Faced with these harsh realities, the rush for renewable energy and the need to reduce carbon emissions are unlikely to be priorities for them.

While it is crucial for the world to shift away from fossil fuels and adopt sustainable and clean energy sources, it is unjust to demand this from people who are struggling to survive the winter or to earn a living. The concept of energy justice entails three principles of justice, and all three must be taken into account to address energy inequalities.

First is distributive justice—the right of all segments of society, regardless of their socio-economic status, to benefit from the energy resources produced by a country. Studies have shown that energy burdens of low-income and impoverished segments of society are disproportionately higher relative to their income than those earning higher income (Lewis et al. 2020). So, when a low-income person or a pensioner receives their salary or pension, they would normally settle their utility bills first, and whatever is left would be used for food and other necessities. In other words, high energy burdens may force them to make undesirable tradeoffs like being unable to afford visits to a doctor or dentist or buying healthy and nutritious foods.

Second is procedural justice—the democratic right of the whole of society, regardless of socio-economic status, to participate in energy-related decision-making. This is particularly important when governments decide to develop new energy plants or undertake any research or excavation in an area that is populated or could affect the daily quality of life of the people in that area. Procedural justice requires that peoples' voices are duly heard and considered relative to the priorities and benefits of the entire nation. Decisions sometimes have to be made that may not please certain parties but must be taken as long as they are driven by the greater public interest, rather than by corporate greed or benefits limited to a specific segment of society.

The third element of energy justice is recognition justice. This principle asserts that everyone has the right to be acknowledged as having specific needs, which the state must address. For example, low-income and impoverished groups may struggle to afford higher gas prices resulting from the nation's efforts to promote renewable energy and reduce access to fossil fuels. Countries must recognise this reality and implement an appropriate energy justice framework, such as tax incentives or other support measures, to assist those in need.

The energy justice framework must be built on two key principles: a prohibitive principle which states that 'energy systems must be designed and constructed in such a way that they do not unduly interfere with the ability of people to acquire those basic goods to which they are justly entitled,' and an affirmative principle which states that 'if any of the basic goods to which people are justly entitled can only be secured by means of energy services, then in that case there is also a derivative entitlement to the energy services' (Jones et al. 2015). These two principles are based on the notion that energy is a material requirement for other fundamental needs to which individuals are entitled under national and international laws. The externalities connected to energy systems frequently obstruct our ability to satisfy basic human needs, and unequal access to energy leads to socio-economic disparities. The organisation and policy-making of energy systems have significant effects on human society; some scholars go further by suggesting that it also affects international politics and impacts state sovereignties (Blondeel et al. 2021).

Energy justice primarily concerns policy-making and its impact on society's broader implications concerning energy distribution and access. The central goal of energy justice is to reduce energy burdens and ensure that all communities, regardless of their socio-economic status, have access to affordable and sustainable energy. The political context also plays a pivotal role in allocating resources, influencing policy formulation and shaping the energy landscape. This intersection uncovers the power dynamics embedded within energy policies and their potential impact on marginalized communities.

As we deal with the foundational principles of energy justice, a fundamental question emerges: why is energy justice essential? Fairness, justice, or equality subjective concepts that evolves over time and vary among individuals. For example, consider the concept of a just transition, initially proposed by coal miners' unions to transition towards sustainable livelihoods, which now as come encompass considerations of gender, intersectionality, agroecology, community resilience, and addressing energy poverty.

Justice in the energy paradigm has begun to focus on addressing energy poverty, which stems from the unequal distribution of energy prices and lack of energy access in certain communities. However, as mentioned earlier, the concept of energy justice is now establishing connections between energy policies and social justice, including economic factors in terms of equalities and inequalities within the production and consumption systems. Essentially, the focus is on identifying those who are left behind. As energy systems go beyond technology and economic development, the concept of energy justice is needed to address not only access to energy inequalities but also the inequalities that persist in energy extraction, production, and transport (Heffron 2022).

Energy extraction, production, and transport constitute vital stages in the energy supply chain, each playing a significant role in shaping the landscape of energy justice. These processes have far-reaching implications on both local and global scales, impacting communities, economies, and the environment. Examining them through the lens of energy justice sheds light on the distribution of benefits and burdens associated with these stages. This is where energy justice is important, as it recognises and rectifies historical and existing inequalities in the energy supply chain. This includes empowering local communities affected by extraction, production, and transport activities, incorporating their perspectives in decision-making processes, and ensuring that the benefits of energy production are distributed equitably. Additionally, promoting environmentally sustainable practices throughout the supply chain contributes to a just and sustainable energy future for all.

1.3 The Link Between Energy Justice and SDG 7

Fostering the Sustainable Development Goal 7 (SDG 7) yields numerous benefits. Goal 7 focuses on guaranteeing access to clean and affordable energy, crucial for the advancement of various sectors like agriculture, business, education, healthcare, and transportation. Despite global strides towards sustainable energy, progress is not swift enough. Projections indicate that by 2030, around 660 million people will still lack access to electricity, and nearly 2 billion will rely on polluting fuels for cooking.

Energy justice goes beyond access to clean energy and talks about addressing persistent challenges related to environmental, social, and governance issues in the sector (Heffron and McCauley 2017; Sovacool et al. 2016). Some scholars characterise today's energy dimensions as faulty due to inadequate attention in addressing fundamental moral and ethical concerns (Sovacool et al. 2016). Hence, is important to consider the intersection between energy justice, sustainable development, and human rights. This recognition underscores the inextricable link between providing fair access to clean energy and upholding fundamental human rights. It becomes evident that energy justice is not just an environmental concern but a matter of social justice and human well-being (Heffron 2021). This acknowledgement is a positive step towards developing energy solutions that prioritise both the planet and its people.

Understanding energy justice requires implementing changes to laws and policies at three interconnected levels: global, national, and local (Heffron and Talus 2016). On a global scale, two significant events occurred in 2015: the establishment of the 17 UN Sustainable Development Goals (SDGs) and the Paris COP21 Agreement. Countries that committed to these agreements are now obliged to incorporate them into their decision-making processes, focusing on future energy and climate goals as they reassess their commitments. The UN SDGs offer a clear framework for achieving sustainability, resolving past confusion stemming from various definitions of sustainability. Consequently, embracing the UN SDGs brings clarity to the pursuit of sustainability objectives.

SDG 7, which aims to 'ensure access to affordable, reliable, sustainable, and modern energy for all,' and the concept of energy justice —are not different but rather complementary. The principles of energy justice can be similarly interpreted when defining policies and laws relating to SDG 7. For example, SDG 7 addresses distributional justice by emphasising universal access to affordable and sustainable energy. This goal seeks to eliminate disparities in energy access, ensuring that everyone, regardless of their location or socio-economic status, has the opportunity to benefit from reliable and modern energy sources. By prioritising equitable energy distribution, SDG 7 aligns with the principle of distributional justice in promoting fair and inclusive access. Similarly, the commitment to incorporating SDG 7 into decision-making processes at global, national, and local levels supports procedural justice. This involvement ensures that diverse voices are heard in shaping energy-related decisions, contributing to fair and inclusive procedures in energy governance. Lastly, the emphasis on sustainability within SDG 7 also addresses the recognition justice by acknowledging the importance of ecological considerations in energy access. By promoting sustainable energy sources, this goal recognises the interconnectedness of environmental concerns with energy justice. It acknowledges the need to balance energy access with environmental preservation, respecting the rights and interests of communities and ecosystems.

In essence, SDG 7 and the concept of energy justice complement each other by sharing common objectives and principles. SDG 7 provides a concrete global goal for promoting universal energy access, while the principles of energy justice offer a conceptual framework that underscores the importance of fairness, inclusivity, and sustainability in achieving that goal. However, the question remains of how the conceptual framework on the international level should be integrated and implemented on national and local levels.

Some global initiatives exert influence at both local and national levels, underscoring the significance of energy justice in facilitating a fair transition within the energy sector. The enhancement of energy justice entails two primary approaches: (a) the introduction of new laws to ensure fairness and (b) the resolution of energy-related conflicts through national courts (Heffron 2022). While the creation of new laws can be intricate and time-consuming in many countries, the resolution of disputes in courts, albeit time-consuming, often results in decisions that shape new laws and policies.

National courts assume a pivotal role in resolving these disputes, identifying key justice issues, and determining their resolutions. Disputes typically arise when an energy sector stakeholder's actions occur in the absence of existing legislation on the matter. These legal institutions play a forward-looking role by interpreting justice based on the law and aligning it with current societal objectives (Heffron 2022). They bear the responsibility of adapting to current public policy, guiding the rules of the game in the energy sector, and defining acceptable behaviour for governments or energy companies (Heffron 2022).

Given the widespread adoption of SDG 7 and the Paris Agreement by countries globally, it is anticipated that case law judgments have evolved since then. Indeed, a discernible trend has emerged, emphasising the protection of human rights (Heffron 2022). Rights such as the right to life, health, minimum subsistence, freedom, human dignity, water, a healthy environment, air, culture, property, adequate housing, security, and a fair trial have garnered increased protection. Energy justice and SDG 7, grounded in environmental justice principles, centre around fairness, affordability, accessibility, and participation in the energy system and its transition, irrespective of factors such as race, nationality, income, or location. Advocates for energy justice push for policy measures to reduce energy costs for low-income customers, prevent disproportionate impacts, and ensure the equitable distribution of energy-related benefits. This involves providing access to reliable and clean energy, engaging communities in decision-making, and fostering development within the energy sector. Therefore, energy justice not only aligns with Sustainable Development Goal 7 but also intersects with various aspects such as energy affordability, infrastructure planning, community renewable energy projects, climate change, energy security, energy economics, and energy finance and taxation.

1.4 The Framework of the Book

The concept of energy justice is poised to gain increased prominence across various disciplines, including the sciences, law, economics, and other social sciences in the future. Therefore, the book aims to delve into the fundamental concept of 'energy justice' within all pertinent disciplines. Its goal is to furnish society with an ethical framework that can guide decision-making in the energy sector.

Specifically, the book aims to:

1. Provide a conceptual analysis of the energy justice framework as it relates to SDG 7,
2. Evaluate the role of the energy justice framework as a decision-support tool for policymakers,
3. Present understandings of energy justice concept across multiple disciplines and see how they can complement each other, and
4. Recommend ways in which energy justice's objectives might be negotiated and implemented among different stakeholders (state, people, and environment).

With the above objectives in mind, the book 'Energy Justice: Ensuring Access to Affordable, Reliable, Sustainable and Modern Energy for All' covers a comprehensive exploration of energy justice from various perspectives. The structure of the book is carefully designed to provide coherence and depth to the topic, avoiding a mere compilation of random chapters.

While this chapter introduces the concept of energy justice, outlining its significance and the areas chosen for detailed exploration, the subsequent chapters deal with specific aspects of energy justice, addressing topics such as the impact of energy burden on low-income groups, regulatory and policy considerations surrounding energy justice, and the relationship between energy justice and environmental justice. The book also explores whether access to affordable, reliable, sustainable, and modern energy can be considered a basic human right, analysing international conventions and the potential implications for countries.

The book further examines energy justice through diverse lenses, including an Islamic perspective, the role of the financial sector, the contributions and challenges of the fossil fuel and nuclear energy sectors, and the impact of energy technology innovation. Additionally, chapters explore the potential of taxation systems in achieving energy justice and private lawsuits against power companies.

The final chapter summarises the main findings and recommendations from the substantive chapters, providing a review coloured by the editors' perspectives. Overall, the book offers a comprehensive examination of energy justice, covering a wide range of topics and perspectives relevant to achieving SDG 7.

References

Blondeel M, Bradshaw MJ, Bridge G, Kuzemko C (2021) The geopolitics of energy system transformation: a review. Geogr Compass 15(7):e12580

Heffron RJ (2021) Energy multinationals challenged by the growth of human rights. Nat Energy 6(9):849–851

Heffron RJ (2022) Applying energy justice into the energy transition. Renew Sustain Energy Rev 156:111936

Heffron RJ, McCauley D (2017) The concept of energy justice across the disciplines. Energy Policy 105:658–667

Heffron RJ, Talus K (2016) The development of energy law in the 21st century: a paradigm shift? J World Energy Law & Business 9(3):189–202

IISD-SDG Knowledge Hub (2019) SDG 7 report warns 650 million people could lack electricity access in 2030 without sustained effort [online]. Available from: https://sdg.iisd.org/news/sdg-7-report-warns-650-million-people-could-lack-electricity-access-in-2030-without-sustained-effort/. Accessed 30 Jan 2024

International Energy agency (IEA) (2019) Energy efficiency 2019 [online]. Available from: https://www.iea.org/reports/energy-efficiency-2019. Accessed 30 Jan 2024

Jones BR, Sovacool BK, Sidortsov RV (2015) Making the ethical and philosophical case for "energy justice." Environ Ethics 37(2):145–168

Küfeoğlu S (2022) SDG-7 Affordable and clean energy. Emerging technologies. Springer, Cham, pp 305–330

Lewis J, Hernández D, Geronimus AT (2020) Energy efficiency as energy justice: addressing racial inequities through investments in people and places. Energ Effi 13:419–432

Minion L, Banerjee A (2024) "I can feel the money going out of the window": how high energy prices evoke negative emotions in people with previous experience of homelessness. Energy Res Soc Sci 108:103387

Oreggioni GD et al (2021) Climate change in a changing world: socio-economic and technological transitions, regulatory frameworks and trends on global greenhouse gas emissions from EDGAR v. 5.0. Glob Environ Chang 189:102350. Available at: https://doi.org/10.1016/j.gloenvcha.2021.102350. Accessed 30 Jan 2024

Prontera A (2024) Winter is coming: Russian gas, Italy and the post-war European politics of energy security. West Eur Polit 47(2):382–407

Qamruzzaman M, Jianguo W (2020) The asymmetric relationship between financial development, trade openness, foreign capital flows, and renewable energy consumption: fresh evidence from panel NARDL investigation. Renew Energy 159:827–842. Available at: https://doi.org/10.1016/j.renene.2020.06.069. Accessed 30 Jan 2024

Sovacool BK, Heffron RJ, McCauley D, Goldthau A (2016) Energy decisions reframed as justice and ethical concerns. Nat Energy 1(5):1–6

The World Bank (2020a) The tracking SDG 7 report: the last decade to leave no one behind [online]. Available at: https://www.worldbank.org/en/topic/energy/publication/the-tracking-sdg-7-report-the-last-decade-to-leave-no-one-behind. Accessed 30 Jan 2024

The World Bank (2020b) Covid-19 intensifies the urgency to expand sustainable energy solutions worldwide. Available from: https://www.worldbank.org/en/news/press-release/2020/05/28/covid-19-intensifies-the-urgency-to-expand-sustainable-energy-solutions-worldwide. Accessed 30 Jan 2024

The World Bank (2021) Report: universal access to sustainable energy will remain elusive without addressing inequalities [online]. Available from: https://www.worldbank.org/en/news/press-release/2021/06/07/report-universal-access-to-sustainable-energy-will-remain-elusive-without-addressing-inequalities. Accessed 30 Jan 2024

Tucho GT, Kumsa DM (2020) Challenges of achieving sustainable development goal 7 from the perspectives of access to modern cooking energy in developing countries. Front Energy Res 8:564104. Available at: https://doi.org/10.3389/fenrg.2020.564104. Accessed 30 Jan 2024

United Nations (2022) Ensure access to affordable, reliable, sustainable and modern energy for all [online]. Available from: https://sdgs.un.org/goals/goal7. Accessed 30 Jan 2024

UN News (2023) Science points to 'climate collapse' as UN chief calls COP28 to action [online]. Available from: https://news.un.org/en/story/2023/11/1144147. Accessed 30 Jan 2024

World Resources Institute (2021) Explaining the exponential growth of renewable energy [online]. Available at: https://www.wri.org/insights/growth-renewable-energy-sector-explained. Accessed 30 Jan 2024

Adnan Trakic is an Associate Professor and Head of Business Law and Taxation Department at the School of Business, Monash University Malaysia. He graduated with an LLB (Hons), a Master of Comparative Laws (MCL), a Doctor of Philosophy (Ph.D.) from the International Islamic University Malaysia (IIUM) and a Graduate Certificate in Higher Education (GCHE) from Monash University. He has taught and researched in the fields of conventional and Islamic finance law, business law, and dispute resolution. A co-author of several books, his published work extends to many articles, book chapters, and other contributions.

Ridoan Karim is a Lecturer at the School of Business, Monash University Malaysia, where he teaches and conducts research in the fields of business and cyber law. He is a passionate academic and researcher who has published widely in peer-reviewed journals and presented papers at numerous national and international conferences. His areas of interest include "energy justice", "science, technology and law", "privacy and data protection law", and "Asian and comparative law", among others. Dr. Karim has served as a Consultant and Fellow on projects funded by the Ministry of Education Malaysia, University of Malaya and the Monash Data Futures Institute (MDFI). He earned his Doctor of Philosophy (PhD) from the University of Malaya, Malaysia, as well as a Master of Business Administration (MBA) from the University of Chichester, UK, a Master of Comparative Laws (MCL) from International Islamic University Malaysia (IIUM), and a Bachelor of Laws (LLB) from BRAC University, Bangladesh.

Pervaiz K. Ahmed is a Distinguished Professor and Director of the Sunway Institute for Global Strategy and Competitiveness, Sunway University. Pervaiz is a Management scholar exploring innovation and strategic dynamics of entrepreneurial change. His outlook on management is colored by moral responsibility, particularly ethics of care and sustainability. Pervaiz has held several senior positions in the UK before coming to Malaysia. He has published extensively in international journals and is a regular keynote speaker with a number of academic awards for his research. In addition, he has served as editor and sits on the editorial boards of several international journals. He has extensive experience working with and advising blue-chip companies and public sector organizations, such as Unilever, Ford, AT&T, NCR, British Telecommunications, and the NHS in Europe. He has also been involved with corporate clients in Asia, as well as government agencies, such as the Singapore National Productivity Council. In recent times Pervaiz has been involved with the Government of Dubai's Public Sector Innovation and Improvement Initiative, and also served on projects commissioned by Economic Planning Unit (EPU), Ministry of International Trade (MITI), and APEC Secretariat.

Energy Justice and Human Rights

2

Muhamad Hassan Ahmad, Ashgar Ali Ali Mohamed, Mohamed Hanipa Maidin, and Mohammad Naqib Eishan Jan

Abstract

The UN aspires to ensure—through Goal 7 of the SDGs—that all major aspects of shared prosperity among the world population are further enhanced by having access to clean, affordable, reliable, and sustainable energy. By 2030, approximately between 2 and 3 billion people will have to rely on unsustainable fuels and technologies. Accordingly, this chapter aims to examine access to affordable, reliable, and sustainable energy, which is the key to the development of all other aspects of having a habitable life guaranteed under most of the constitutions as a 'right to life'. It argues that although there is no explicit mention of 'energy justice' as 'human rights' in international human rights conventions, there exists a 'subtle connection' between the 'right to life'—guaranteed under numerous international conventions as well as most of the constitutions—and 'access to energy' which requires special attention from all stakeholders.

M. H. Ahmad (✉) · A. A. A. Mohamed · M. H. Maidin · M. N. E. Jan
International Islamic University Malaysia, Kuala Lumpur, Malaysia
e-mail: mdhassan@iium.edu.my

A. A. A. Mohamed
e-mail: ashgar@iium.edu.my

M. N. E. Jan
e-mail: naqib@iium.edu.my

Keywords

Energy justice · Clean energy · Affordable energy · Reliable energy · Sustainable energy · Human rights

2.1 Introduction

In 2015, the United Nations (UN) adopted the 'Sustainable Development Goals' (SDGs) as 'a universal call to take action to end poverty, protect the planet, and ensure that all people enjoy peace and prosperity by 2030'. There are 17 integrated SDGs which 'recognise that action in one area will affect outcomes in others, and that development must balance social, economic, and environmental sustainability' (UNDP n.d.). In general, access to affordable, reliable, and sustainable energy is captured under Goal 7 of the SDGs—bearing the theme: 'access to clean and affordable energy'—which set the targets mentioned in Chap. 1 of the book.

The UN envisages ensuring, through Goal 7, that all major aspects of shared prosperity among the world population, such as in health care, agriculture, transportation, communications, education and business, among others, are further developed by having access to clean and affordable energy (United Nations n.d.).

In modern times, it has become more apparent that energy has to be affordable for the general public in order to enjoy essential services at an

A. Trakic et al. (eds.), *Energy Justice*, Sustainable Development Goals Series,
https://doi.org/10.1007/978-981-97-6059-6_2

optimum level for their survival, especially in health care, transportation, and communications. Besides, it has to be reliable in the sense that those essential services are rendered to the general public without any interruption. For example, a power outage during a surgical operation in the hospital will significantly increase the risks of complications or may even cause the loss of life (see also Foster Fuels n.d.).

On top of everything, in the contemporary world, any kind of energy offered to the general public has to be sustainable so that the environment is protected from by-products and side effects of the usage of such energy. The notion of sustainable energy was developed mainly to preserve the environment by reducing the pollution associated with unsustainable energy and conserving the natural resources of the planet. It can break free from the reliance on fossil fuels, which not only have adverse effects on the planet's environment but also become more expensive and less accessible to the general public (Routledge n.d.). Hence, sustainable energy is defined as the 'most reliable form of energy' in the sense that it will never deplete, such as solar power, wind power, and hydroelectric power (see also Nationalgrid n.d.).

Nowadays, as much as energy is vital to all aspects of life, it is equally important to make it affordable, reliable, and sustainable. Although the world is marching towards sustainable energy targets, it is still not at the desirable speed. It is predicted that, even by 2030, approximately 2 and 3 billion people will have to rely on polluting fuels and technologies to access energy, which is unsustainable. In order to ensure access to sustainable energy for all by 2030, the UN urges accelerating electrification through clean energy sources such as solar, wind, and thermal. It further encourages upgrading technology in providing clean energy and expansion of infrastructure by increasing investments in renewable energy. Above all, it notably recommends that the member states develop 'enabling policies and regulatory frameworks' to guarantee the general public access to affordable, reliable, and sustainable energy (United Nations n.d.). Accordingly, this chapter aims to explore whether access to affordable, reliable, and sustainable energy is, all over again, the key to the development of all other aspects of having a habitable life guaranteed under most of the constitutions as 'right to life'.

2.2 Access to Energy as a Basic Human Right Under International Law

2.2.1 International Human Rights Conventions and Instruments

The contemporary world marks the end of the Second World War as the beginning of the realisation of human rights. In 1945, the Charter of the UN was signed, and the promotion and protection of human rights became one of the main goals of the organisation (see Articles 1(3), 55 and 56 of the UN Charter). In its preamble, the Charter 'reaffirms faith in fundamental human rights, in the dignity and worth of the human person, in the equal rights of men and women and of nations large and small'. Article 1 (2) of the Charter provides that: 'The achievement of international co-operation in resolving international problems of an economic, social, cultural, or humanitarian nature, and in promoting and promoting respect for human rights and fundamental freedoms for all without distinction, in terms of race, gender, language, or religion' (Ahmad et al. 2022: 209–210).

With regard to the establishment of 'the conditions of stability and prosperity necessary for peaceful and friendly relations between nations on the basis of respect for the principle of equal rights and the determination of the destiny of the people', Article 55 of the Charter provides that the UN 'promotes universal respect for human rights and fundamental freedoms for all without distinction as to race, sex, language, or religion'. Again, Article 56 of the Charter states: 'All members pledge themselves to take joint and separate action in co-operation with the Organisation (UN) for the achievement of the purposes set forth in Article 55'. The term 'pledge' implies a legal obligation on the part of member states (Ahmad et al. 2022: 210).

However, the obligation is said not to observe human rights immediately but to work towards their fulfilment in the future. Nevertheless, States must not go backwards and deliberately violate human rights. In its Advisory Opinion in the *Namibia case* [1971] ICJ Rep 16, the International Court of Justice (ICJ) has pointed out that, under Article 2(7) of the Charter, the UN cannot interfere in internal matters of a sovereign State on the pretext of protecting human rights, but it can do so if human rights violation in the State is such that endangers world peace and security (Ahmad et al. 2022: 210).

In 1948, the Universal Declaration of Human Rights (UDHR), which consists of 30 articles, was adopted by the UN General Assembly on 10 December 1948 and declared that all human beings are free and equal, regardless of gender, colour, religion, or other characteristics. As a turning point, the UDHR sets out rights and freedoms for all, including the right to freedom from torture; the right to freedom of expression; the right to education; the right to asylum; other civil and political rights such as the right to life, liberty, and privacy; economic, social, and cultural rights such as social security, health, and housing rights (Ahmad et al. 2022: 210).

Although the rights and freedoms mentioned in the UDHR apply to all people in all States around the world, they are restricted in the sense that they have to be exercised in a way that does not violate the rights of others, must not adversely affect public order, welfare, or moral of the society. The UDHR is not a treaty but is based on the UN General Assembly resolution, which is a recommendation and thus is not binding. However, 'human rights are foreign to no culture and intrinsic to all nations. They belong not to a chosen few, but to all people. It is this universality that endows human rights with the power to cross any border and defy any force' (Annan 1999). Hence, it is generally agreed that some, although not all, provisions of the UDHR are crystalised into norms of customary international law binding on all States. This may be true with regard to certain civil rights, but the same is doubtful with regard to economic rights. Nevertheless, the UDHR has served as a yardstick and base for the development of various international human rights laws (Ahmad et al. 2022: 210–211).

The UDHR also inspired the adoption of international human rights treaties, including the International Covenant on Civil and Political Rights (ICCPR) (adopted and opened for signature, ratification and accession by General Assembly resolution 2200 A (XXI) of 16 December 1966. Entry into force: 23 March 1976, in Accordance with Article 49) and the International Covenant on Economic, Social and Cultural Rights (ICESCR) (adopted and opened for signature, ratification and accession by General Assembly resolution 2200 A (XXI) of 16 December 1966. Entry into force: 3 January 1976, in Accordance with Article 27). The ICCPR, which was adopted in 1966 and came into force in 1976, is a treaty that incorporates the civil and political rights provisions of the UDHR. As a treaty, ICCPR is binding upon its parties. It requires each State party 'to take the necessary steps to adopt such legislative or other measures as may be necessary to give effect to the rights recognised in the present Covenant'. There are two Optional Protocols to the ICCPR granting additional rights *vis-à-vis* the ratifying States. The First Optional Protocol (adopted and opened for signature, ratification and accession by General Assembly resolution 2200 A (XXI) of 16 December 196. Entry into force: 23 March 1976, in Accordance with Article 9) allows individuals to directly bring complaints before the Human Rights Committee, which is composed of 18 members elected by the States parties to the ICCPR, and all that this Committee can do is to call upon the State concerned for explanation and make recommendations. The Second Optional Protocol (adopted and proclaimed by General Assembly resolution 44/128 of 15 December 1989. Entry into force: 11 July 1991, in Accordance with Article 8(1)) requires the ratifying States to abolish the death penalty, but few states have ratified it (Ahmad et al. 2022: 211).

Meanwhile, the ICESCR, which was adopted in 1966 and came into force in 1976, generally incorporates the economic, social, and cultural rights provisions of the UDHR. A multilateral treaty, but unlike ICCPR, is not directly binding, as its implementation depends on the goodwill and resources of States. Article 2 of the ICESCR clarifies that the State party undertakes 'steps to the maximum of its available resources, with a view to progressively achieving the full realisation of the rights recognised in the present Covenant'. This provision makes the legal obligation rather weak, and it could be for this reason that the Covenant does not make any provision for individual petitions (Ahmad et al. 2022: 211–212).

There are a number of other international human rights treaties that have been adopted under the auspices of the UN. These treaties include the Convention on the Prevention and Punishment of the Crime of Genocide, the International Convention on the Elimination of All Forms of Racial Discrimination (ICERD) (adopted and opened for signature and ratification by General Assembly resolution 2106 A (XX) of 21 December 1965. Entry into force: 4 January 1969, in Accordance with Article 19), the Convention on the Elimination of All Forms of Discrimination Against Women (CEDAW) (adopted and opened for signature, ratification and accession by General Assembly resolution 34/180 of 18 December 1979. Entry into force: 3 September 1981, in Accordance with Article 27(1)), Optional Protocol to the Convention on the Elimination of All Forms of Discrimination against Women (adopted by General Assembly resolution 54/4 of 6 October 1999 and opened for signature, ratification and accession on 10 December 1999. Entry into force: 22 December 2000, in Accordance with Article 16), the Convention on the Rights of the Child (CRC) (adopted and opened for signature, ratification and accession by General Assembly resolution 44/25 of 20 November 1989. Entry into force: 2 September 1990, in Accordance with Article 49), Optional Protocol to the Convention on the Rights of the Child on the sale of children, child prostitution and child pornography (adopted and opened for signature, ratification and accession by General Assembly resolution 54/263 of 25 May 2000. Entry into force: 18 January 2002, in Accordance with Article 14.), Optional Protocol to the Convention on the Rights of the Child on the involvement of children in armed conflicts (adopted and opened for signature, ratification and accession by General Assembly resolution 54/263 of 25 May 2000. Entry into force: 12 February 2002, in Accordance with Article 10), Convention against Torture and Other Cruel, Inhuman or Degrading Treatment or Punishment (CAT) (adopted and opened for signature, ratification and accession by General Assembly resolution 39/46 of 10 December 198. Entry into force: 26 June 1987, in Accordance with Article 27(1)), Optional Protocol to the Convention against Torture and Other Cruel, Inhuman or Degrading Treatment or Punishment (adopted by General Assembly resolution 57/199 of 18 December 2002. Entry into force: 22 June 2006, in Accordance with Article 28 (1)), and the International Convention on Protection of the Rights of All Migrant Workers and Members of Their Families (ICMRW) (adopted by General Assembly resolution 45/158 of 18 December 1990. Entry into force: 1 July 2003, in Accordance with Article 87(1), among others (Ahmad et al. 2022: 212).

All the above-cited treaties, which have been ratified by the majority of the UN Member States, are designed to provide universal respect for human rights. However, the gap between the laudable rhetoric of international human rights treaties and the reality of human rights action or implementation remains wide. Violations of human rights are rampant, but at the international level, there is no proper human rights enforcement body. For instance, the ICJ is inactive in the area of human rights simply because of the limitations of its jurisdiction and the firm desire of most States to preclude it from hearing human rights cases. Further, although all the core international human rights treaties have monitoring bodies to oversee the implementation of the treaty provisions, but these bodies have no teeth to bite, for they merely consider State parties' yearly reports as well as individual complaints or communications, and make recommendations reminding states to be observant of their human rights obligations (Ahmad et al. 2022: 212–213).

2.2.2 Access to Energy as a Basic Human Right

Access to affordable, reliable, and sustainable energy not only to support all sectors, including business, medicine, education, agriculture, infrastructure, communication, and advanced technology, but also to realise various human rights, including the right to life, health, and development, is vital. Access to energy as one of the basic human rights is so fundamental that if it is guaranteed, it helps to effectively realise other basic human rights guaranteed by international human rights conventions and other relevant international instruments. However, mere acceptance of human rights treaties is not enough in itself; they must be implemented in good faith, and people must be able to achieve their rights, including the right to access energy.

The commitment of States around the world requires taking positive steps to ensure that people everywhere have access to clean energy so that they can effectively enjoy their other basic rights. It is a legal obligation that member states of human rights treaties must commit themselves to respect, protect, and implement. In relation to energy access, the obligation to respect means that State action must not undermine existing access to energy. States may create a policy that ensures that all citizens have access to an optimum amount of energy at an affordable cost.

Fulfilling human rights, including access to affordable energy, may be difficult but not impossible. Typically, the obligation to fulfil involves a positive duty whereby the government helps people to exercise their rights. Actions taken on behalf of individuals, communities, and entire countries have led to a growing number of precedents in some countries in which courts have recognised the right to energy. In some cases, courts have found large players—such as utilities and governments—guilty of failing to protect or complement these rights.

Also, as part of their commitment to human rights, States should enact laws to promote energy justice. Legislating laws to promote energy accessibility to all is crucial, particularly at a time when hundreds of millions of people around the world still do not have access to electricity and billions of people still rely on polluting fuels, such as animal dung, wood, and crop residues for cooking (Sagar 2005). Access to clean energy is necessary as one of the basic human rights, and efforts should be made to realise it because its realisation helps realise other basic human rights.

Realisation of basic human rights, including food, shelter, clothing, water, sanitation, medical care, schooling, and access to information without having access to energy, is difficult, if not impossible. Access to energy helps provide basic human needs such as cooked food, comfortable living temperature, lighting, use of electrical appliances, running water or sewage, essential health care (refrigerated vaccines, emergency, and special care), and transportation. Access to energy also contributes to productive activities, including agriculture, commerce, manufacturing, industry, and mining.

On the contrary, lack of access to energy not only contributes to poverty, deprivation, and economic decline but is especially harmful to women and children in developing countries (Sagar 2005). For example, in least-developed countries, it is traditionally women who are responsible for preparing and cooking food. As a result, without access to energy services, they have to spend considerable time searching for firewood, cooking, and other needs.

In a world with widespread poverty, unprecedented levels of inequality, and the culmination of climate change and biodiversity loss, energy has a key role to play in correcting and reversing these trends. Access to energy has been a catalyst of well-being for billions of people and is essential to modern life (Energy Access Goals Working Group 2016). Having realised the importance of energy and access to it, Goal 7 of the SDGs explicitly aims to ensure access to affordable, reliable, sustainable, and modern energy for all by 2030 (UN Doc A/RES/70/1).

The 2030 Agenda represents a global commitment to address energy-related challenges. It explicitly states that the SDGs are 'grounded' in the UDHR and 'international human rights treaties' and 'informed' by the 'Declaration on the Right to Development' (UN Doc A/RES/70/1). This basis of the 2030 Agenda in human rights, at least in principle, should enable integrated and comprehensive action to achieve a just energy transition in line with the SDGs (Winkler and Williams 2017).

Binding international human rights treaties, such as the ICESCR, recognise the right to an adequate standard of living, including the right to enjoy the benefits of scientific progress and its applications, which encompasses access to energy. The Committee on Economic, Social and Cultural Rights, the body responsible for monitoring the ICESCR, has highlighted the importance of energy in realising several economic, social, and cultural rights (Ahmad et al. 2022: 211).

States are bound by international human rights obligations and the realisation of the right to energy, and this obligation may include the obligation to enact laws that promote energy justice. States should take positive steps to ensure access to energy for all, especially for vulnerable and marginalised groups. This task may include implementing policies and regulatory frameworks that address the affordability, reliability, sustainability, and modernisation of energy systems. Hence, States have a duty to address energy poverty, given that lack of access to energy disproportionately affects certain communities and individuals. Providing laws to eliminate discriminatory practices and guarantee equal access to energy resources can be seen as a step towards the implementation of international human rights obligations. If countries fail to take sufficient steps to ensure access to energy, they can be held accountable under international law. This could lead to legal challenges at the domestic and international levels, potentially leading to sanctions or other consequences.

The recognition of access to energy as a fundamental human right would strengthen international co-operation in addressing global energy challenges. Countries would be encouraged to collaborate on research, technology transfer, and financial assistance to ensure sustainable and inclusive energy solutions. States would be obliged to review and modify their energy policies. Governments would need to prioritise investments in renewable energy sources, improve energy infrastructure, and implement measures to address energy inequality. This recognition could also empower civil society organisations and individuals to defend their rights to access energy. This may lead to increased public awareness, participation, and pressure on governments to take meaningful actions.

2.3 Access to Energy as a Basic Human Right Under Domestic Laws

2.3.1 Human Rights Under Domestic Laws

As mentioned above, the notion of the protection of human rights perceived today is relatively contemporary, as it was formally acknowledged only after the formation of the UN in 1945. It began with the recognition of some principles of human rights under the UDHR, which, as stated earlier, contains thirty articles of rights of universal application to all people (UN 2021). Nevertheless, since it is not a treaty signed and ratified by the Members of the UN but merely a declaration, it does not create any legal obligation on the State Members to adhere to (Ahmad et al. 2022: 213).

Although that being the case, UDHR still paves the way for the adoption of more than seventy human rights treaties at global as well as regional levels, and all of those human rights treaties make references to the UDHR in their preambles (UN n.d.). These include the Convention relating to the Status of Refugees 1951, the ICERD, the ICCPR, the ICESCR, the CEDAW, the CRC, among others (see Jan et al. 2017, Albar 2021). Of course, all of the above human rights conventions are binding upon the respective parties. Some commentators on international law even argue that principles of fundamental and universal human rights enshrined in the UDHR could be construed as well-established customary international laws, as many countries have invoked those principles in various international treaties and legal instruments since the establishment of the UN (Jan 2011).

With regard to the implementation of international human rights standards in the national jurisdiction, parties to the respective international treaties and legal instruments are obliged to incorporate those standards into their relevant domestic laws. Article 26 of the Vienna Convention on the Law of Treaties 1969 (VCLT) reflects the notion of '*pacta sunt servanda*' by stating that: 'Every treaty in force is binding upon the parties to it and must be performed by them in good faith'. The subsequent Article 27 provides: 'A party may not invoke the provisions of its internal law as justification for its failure to perform a treaty'. However, there is no prohibition on a state that is not a party to a particular convention from adopting the principles of human rights mentioned in that convention in its national legislation. Therefore, the national laws of the States pertaining to human rights are varied from one jurisdiction to another (Ahmad et al. 2022: 213–214).

2.3.2 Access to Energy as a Basic Human Right in Selected Jurisdictions

The initial idea of linking energy justice to human rights emanated from the Brundtland Report of 1987. The 'Brundtland Report' was duly named after the Commission's chairwoman, Gro Harlem Brundtland. The Report which was entitled 'Our Common Future' came into being due to a splendid initiative by the World Commission on Environment and Development (WCED), which had been set up in 1983 (Federal Office for Spatial Development Area n.d.). The Report plainly connected human rights, sustainable use of natural resources, access and due process, poverty and equality, intergenerational equity, and social justice to energy issues. The right to life is one of the core human rights which is duly embedded in a slew of international human rights instruments and treaties such as the UDHR (see Article 3 of the UDHR which clearly states that: 'Everyone has the right to life, liberty and the security of person'.), the ICCPR (see Article 6(1) of the ICCPR which categorically enshrines this basic right by stating that: 'Every human being has the inherent right to life. This right shall be protected by law. No one shall be arbitrarily deprived of his life'), and the ICESRS. The nexus between human rights to life and energy justice becomes clearer when there is an obvious trade-off between the protection of human health and the costs of emissions control (see Sovacool and Dworkin 2014).

In Spain, for instance, more people are said to have died prematurely due to fuel poverty than from car accidents, and worse still, up to 100,000 Europeans die each year as a result of a cold home (see Right to Energy Coalition 2020). Margaretha Wewerinke-Singh has shared these startling statistics with us. There are still 790 million people around the globe who remain without access to electricity and clean cooking facilities, and 2.6 billion people still depend on polluting fuels for cooking. To rub salt to the wound, the established global energy system has not done much to improve the living standards of the most vulnerable. Be that as it may, it goes without saying that acquiring, guaranteeing, and securing a rapid and equitable energy transition globally to the realisation of the human rights of billions of people is of paramount importance (see Wewerinke-Singh 2021).

2.3.2.1 The European Union

Being a group of developed nations, Europe is relatively robust in its energy transition towards a low-carbon energy system. Apparently, a low-carbon energy system is one of the goals in the 17 SDGs introduced by the UN. To smooth the way for such a transition, the European Commission has presented a new energy policy bundle known as the Clean Energy for all Europeans packages, or, in short, it is often called as 'Clean Energy Package' (CEP). The proposal of this package was initially mooted by the European Commission in November 2016 and nicknamed as the Winter Package at that time. Basically, the package contains eight legislative acts on the energy performance of buildings, renewable energy, energy efficiency, governance, and electricity market design (Carella 2021).

Since 2021, household electricity prices in the European Union (EU) have shown a sharp increase, and as a result of these bewildering soaring prices, the European Commission decided to launch the 'Energy Poverty Advisory Hub' (EPAH). The European Parliament—being the EU's law-making body—was also instrumental in the establishment of EPAH. Perhaps it is no exaggeration to say that EPAH may be considered to be the leading EU initiative to get rid of energy poverty and, in turn, snowball the just and equitable energy transition of European local governments (Carella 2021).

EPAH is a living testament to the importance of the government's intervention in drawing up a good, sound, and workable policy in order to stabilise the electricity and gas prices and, in turn, to actualise the notion of energy justice. Having been in a significant surge in electricity and gas prices—which kicked off even prior to the Russian invasion of Ukraine and, in turn, skyrocketed up to the second semester of 2022—the EU governments, nonetheless, managed to stabilise the prices of both essential items by devising and drawing up good, sound, practical, and workable policies. Via EPAH, this important energy advisory hub is expected to play vital roles in providing direct support, online training, and research to local authorities and civil society organisations and by building a collaborative network of all stakeholders interested in taking action to combat energy poverty in Europe (Carella 2021).

Apart from establishing EPAH and introducing usual broad-based policies such as price-suppressing measures—including subsidies, tax cuts and price controls—Europe has also taken legislative steps, for instance, by setting up 'Social Climate Fund' (SCF), which is slated to financially assist and support those who are relatively fragile in society, i.e., the most affected by energy and mobility poverty (Brun 2023). Many countries in Europe also introduced other measures, such as a 'Fuel Poverty Emergency Fund', which is akin to COVID-19 relief packages, in order to reduce the burdens on poorer households.

It is interesting to note that the original idea of introducing this particular measure emanated from civil society. The 'EU's Renovation Wave Strategy', which seeks to tackle fuel poverty and worst-performing buildings, is also part of Europe's grand strategies in combating the scourge of energy poverty. This strategy has been specifically designed due to the fact that buildings are often said to have been responsible for about 40% of the EU's energy consumption and 36% of greenhouse gas emissions from energy, yet only 1% of buildings that have undergone energy efficient renovation every year. It has a humble beginning when it was formed in 2014 by neighbourhood associations, workers' assemblies as well as water and housing rights platforms (Press Corner n.d.).

It seems that the role of pressure groups and Non-Governmental Organisations (NGOs) in Europe are equally crucial and significant in redressing the grievances of vulnerable groups resulting from the scourge of energy poverty or energy injustice. Hence their contributions should never be discounted. One such NGO is the 'Alliance Against Energy Poverty' (also known in Spanish as 'Alianza Contra La Pobreza Energética' [APE])—a grassroots organisation—which has actively involved in challenging the dominant energy system and addressing the root causes of energy poverty (European Commission n.d.). In 2021, for example, APE was at the forefront of leading a campaign that freed 35,000 families of their energy debts. When the Catalonian Parliament decided to ban electricity disconnections for vulnerable households in 2015, such a parliamentary action was, in fact, motivated by a tireless and successful campaign by APE (Cataluña Transnational Institute 2023). Due to its victorious mobilisation, the city of Barcelona, in turn, reacted positively by agreeing to establish advice centres and launch a public campaign informing people of their rights (Right to Energy Coalition 2020).

In addition, the 'Right to Energy' (the Coalition) is an effective pressure group which manages to assemble and unite multiple bodies with their respective different agenda, such as trade unions, anti-poverty groups, social housing providers, environmental campaigners, NGOs, health organisations, and energy cooperatives across Europe whose primary function is to campaign to tackle energy poverty at an EU, national and local level. The Coalition' designated aim is also for an energy system that puts people and the planet before profit (Right to Energy Coalition 2020).

2.3.2.2 The United States

Since greenhouse gas (GHG) emissions are proven to have endangered public health, bringing out vulnerable groups from this health risk is also part of energy justice initiatives. As far as the United States is concerned, the most affected areas that are caused by the endangered greenhouse gases may be found mostly across the Gulf South, in states where some of the poorest black, Latino, Asian, and Native American communities are located. Many of these places have already sustained huge tolls from climate-related catastrophes (Brentin Mock 2019).

Hence, in response to the need to incorporate distributional considerations into energy and climate policy, the United States has come up with a proposal known as a 'Green New Deal Resolution' introduced by Representative Alexandria Ocasio-Cortez and Senator Ed Markey on 7 February 2019. The Green New Deal Resolution essentially seeks to ingrain equity and social justice within climate change mitigation goals. In addition, it also commits to job training and economic development support for individuals and communities that might be adversely affected by a transition to new sources of energy. Though this new deal is generally welcomed, some environmental justice activists and lobbyists are particularly concerned with the inefficacy of this new deal, as it is only a resolution with no power teeth to bite since it has no binding power to force Congress to act on anything (Alaina Boyle et al. 2021).

2.4 Right to Life and Access to Energy

Although there are several international human rights conventions, none of these conventions explicitly mention 'energy justice' as 'human rights'. However, it does not mean that energy justice is not totally within the purview of those recognised human rights. Rather, it is mainly associated with several other human rights, especially the 'right to life' guaranteed not only under numerous international conventions but also under most constitutions.

In this context, the expression 'life' does not refer to mere existence, but it incorporates all those facets that are an integral part of life itself and those matters that form the quality of life. These include rights to seek and be engaged in lawful and gainful employment, to receive benefits that the society has to offer to its members, to live in a reasonably healthy and pollution-free environment, to continue in public service subject to removal for good cause by resort to a fair procedure, among others (Manikam et al. 2022: 99–104).

2.4.1 Right to Life Under the Constitution of India

In India, the Supreme Court, in the case of *Olga Tellis v. Bombay Municipal Corp* [1986] AIR SC 180, observed that: 'Deprive a person of his right to livelihood and you shall have deprived him of his life'. Again, in *Delhi Transport Corp v. DTC Mazdoor Congress & Ors* [1991] AIR SC 101, Satwant J, when delivering the majority judgment of the Supreme Court, said that: 'The right to life includes right to livelihood. The right to livelihood therefore cannot hang on to the fancies of individuals in authority'. Moreover, the Supreme Court of India in *Shantisar Builders v. Narayan Khimlal Totame* [1990] AIR SC 630 concluded that: 'The right to life includes among others, the right to reasonable accommodation' (Manikam et al. 2022: 99–103).

Access to modern health care is crucial in preserving life. In *Virendra Gaur & Ors v. State of Haryana* [1995] 2 Supreme Court Cases 577, the petitioner was seriously injured and was seeking medical care. Several government hospitals could not treat him due to a lack of facilities, and he was forced to seek medical care in a private hospital. The issue here was whether the lack of such facilities could amount to a denial of his fundamental rights under Article 21 of the Constitution of India. The Supreme Court opined such a lack of medical facilities amounted to a violation of Article 21. The government is under a duty to provide adequate medical facilities to ensure the welfare of the people if taken care of. Access to medical care is crucial to ensure a decent quality of life (Manikam et al. 2022: 103–104).

Any act that damages the environment would also affect a person's quality of life. Recognising a healthy environment as a part of fundamental rights would be crucial to preserving and conserving the environment. In *Virendra Gaur & Ors v. State of Haryana* [1995] 2 Supreme Court Cases 577, the courts were of the opinion that 'life' in Article 21 also means the right to live with human dignity, which includes living in an environment that is free from pollution. The term environment was interpreted in a liberal manner, which includes a hygienic atmosphere and ecological environment. Therefore, the state has an obligation to maintain ecological balance and a hygienic environment. This decision is an example of a proactive approach taken by the courts to recognise the right to a healthy environment as a fundamental right. Such an approach will protect the environment and enable a person to enjoy a decent quality of life (Manikam et al. 2022: 103).

It is worthwhile noting that the courts in India have given a broad and liberal interpretation to the term 'right to life'. The above term is not restricted to the right to livelihood alone but has been extended to all other rights associated with the right to life. For example, the right to health and medical care (see *Consumer Education & Research Centre and Others v. Union of India & Others* (1995) AIR 922, SC (Ind); *Lambodar Panigrahi v. Civil Judge, Athagarh* (1996) Lab IC 2790 (Orissa HC); *Suzana bt Md Aris (claiming as administrator of the estate and a dependant of Mohd Anuar bin Sharip, deceased) v. DSP Ishak bin Hussain & Ors* [2011] 1 MLJ 107, [2011] 1 CLJ 226), opportunities to eliminate sickness and physical disability of the workman (see *Kirloskar Brothers Ltd v. Employees' State Insurance Corpn* [1996] Lab IC 1718 at 1721, SC (Ind)), the right to receive wages for minimum subsistence (see *Radhey Shyam v. District Judge, Moradabad* (1997) Lab IC 743 (Alahabad, HC); *v. Sattamma v. Maize Breeder and Head Agricultural Research Station* (1996) Lab IC 1492 (Andhra Pradesh, HC)) which is regarded as a minimum requirement to enable a person to live with human dignity (see *Charles Sobraj through Marie Andre'o v. The Superintendent, Central Jail, Tihar, New Delhi* (1978) AIR 1514, SC (Ind)), and the right to a person's reputation (see *Sowmithri Vishnu v. Union of India and Another* [1985] AIR 1618 at 1621–1622, SC (Ind)), among others (Manikam et al. 2022: 100–101).

2.4.2 Right to Life Under the Federal Constitution of Malaysia

Malaysian courts, in some cases, have adopted a similar view to that of Indian courts regarding the term 'right to life' under Article 5(1) of the Federal Constitution of Malaysia which sets out the general notion that no person shall be deprived of his or her life, liberty and property without due process of law. It is worthwhile noting that Article 5(1) of the Federal Constitution is *pari materia* with Article 21 of the Indian Constitution, and therefore, the interpretation of the Indian cases is relevant and often quoted by Malaysian judges when determining cases on Article 5. It was stated that the Federal Constitution, being a 'living piece of legislation', did not spell out each and every aspect of the fundamental liberties of the citizen (see *Dato Menteri Osman bin Baginda & Anor v. Dato Ombi Syed Alwi bin Syed Idrus* [1981] 1 MLJ 29, FC, per Raja Azlan Shah). Judicial creativity in implementing 'the true brooding spirit of the framers of the Federal Constitution' is, therefore, 'necessary and timely' (see *Utra Badi a/l K Perumal v. Lembaga Tatatertib Perkhidmatan Awam & Anor* [1998] 3 MLJ 676 at 691; *Krishnan Kutty a/l Sanguni Nair v. Telekom Malaysia Kawasan Utara & Anor* [1999] 4 MLJ 393 at 408 per Kang Hwee Gee J). The contemplation is not only of what has been but also of what may be (See *Weems v. United States* 54 L Fd 793 at 801 (1909) (US SC)). Therefore, the 'right to life' in Article 5(1) is not merely confined to physical existence by declaring that it is unlawful to extinguish or take away another's life except by due process of law but includes the right to quality of life which necessarily includes right to 'livelihood'. Interestingly, it was further extended to the right to 'reputation' (See *Lembaga Tatatertib Perkhidmatant Awam, Hospital Besar Pulau Pinang v. Utra Badi a/l K Perumal* [2000] 3 MLJ 281 at 296).

Gopal Sri Ram in the Malaysian case of *Tan Tek Seng v. Suruhanjaya Perkhidmatan Pendidikan & Anor* [1998] 1 MLJ 481 adopted a liberal approach to Article 5(1) and concluded that 'life' under Article 5(1) is more than mere existence and includes all aspects of life that are crucial to the quality of life. His lordship stated:

> In my judgement, the courts should keep in tandem with the national ethos when interpreting provisions of a living document like the Federal Constitution. Lest they be left behind while the winds of modern and progressive change passes them by. Judges must not be blind to the realities of life, neither should they wear blinkers when approaching a question of constitutional interpretation. They (Judges) should, when discharging their duties as interpreters of the supreme law, adopt a liberal approach in order to implement the true intention of the framers of the Federal Constitution. Such an objective may only be achieved if the expression 'life' in Article 5(1) is given a broad and liberal meaning. Adopting the approach that commends itself to me, I have reached the conclusion that the expression 'life' appearing in Article 5(1) does not refer to mere existence. It incorporates all those facets that are an integral part of life itself and those matters which go to form the quality of life. Of these are the rights to seek and be engaged in lawful and gainful employment and to receive those benefits that our society has to offer to its members. It includes the right to live in a reasonably healthy and pollution-free environment. For the purposes of this case, it encompasses the right to continue in public service subject to removal for good cause by resort to a fair procedure.

In the arriving at the said conclusion, his lordship Gopal Sri Ram relied on two Indian cases to substantiate his argument, the case of *Olga Tellis v. Bombay Municipal Corp* [1986] AIR SC 180 and the case of *Dehli Transport Corp v. STC Mazdoor Congress & Ors* [1991] AIR SC 101, which seem to suggest that right to life includes the right to livelihood.

In *Olga Tellis v. Bombay Municipal Corp* [1986] AIR SC 180, Chandrachud CJ said the following:

> For purposes of argument, we will assume the factual correctness of the premise that if the petitioners are evicted from their dwellings, they will be deprived of their livelihood. Upon that assumption, the question which we have to consider is whether the right to life includes the right to livelihood. We see only one answer to that question, namely, that it does. The sweep of the right to life conferred by Article 21 is wide and far-reaching. It does not mean merely that life cannot be extinguished or taken away as, for example, by the imposition and execution of the death sentence, except according to procedure established by law. That is but one aspect of the right to life. An equally important facet of that right is the right to livelihood because, no person can live without the means of living, that is, the means of livelihood. If the right to livelihood is not treated as a part of the constitutional right to life, the easiest way of depriving a person of his right to life would be to deprive him of his means of livelihood to the point of abrogation. Such deprivation would not only denude the life of its elective content and meaningfulness but it would make life impossible to live. And yet, such deprivation would not have to be in accordance with the procedure established by law, if the right to livelihood is not regarded as a part of the right to life. That, which alone makes it possible to live, leave aside what makes life livable, must be deemed to be an integral component of the right to life. Deprive a person of his right to livelihood and you shall have deprived him of his life.

Furthermore, in *Delhi Transport Corp v. DTC Mazdoor Congress & Ors* [1991] AIR SC 101, Satwant J, when delivering the majority judgment of the Supreme Court, said the following: 'The right to life includes right to livelihood. The right to livelihood therefore cannot hang on to the fancies of individuals in authority. The employment is not a bounty from them nor can its survival be at their mercy. Income is the foundation of many fundamental rights and when work is the sole source of income, the right to work becomes as much fundamental. Fundamental rights can ill-afford to be consigned to the limbo of undefined premises and uncertain applications. That will be a mockery of them'. As noted earlier, the courts in India have given a broad and liberal interpretation to the term 'right to life'.

Reverting back to the Malaysian authorities, in the case of *Ketua Pengarah Jabatan Alam Sekitar v. Kajing Tubek* [1997] 3 MLJ 23, the court recognised that the building of a hydro-electric dam would displace more than 10,000 natives who occupy the land under the customary right and such deprivation is a deprivation of their life under Article 5(1). Similarly, in *Nor Anak Nyawai v. Borneo Pulp Plantation Sdn Bhd & Ors* [2001] 6 MLJ 241, Ian Chin J was of the opinion that if there is a longhouse with a sizable number of families in a remote area, it is expected that activities such as hunting, fishing, the gathering of forest produce, and farming have been carried out. These activities can be expected to have been carried out within an area surrounding the longhouse that can be covered by a half-day foot journey unless it is limited by the presence of another longhouse in the vicinity. As such, anything that affects their livelihood affects their rights under Article 5 (Manikam et al. 2022: 101–102).

In *Maria Chin Abdullah v. Ketua Pengarah Imigresen & Anor* [2021] 1 MLJ 750, FC, Nallini Pathmanathan, in delivering the dissenting judgment of the Federal Court, stated that 'the right to travel abroad is now added to the list of essential rights comprising a part of the right to life. The list is by no means exhaustive. It has and will continue to evolve to meet the changing needs of our society, country and the world'. Her Ladyship cited the views of Raja Azlan Shah LP in *Dato Menteri Othman bin Baginda & Anor v. Dato Ombi Syed Alwi vin Syed Idrus* [1981] 1 MLJ 29, and the views of Abdoolcader J in *Merdeka University Berhad v. Government of Malaysia* [1981] 2 MLJ 356, that the Constitution is a living piece of legislation, and accordingly its provisions should be construed broadly and generally, not narrowly nor restrictively.

Hence, the right to life guaranteed under most of the constitutions should not only be interpreted from the rule of law perspective but should extend its interpretation to include broader aspects of life by counting other basic necessities in supporting the survival of humans in modern times which would then well include access to affordable, reliable, and sustainable energy. In this way, the nexus between the 'right to life' and the 'access to energy' as a fundamental human right becomes more apparent.

2.5 Conclusion

Unquestionably, access to affordable, reliable, and sustainable energy becomes an extremely crucial and fundamental right of a person to survive in modern times because modern societies would not be able to function without having access to sustainable energy, which is dependable and less expensive to consume. This situation demands all forms of governments at international, regional, national, and domestic levels to guarantee such a right to have access to affordable, reliable, and sustainable energy to the world population. Accordingly, the UN responds to this call in the 2030 Agenda via Goal 7 of the SDGs and urges member states to develop 'enabling policies and regulatory frameworks' to ensure access to affordable, reliable, and sustainable energy.

This chapter presents that even though there is no explicit mention of 'energy justice' as 'human rights' in international human rights conventions, there exists a 'subtle connection' between the 'right to life'—guaranteed under numerous international conventions as well as most of the constitutions—and 'access to energy' which requires special attention from all stakeholders.

In a nutshell, access to affordable, reliable, and sustainable energy is gradually more recognised as a critical element of the right to an adequate standard of living. While international conventions and instruments provide a basis for realising this right, the duty of States to introduce legislation to promote energy justice is an evolving concept. Implications for countries include international cooperation, policy reform, legal accountability, and increased public awareness. As the world seeks to address the challenges of climate change and sustainable development, the recognition of access to affordable, reliable, and sustainable energy as a basic human right becomes more imperative.

References

Ahmad MH, Jan MNE, Mohamed AAA (2022) Human rights and the rule of law: the common law and Shari'ah perspective. In: Trakic A (ed) Shari'ah and common law: the challenge of harmonisation. De Gruyter.

Albar SH (2021) Human rights: high time to enforce seriously. MLJ 3: ccclxix- ccclxxii

Annan KA (1999) 'Foreword' in the UDHR: fifty years and beyond. In: Y Danieli, E Stamatopoulou, CJ Dias (eds). Baywood Publishing Company

Boyle A et al (2021) Green new deal proposals: comparing emerging transformational climate policies at multiple scales. Energy Res Soc Sci 81:102259

Brun M-A (2023) Why adopting the social climate fund is the right choice for the EU, can Europe. https://caneurope.org/why-adopting-the-social-climate-fund-is-the-right-choice-for-the-eu/. Accessed 29 Dec 2023

Carella C (2021) The clean energy for all Europeans package, Florence school of regulation. https://fsr.eui.eu/the-clean-energy-for-all-europeans-package/. Accessed 29 Dec 2023

Cataluña Transnational Institute (2023) Alliance against energy poverty. https://www.tni.org/en/article/alliance-against-energy-poverty-cataluna. Accessed 29 Dec 2023

Corner P (n.d.) European Commission, European Commission. https://ec.europa.eu/commission/presscorner/detail/en/ip_20_1835. Accessed 29 Dec 2023

Energy Access Goals Working Group (2016) More than a light bulb: five recommendations for making modern energy access meaningful for people and well-being, centre for global development. https://www.cgdev.org/node/3124016. Accessed 29 Dec 2023

UN Environment Programme (n.d.) GOAL 7: affordable and clean energy. https://www.unep.org/explore-topics/sustainable-development-goals/why-do-sustainable-development-goals-matter/goal-7. Accessed 29 Dec 2023

European Commission (n.d.) Energy poverty advisory hub. https://energy-poverty.ec.europa.eu/alliance-against-energy-poverty-alianza-contra-la-pobreza-energetica_en#: ~ :text=non%2Dprofit%20organisation-,Alliance%20against%20energy%20poverty%20(Alianza%20Contra%20la%20Pobreza%20Energ%C3%A9tica),and%20to%20defend%20human%20rights. Accessed 29 Dec 2023

Federal Office for Spatial Development Area (n.d.) 1987: Brundtland report. https://www.are.admin.ch/are/en/home/media/publications/sustainable-development/brundtland-report.html. Accessed 29 Dec 2023

Foster F (n.d.) What happens when a hospital loses power? https://fosterfuels.com/blog/happens-hospital-loses-power/#: ~ :text=If%20the%20lights%20go%20out,risks%20and%20complications%20of%20surgery. Accessed 29 Dec 2023

Jan MNI, Mohamed AAA, Ahmad MH (2017) International refugee law practice and procedure. IIUM Press

Jan MNI (2011) Principles of public international law: a modern approach, 2nd edn. IIUM Press

Manikam V, Ramalingam CL, Panchalingam L, Mohamed AAA (2022) Right to life and personal liberty. In: Mohamed AAA, Ahmad MH (eds) Constitutional law in Malaysia. LexisNexis

Mock B (2019) The green new deal vs. the old green deals. https://www.bloomberg.com/news/articles/2019-02-12/the-green-new-deal-vs-the-old-green-deals. Accessed 29 Dec 2023

Nationalgrid (n.d.) what are the different types of renewable energy? https://www.nationalgrid.com/stories/energy-explained/what-are-different-types-renewable-energy#: ~ :text=Renewable%20energy%20is%20energy%20that,and%20hydroelectric%2C%20including%20tidal%20energy. Accessed 29 Dec 2023

Right to Energy Coalition (2020) About energy poverty. https://righttoenergy.org/about-energy-poverty/. Accessed 29 Dec 2023

Routledge (n.d.) What is sustainable energy and why do we need it? https://www.routledge.com/blog/article/what-is-sustainable-energy-and-why-do-we-need-it#: ~ :text=It%20will%20never%20deplete%20and,pollution%20associated%20with%20unsustainable%20energy. Accessed 29 Dec 2023

Sagar AD (2005) Alleviating energy poverty for the world's poor. Energy Policy 33:1367

Sovacool BK, Dworkin MH (2014) Global energy justice: problems, principles, and practices. Cambridge University Press

UN (n.d.) Universal declaration of human rights. https://www.un.org/en/about-us/universal-declaration-of-human-rights. Accessed 4 Apr 2022

UN (2021) Where can I find UN information on the 'Cylinder of Cyrus' a.k.a. 'Edict of Cyrus'? https://ask.un.org/faq/194027. Accessed 4 Apr 2022

UNDP (n.d.) The SDGs in action: what are the sustainable development goals? https://www.undp.org/sustainable-development-goals?utm_source=EN&utm_medium=GSR&utm_content=US_UNDP_PaidSearch_Brand_English&utm_campaign=CENTRAL&c_src=CENTRAL&c_src2=GSR&gclid=CjwKCAjwxZqSBhAHEiwASr9n9HY-lectv0IpRjOlUCdxTJqNnivCAbLgXwLNrpY-P0RatEZhtuHS7hoC6eQQAvD_BwE. Accessed 4 Apr 2022

United Nations (n.d.) Ensure access to affordable, reliable, sustainable and modern energy. https://www.un.org/sustainabledevelopment/energy/. Accessed 29 Dec 2023

Wewerinke-Singh M (2021) A human rights approach to energy: realizing the rights of billions within ecological limits. Rev Eur Comp Int Environ Law 31 (1):16–26

Winkler IT, Williams C (2017) The sustainable development goals and human rights: a critical early review. Int J Human Rights 21:1024

Muhamad Hassan Ahmad is an Assistant Professor at the Civil Law Department, Ahmad Ibrahim Kulliyyah of Laws, International Islamic University Malaysia (IIUM). He had been the 'External Examiner' for 'Master of Laws (LLM) (International Business and Trade Law)' programme from 2018 to 2022 at the Taylor's Law School, Taylor's University which currently ranks as the Top 1 Private University in Southeast Asia in the QS World University Rankings. He also served as the 'Adjunct Lecturer' for 'Master of Laws (LLM) (International Law)' programme at the Faculty of Shariah and Law, Universiti Islam Sultan Sharif Ali (UNISSA), Brunei. He graduated with a Bachelor of Laws (LLB (Hons)), Master of Comparative Laws (MCL), and Doctor of Philosophy (PhD) from IIUM. He has a special research interest in Public International Law, Private International Law, Alternative Dispute Resolution, Comparative Law, Jurisprudence, and Online Legal Education. He is the Editor-in-Chief of the 'Journal of Islamic Law Review' indexed in 'Index to Foreign Legal Periodicals (IFLP)' produced by the American Association of Law Libraries at the University of California, Berkeley, School of Law, the United States. He published a number of research books, textbooks, and research articles in various academic journals; presented several research papers at numerous academic conferences as well as symposiums; and talked in many public forums.

Ashgar Ali Ali Mohamed (Dato' Sri) is a Professor at Ahmad Ibrahim Kulliyyah of Laws, International Islamic University Malaysia (IIUM). He graduated with a Bachelor of Laws (LLB (Hons)) and Master of Comparative Laws (MCL) from IIUM, a Master of Laws (LLM (Hons)) from the University of Auckland, New Zealand and a Doctor of Philosophy (PhD) from the Graduate School of Management, University Putra Malaysia. He was admitted to the Malaysian Bar as an advocate and solicitor by the High Court of Malaya in 1992. His areas of interest include employment law, civil litigation, and alternative dispute resolution. He has authored several books on the above subjects, published numerous articles both in local and international journals/reports, and presented papers on the abovementioned areas at various international and local conferences.

Mohamed Hanipa Maidin is a former Deputy Minister of Law and ex-member of Parliament for two terms. In 1994, he was called to the Malaysian Bar to be duly licensed as an advocate and solicitor. In 1997, he set up his own legal firm under the name of Mohamed Hanipa & Associates. He was an active litigation lawyer for almost 25 years in practice and being a lawyer, he had appeared in various public interest litigations in Malaysian courts including the Federal Court. Several of his cases were reported in media as well as in various legal journals in Malaysia. He obtained his Bachelor of Laws (LLB (Hons)) and Bachelor of Laws (LLB (Shariah)) from the International Islamic University Malaysia (IIUM). He is also a passionate writer and many of his articles have been published in media in Malaysia and abroad.

Mohammad Naqib Eishan Jan is a Professor at Ahmad Ibrahim Kulliyyah of Laws, International Islamic University Malaysia (IIUM). He graduated with Bachelor of Laws (LLB (Hons)), Master of Comparative Laws (MCL) and Doctor of Philosophy (PhD) from IIUM. His areas of expertise include Public International Law, International Humanitarian Law, International Human Rights Law, International Dispute Resolution, International Commercial Arbitration, Use of Force in International Relations, Refugee Law and the Law of the United Nations. He has taught and published extensively in these specialised areas. Further, he has published a number of books and numerous articles in refereed law journals, and presented papers at various conferences.

3 Does Energy Justice Trump Environmental Justice?

Maizatun Mustafa and Farahdilah Ghazali

Abstract

Transitioning towards renewable energy sources is imperative to combat climate change and promote sustainable development. This transition requires policies focusing on reducing carbon emissions and enhancing energy efficiency. Intertwined with environmental justice, energy justice emphasises equitable access to affordable and reliable energy. However, achieving this balance poses challenges, especially in reconciling environmental preservation with societal energy needs. This chapter explores the nexus between energy justice and environmental justice, elucidating their shared objectives and distinct focuses. Through discussions on clean energy development, it emphasises the importance of incorporating both perspectives into energy system strategies. Although renewable energy sources offer significant potential for combating climate change and its effects, addressing distinct environmental concerns associated with different types of renewable energy projects is crucial to ensure their sustainable and responsible deployment to avoid compromising environmental justice in meeting energy needs.

Keywords

Energy justice · Environmental justice · Renewable energy · Climate change · Sustainable development

3.1 Introduction

The energy sector has evolved and experienced major changes in addition to developments in technology, businesses, and property rights. State-owned energy businesses previously dominated the market with monopoly rights in energy production, transmission, and distribution (Milčiuvienė et al. 2019; IDEAS 2023). To prevent the abuse of monopoly power in the energy market, private players have been invited by the governments in the operations of generation, transmission, distribution, and supply where

The original version of the chapter has been revised: The affiliation of the second author has been updated. A correction to this chapter can be found at https://doi.org/10.1007/978-981-97-6059-6_17

M. Mustafa (✉)
International Islamic University Malaysia, Kuala Lumpur, Malaysia
e-mail: maizatun@iium.edu.my

F. Ghazali
School of Business and Economics, Universiti Putra Malaysia, Serdang, Selangor, Malaysia
e-mail: farahdilah@upm.edu.my

A. Trakic et al. (eds.), *Energy Justice*, Sustainable Development Goals Series,
https://doi.org/10.1007/978-981-97-6059-6_3

energy corporations were privatised (Milčiuvienė et al. 2019). With emerging new ideas of economic regulation, particularly on the distributed economy and energy justice, many have interests and invested in energy generating based on self-consumption, especially in the solar photovoltaic and batteries industries (Burke and Stephens 2017). The introduction of the Single European Act (1987) on the European Economic Community (EEC) members was to facilitate the free movement of products, people, services, and capital across European borders. Through this liberalisation movement, the local barriers to the electricity and natural gas market were lifted, resulting in a competitive energy market (Diaz-Rainey et al., 2021). This energy liberalisation has culminated in many advantages and has enhanced the policy for environmental emissions management by creating trading mechanisms (Pollit 2012; Ali et al. 2022).

On the other hand, to ensure long-term economic growth, governments are required to implement policies to minimise carbon dioxide emissions as a result of the global issue of climate change, which has significant effects on economies and society. In emerging nations, production and consumption are growing quickly and becoming more diverse as a result of globalisation (Ristovska 2014). Since the utilisation of fossil fuels in commodity manufacturing causes environmental degradation and presents health hazards to individuals and societies, nations have begun to transition to renewable energy alternatives. (UNDP 2016; Pata 2021). The volume of energy used for business activities exerts an immense effect on the emissions of pollutants (Shahzad 2020; UNEP 2022), while a heavy reliance on natural gas might make it more difficult to move to an economy with lower carbon emissions (Najm and Matsumoto 2020). Therefore, technological developments, higher energy efficiency, competitive economies, improved energy security, and a focus on tackling climate change are all necessary for the development of inexpensive and sustainable energy solutions and measures (Kabeyi and Olanrewaju 2022).

The global economy has traditionally revolved around the energy sector. Before industrialisation, energy was essential to growth, but due to industrialisation, access to energy sources became a crucial concern for governments and corporations worldwide (Heffron and Talus 2016). In the same vein, energy justice has become an emerging inter and transdisciplinary research area and has drawn attention together with environmental justice and climate justice. Through this 'justice' concept, researchers are now considering how their findings may be put into practice, influencing the future energy mix that people use and fostering a more sustainable economy where policies of the environment and reducing climate change are crucial. (Heffron and McCauley 2017). According to Wood and Roelich (2019), less consumption of fossil fuels could minimise the negative effects of climate change. They further argued that reducing climate change and using fossil fuels should be seen in the context of their respective contributions to well-being rather than as distinct concepts. When energy justice is implemented in reality, it will be compatible with society's objective of accomplishing its environmental and climate change goals. Therefore, it should help achieve the 1.5–2 °C temperature increase goal outlined in the 2015 Paris Agreement. This is vital because energy justice can potentially reverse these environmental damages (Heffron 2022).

This chapter explores how energy justice relates to environmental justice. There is no doubt that the two concepts do overlap to a large degree. However, they could also have some astonishing differences, particularly concerning their objectives. While environmental justice focuses on the preservation of the environment and the advancement of a clean source of energy, the emphasis in energy justice is predominantly on people and their right to an affordable and reliable source of energy (Opal and Nathwani 2023). The two seemingly competing interests are reconcilable. This chapter addresses the implementation of the concept of 'Affordable and Accessible Energy for All' while considering the framework related to the environmental perspective. Through the clean energy development in the energy system, the chapter provides a discussion on the development of renewable energy and nuclear as well as problems relating to the alternative sources research. The chapter goes into additional detail about the primary components of this new agenda, including its normative scope, demonstrating how energy justice impacts and delivers benefits, especially for the environment. Lastly, the chapter shows how energy systems strategies are incorporated to balance energy justice and environmental justice, which might offer a clean, dynamic framework for energy production and consumption.

3.2 The Concept of Energy Justice and Environmental Justice

In pursuit of justice research, energy is a new focal point. There is, of course, ample documentation on the 'energy challenge' particularly with increasing population and limited resources (Jenkins et al. 2016). According to Heffron et al. (2018), energy justice is an emerging ideology that prescribes moral, philosophical, and ethical movements that share environmental and climate justice theories. It historically developed from environmental justice (McCauley et al. 2013). One of the seven defining principles of energy law theory is said to be rapidly changing in light of the theory's applicability. The other six principles are: human health, climate change, sovereignty over natural resources, prudent, rational, and sustainable resource use, and access to modern energy services (Heffron et al. 2018). Energy justice hinges on the equitable treatment of disadvantaged and privileged groups in addressing moral intuition (Pesch et al. 2017). Energy law can significantly improve a country's economy and environmental conditions when it is implemented effectively, through which energy justice can be preserved.

Setyowati (2021) stated that equitable sharing of the advantages and disadvantages associated with energy production and consumption is known as distributive energy justice. Distributive justice necessitates an equitable distribution of opportunities and hazards; procedural justice entails access to decision-making authority; and recognition justice entails respect for all peoples and worldviews (McCauley et al. 2013). Ensuring that all parties involved in energy decisions have equal and meaningful participation is the main goal of procedural justice (Lee and Byrne 2019). In order to end the socio-cultural dominance of some groups over others, recognition justice recognises the unique identities of individuals with respect to the energy system (Setyowati 2021) and (Van Uffelen 2022). Heffron (2022) has incorporated restorative and cosmopolitan justice as an additional component of energy justice. According to restorative justice, any injustice that the energy industry does should be made right; decommissioning procedures are a prime example (Hazrati and Heffron 2021). Fundamentally, cosmopolitanism justice is based on the idea that, when it comes to energy, we are all citizens of the same globe, and as such, the impacts of energy-related cross-border activities must be taken into account (Heffron 2022).

In their study on energy justice, Jenkins et al. (2016) focused on assessing the areas where injustices occur, the reasons the impacted communities receive less attention, and how the existing systems address these issues to reduce these injustices. Martin acknowledged that environmental catastrophes and social inequality go hand in hand. Many believe that incremental adjustments are running out of time to be effective and that a fundamental, transformative shift that balances justice and sustainability is unavoidably needed. Therefore, many countries, especially in Europe, have initiated a clean energy development programme. Nevertheless, these programmes also face challenges as their scope and ambition increase. These include who bears the cost of the transition and location of large-scale solar and wind farms (Dupré la Tour 2023). Another concern is whether the new 'green' economy is more equitable and offers more opportunities than fossil fuels. Welton and Eisen alleged that the quality of the local air and the climate greatly benefit from clean energy. However, a cleaner energy economy does not inevitably result in a more equitable economy. How to pay for the transformation is the first issue with clean energy justice. Currently, most new energy infrastructure is financed by electricity prices (Feldman et al. 2020). Funding a clean energy buildout through these volumetric bills, however, may not feel as equitable as it once did when some power customers turned producers by implementing rooftop solar, electricity storage, or other cutting-edge technology (Kim et al. 2023). There may be a class disparity in how clean energy is funded if wealthy users can reduce their electricity costs the most through clean energy expenditures.

Transitions to clean energy, such as renewable energy, are crucial for reducing global warming and decarbonising the global economy. However, several energy sources that are categorised as renewable pose risks to human health, endangering the well-being of those who are already the most vulnerable to the effects of climate change (Hanke et al. 2021). Keady et al. (2021) studied how renewable energy policies have addressed energy vulnerability and energy justice. They found that existing policies do not equitably dispense household transition benefits to address vulnerability. Hanke et al. (2021) argued that policies that promote the use of renewable energy have increased significantly. While research on energy justice is rapidly expanding, a deeper comprehension of this concept and how energy injustice is being committed during low-carbon energy transition would be possible by using an energy justice lens to the analysis of the energy transition process (Setyowati 2021). On the other edge, over the past forty years, the concept of environmental justice has emerged to highlight the unequal impacts of environmental burdens on populations who are impoverished, marginalised by race, or otherwise excluded (Hanke et al. 2021). The term environmental justice refers to the equitable treatment and significant participation of all individuals, irrespective of their race, colour, national origin, or income, in the formulation, execution, and observance of laws, regulations, and policies that impact the environment and/or public health. Environmental justice movements and the academic literature that have supported it since the field's inception in the late 1980s and early 1990s are dedicated to proving that environmental burdens are 'inequitably distributed both spatially and socially' and to demanding solutions to this fact (Hanke et al. 2021).

Environmental justice is a wide theory and encompasses three fundamental elements: distributive justice, recognition, and procedural justice (Enssle and Kabisch 2020). The division of costs and rewards associated with environmental actions is referred to as distributive justice. Justice as recognition is about recognising and taking into account the interests, values, and viewpoints of those who are accorded respect and those who are not. While capacities theory focuses on people's ability to live in a valuable way, procedural justice is concerned with who is involved and has influence in decision-making (Svarstad and Benjaminsen 2020). The unequal distribution of environmental costs, benefits, and related well-being outcomes is highlighted by environmental justice analysis, which also aims to identify the underlying and proximal causes of this imbalance (Martin et al. 2020). The key takeaway from environmental justice is that distribution disputes and current social inequities are linked to the patterns of environmental inequality (Martin et al. 2020).

Thus far, there has been a dearth of review work that demonstrates the current status of energy justice research in developing nations' contexts and evaluates whether or not energy justice scholarship is beginning to produce relevant insights. Furthermore, there are not many analyses of specific energy justice issues that relate to the environment and climate issues that emerge in developing nations. Table 1 below shows the examples of research that explicitly links energy and environmental justice.

	Authors/year	Objectives and the outcomes of the study
1	Hess et al. (2022)	This study develops the understanding of opposition to energy infrastructure from a justice perspective by drawing on studies on environmental justice and energy justice. A comprehensive set of records containing 70 incidents of protests against the socio-environmental impacts of proposed power lines was assembled. The examination of opposition parties' strategic frameworks serves as the foundation for assessing theoretical conceptions of justice, their practical application, and their limitations
2	Wood and Roelich (2020)	The article examines the relationships between the strategic objectives and moral philosophical underpinnings of two notions of energy justice: the 'principled approach' and the 'triumvirate conception.' The triumvirate approach's embrace of a trivalent vision of justice today conflicts with its overall top-down approach despite its initial development to capture and represent the values of activist-led environmental justice movements
3	Jenkins (2018)	This paper discusses the prospects for energy justice compared to environmental and climatic justice by outlining the history, achievements, and shortcomings of the principles of environmental and climate justice
4	Finley-Brook and Holloman (2016)	This paper highlights how the US energy transitions build from and contribute to environmental injustices. Energy justice requires not only ending disproportionate harm but also involvement in the design of solutions and fair distribution of benefits, such as green jobs and clean air. The researchers study the extent to which the confluence of state, civic, and market processes assure 'just' transitions to clean, low-carbon energy production involving equitable distribution of costs, benefits, and decision-making power. They also explore fossil fuel divestment, carbon taxes, the social cost of carbon measurements, cap-and-trade, renewable energy, and energy efficiency
5	Hess and Ribeiro (2016)	By offering a qualitative approach based on energy justice protocols to systematically assess energy justice in energy projects, this paper seeks to close this gap and expand on the idea of energy justice, and create a framework for actually implementing the concept in energy projects
6	Welton and Eisen	This study presents the difficulties in developing a 'clean energy justice' agenda. This agenda's mapping exposes the equity issues that may arise throughout the switch to sustainable energy and enables more thorough, innovative approaches to legal and regulatory remedies
7	Setyowati (2021)	This article explores Indonesia's efforts to mobilise private capital for renewable rural electrification to fulfil its vision of energy justice. It specifically looks at how and to what extent Indonesia has dealt with issues of energy justice and their societal ramifications. Based on interviews and document analysis, policies and efforts that specifically target distributive energy justice in terms of energy affordability and accessibility are how Indonesia has embodied its goal for energy justice
8	Hanke et al. (2021)	This exploratory research examines how RECs fulfil this social purpose by providing affordable energy and energy efficiency to vulnerable households and enhancing participatory procedures to facilitate involvement from vulnerable groups based on data gathered from 71 European RECs
9	Levenda et al. (2021)	This study aims to provide an overview of the established effects on environmental justice (EJ) related to renewable energy technologies incorporated into numerous renewable energy programmes worldwide. This research is to make sure that justice considerations are taken into account when developing and implementing renewable energy policy
10	Wang and Lo (2023)	This research provides a thorough analysis of scholarly studies on EEJ in Chinese contexts. They found that Chinese EEJ scholarship has pointed out instances of injustice in China's ecological and energy transitions and emphasised certain features, such as regional differences, the disparity between urban and rural areas, and the disproportionate effect on migrant communities. However, the awareness

(continued)

	Authors/year	Objectives and the outcomes of the study
		of the many and local interpretations of justice-related research in China is undermined by the desire to adopt Western justice principles. The review recommends that cross-cultural studies and a more critical approach that emphasises the role of power in perpetuating injustice are two ways that future research can promote theoretical progress
11	Lai (2022)	This article explores how geohistorically produced worries about the village's socioeconomic marginalisation and suffering from petrochemical pollution are entwined with the emergence, development, and framing of the Taihsi Green Energy and Health Community Initiative in Taiwan. To better assist advocates and policymakers in enhancing energy justice in underprivileged communities, the paper makes the case that a relational place-based analysis is essential to understanding the grounded meanings and values of a community energy initiative. Their findings show that community energy was a proactive continuation of place-based activism for environmental justice in this particular context

3.3 Affordable and Accessible Energy for All: An Environmental Perspective

To accomplish the United Nations Sustainable Development Goals (SDGs) in the context of energy justice, it is imperative to understand the trade-offs and synergies between sustainable transitions and challenges to energy access. Specifically, the concept of affordable and accessible energy for all is a critical component of sustainable development, as recognised by SDG 7, which aims to ensure access to affordable, reliable, sustainable, and modern energy for all by 2030 (Tomei et al. 2020). This notion is consistent with the concept of energy justice, which is concerned with the equitable distribution of energy resources, including affordability, which is a critical component of guaranteeing equitable access to energy (Wallsgrove 2022). Universal access to energy encompasses the provision of adequate, affordable, reliable, safe, and environmentally sound energy and energy services for personal, domestic, public, and commercial use for all (Savić 2019). It is growing more evident that energy access is not an objective in itself but is linked to various other development goals (Tomei et al. 2020). In many countries around the world, affordable energy is regarded as a critical component of energy policy (Demski et al. 2017), and the integration of energy justice and access to affordable energy is critical in addressing disparities in energy access and ensuring equitable distribution of energy resources. It emphasises the significance of combating energy poverty and ensuring that everyone has access to affordable energy (Stojilovska 2021). Furthermore, energy justice goes beyond energy accessibility to include the full energy utilisation chain, ensuring that all players benefit equitably from energy consumption (Wang et al. 2019). Figure 3.1 below explains how the use of conventional energy sources can address several issues, including energy poverty and energy insecurity, consequently achieving energy justice.

Access to affordable, reliable, sustainable, and modern energy is inextricably linked to energy poverty and energy security. Energy security is critical for economic growth and human well-being, particularly in emerging countries with high levels of energy poverty (Ullah et al. 2021; Thomson et al. 2017). Energy access is critical to addressing energy poverty and improving living standards, as it is essential to various aspects of human well-being, including health care, education, and economic productivity (Bousnina and Gabsi 2023) and (Marchand et al. 2019). Energy poverty might exacerbate environmental stress and raise carbon emissions, emphasising the

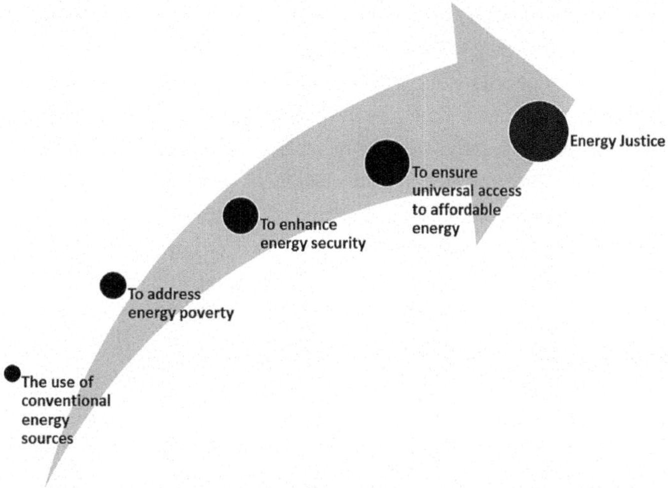

Fig. 3.1 Achievement of energy justice through the use of conventional energy sources. *Source* Authors

necessity of addressing energy poverty in the transition to sustainable energy systems (Burlinson et al. 2018) and (Bousnina and Gabsi 2023). Energy poverty, which is influenced by factors such as inefficient residential stocks, affordability issues, and spatial and technical limitations, plays a significant role in determining energy deprivation (Bouzarovski 2013). According to research by Mangula et al. (2018), economical and readily available energy can contribute to social and economic growth (Mangula et al. 2018). Furthermore, energy security is frequently described by perceptions of availability, acceptability, price, and accessibility (Tuğrul 2017). The dynamic features and diversity of energy access are crucial for ensuring long-term energy access, particularly in terms of enhancing security, autonomy, and protecting valued livelihoods (Smith 2021). Measuring energy availability and affordability is critical for policymakers, and more accurate, faster, and less expensive methods of assessing energy access as an indicator of well-being are needed (Seuret-Jiménez et al. 2020).

The use of oil, natural gas, and coal is critical in providing affordable power access, especially in tackling energy poverty and satisfying the energy needs of various populations. These traditional energy sources have played an important role in guaranteeing affordable energy availability through a variety of processes and applications. Historically, oil, natural gas, and coal have been the dominant sources of energy for electricity generation, heating, and transportation, providing reliable and cost-effective energy to fulfil the demands of rising populations (Mittal et al. 2021). Existing infrastructure and technological advances in the extraction, processing, and utilisation of these resources have contributed to their affordability and widespread availability, particularly in areas where alternative energy sources may be less accessible or economically viable (Mittal et al. 2021).

Nonetheless, conventional energy sources have been linked to several environmental issues, such as air and water pollution, greenhouse gas emissions, and habitat destruction (Ristic et al. 2021) (Hoque and Rashid 2021) (Neelakandan et al. 2021) and (Vaish et al. 2022). These conventional energy sources' extraction, production, and combustion have been linked to negative environmental effects, including climate change, air and water pollution, and ecosystem degradation (Addy et al. 2019) (Qin et al. 2021) (Zhang et al. 2023). The life-cycle greenhouse gas emissions of natural gas, coal, and petroleum have been a source of contention, with concerns raised about the effects on water and air quality, as well as global climate change (Burnham et al. 2011). Furthermore, conventional energy sources such as oil, coal, and natural gas have been highlighted as carrying greater environmental hazards than renewable energy sources, emphasising the importance of addressing their environmental repercussions (Addy et al. 2019). Coal extraction and use have been connected with significant adverse environmental effects, including interference with surface water, groundwater, air, and land, resulting in a slew of environmental issues (Jiang et al. 2016). Whereas the environmental effect potential of coal-fired power plants has been studied, emphasising the environmental costs connected with coal use.

Other energy sources, such as oil and gas, have been associated with major environmental and social implications, such as air and water pollution, biodiversity issues, and detrimental impacts on indigenous people (Finer et al. 2008) and (Nriagu et al. 2016). The distribution of adverse consequences from oil extraction and consumption has been examined through the lens of environmental justice, showing the disproportionate burden on particular populations (O'Rourke and Connolly 2003) and (Fry and Hilburn 2020). Compliance, production costs, and community awareness are all challenges for initiatives to mitigate the adverse environmental impacts of oil and gas extraction (Arthur 2020). The application of innovative environmental taxation as a means of promoting sustainable growth within the oil and gas business has been investigated (Huseynov et al. 2021). Environmental management systems and environmental tax enforcement have been identified as critical variables in improving compliance with environmental standards in the oil and gas industry (Mas'ud et al. 2020) and (Muzanni et al. 2022).

Similar to other energy sources, coal mining activities have been shown in studies to contribute to environmental degradation, air and water pollution, and adverse health impacts in nearby communities, leading to increased risks of respiratory and cardiovascular diseases, as well as higher mortality rates (Hendryx et al. 2011) and (Shi et al. 2019). While coal use has been linked to severe environmental and health consequences, particularly in coal mining regions, this has resulted in discrepancies in the distribution of environmental burdens and benefits, which is important to the concept of environmental justice (Hendryx et al. 2011). The environmental and health impacts of coal mining have been correlated with socioeconomic and health inequities in coal mining regions, emphasising the need to deal with these concerns within an environmental justice framework (Hendryx et al. 2011) and (Shi et al. 2019). The environmental and health consequences of coal combustion have implications for environmental justice, considering disadvantaged communities constantly endure a disproportionate burden of air pollution from coal-fired power plants, exacerbating existing environmental inequalities (Anderson et al. 2018) (Kim et al. 2018) and (Tang et al. 2017).

Disparities in coal consumption between urban and rural areas have been observed, underlining the need to address the unequal distribution of environmental repercussions connected with coal usage (Chen et al. 2018) and (Shao et al. 2020). Efforts to mitigate the environmental and health consequences of coal consumption must be undertaken within the context of environmental justice. The development of clean coal technology, efficient coal-related air pollution control policies, sustainable coal supply chain management, and other measures have been promoted as potential solutions to the environmental and health disparities linked to the use of coal. (Naderi et al. 2020) and (Córdoba 2019). Likewise, the assessment of air pollution indices and the characterisation of gas-phase organics emitted by residential coal combustion have provided a significant understanding into the environmental and health consequences of coal use, particularly in disadvantaged communities (Klein et al. 2018) and (Ghose and Banerjee 2004). Consequently, initiatives to reduce the environmental and health consequences of coal use within an environmental justice framework are critical. Sustainable coal supply chain management, effective coal-related air pollution control regulations, and the development of clean coal technology have been offered as viable solutions to the environmental and health inequities associated with coal use (Naderi et al. 2020).

The above discussions have shown that the use of conventional energy sources has significant implications for environmental justice on a range of levels, including distributive, procedural, and recognition. Conventional energy extraction, such as oil, gas, and coal, frequently contributes to increased environmental injustices. Affected populations bear a disproportionate share of the environmental and health risks associated with traditional energy sources, highlighting the distributive inequality inherent in their use. As a result, addressing these gaps requires comprehensive methods that include environmental justice concerns in policies and regulatory frameworks. Renewable energy sources, on the other hand, have been recognised for their great environmental benefits, contributing to the long-term expansion of human activities while reducing the nasty environmental consequences of conventional fuels (Neelakandan et al. 2021) and (Vaish et al. 2022).

Renewable energy technologies offer several advantages that align with the principles of energy and environmental justice (Levenda et al. 2021). The deployment of renewable energy sources can contribute to mitigating environmental injustices by reducing air and water pollution, as well as minimising the negative impacts associated with fossil fuel extraction and combustion (Kumar 2020). This transition to cleaner energy sources can lead to improved air quality and public health. The expansion of renewable energy infrastructure has the potential to create new employment opportunities and stimulate local economic development (IRENA and ILO 2022). This can contribute to addressing economic injustices by providing job opportunities and supporting the transition to more sustainable and equitable economies (Farghali et al. 2023). In addition, renewable energy projects can enhance energy access and affordability by decentralising energy production and simultaneously assist in addressing energy poverty and energy insecurity, thereby promoting energy justice. (McGee and Greiner 2019). With the principles of distributive justice, the benefits and burdens of renewable energy development are equitably distributed.

3.4 Clean Energy Development: Issues and Challenges

Renewable energy and alternative energy have essential roles to play in advancing energy justice. Wind and solar power, for example, contribute to energy justice by offering clean and sustainable energy solutions that can help alleviate energy poverty and minimise environmental injustice. The expansion of renewable energy can address distributive and procedural justice problems by ensuring equitable access to energy resources and including the community in energy transition decision-making processes (Liebe and Dobers 2020). Nuclear energy, on the other hand, may help energy justice by acting as a stable and low-carbon energy source, particularly in instances when renewable energy may be insufficient (Chen and Li 2022).In addition, nuclear power can alternatively supplement energy for renewable energy power plants, ensuring a more stable and resilient energy supply (Chen and Li 2022).

The German nuclear phase-out is an example of how nuclear energy policy changes might be analysed from an energy justice standpoint. The phase-out represents a change towards environmental justice and is consistent with the aims of encouraging cleaner, safer energy choices (McCauley et al. 2018) and (Rehner and McCauley 2016). It is crucial to emphasise, however, that there are diverse perspectives on the role of nuclear power in attaining energy justice. While some suggest that nuclear power should be abandoned in order to achieve energy justice (McCauley et al. 2018), others emphasise the importance of a balanced strategy that takes into account the specific benefits and constraints connected with both renewable and nuclear energy sources (Chen and Li 2022).

Figure 3.2 shows how the utilisation of renewable and alternative energy, like nuclear power, links to both energy and environmental justice. By addressing equity, access, and sustainability concerns in energy systems, both renewable energy and nuclear power can potentially contribute to energy justice. A holistic approach to energy justice should take into

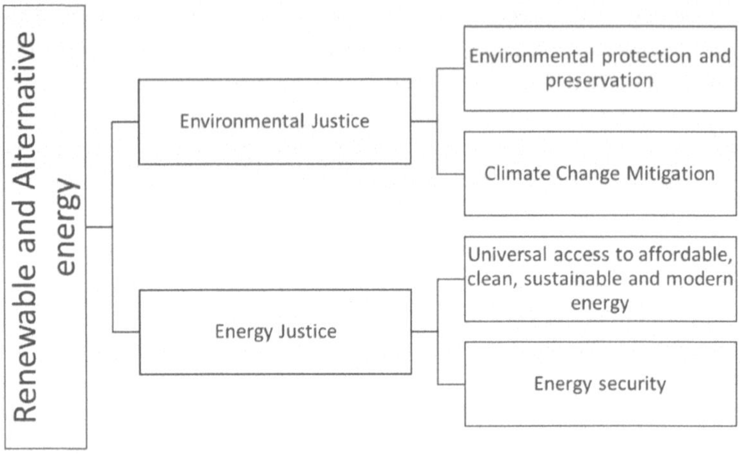

Fig. 3.2 Impacts of renewable and alternative energy power on energy and environmental justice. *Source* Authors

account the distinct characteristics of each energy source and try to establish a balanced and inclusive energy transition that benefits all members of society. Renewable energy and nuclear power play crucial roles in environmental protection and climate change mitigation. The increase in renewable energy sources, such as wind and solar power, makes a substantial contribution to lowering greenhouse gas emissions and mitigating climate change (Luderer et al. 2011) and (Chen et al. 2023). It has been agreed that when renewable energy technologies are incorporated into the energy system, they can accelerate the shift of investments away from high-emission energy sources and towards cleaner alternatives, thus helping climate change mitigation (Luderer et al. 2011) and (Chen et al. 2023).

Nuclear power has the ability to help mitigate climate change due to its low carbon dioxide emissions (Pidgeon et al. 2008). Nuclear power has been identified as an important contribution to climate change mitigation methods, with studies emphasising the need to address financial restrictions, waste disposal, and public acceptability in order to renew its place in the energy mix (Pidgeon et al. 2008). The development of novel nuclear power plants is viewed as a means to contribute to long-term development, environmental protection, and the readiness for rapidly expanding energy demand (Heo et al. 2019). However, the function of nuclear power in climate change mitigation is related to risk perception and confidence in various information sources (Vainio et al. 2016). Additionally, the safe disposal of high-level radioactive waste is critical for ensuring environmental protection and the sustainable development of the nuclear energy industry (Liu et al. 2020).

Many studies revealed that the use of renewable energy resources such as wind and solar power can reduce the adverse impacts on the environment by replacing traditional polluting sources such as coal, gas, and oil (Chen et al. 2023). Nevertheless, solar farms have been the subject of substantial research into their environmental implications (Torres and Petrakopoulou 2022). Large-scale solar farms have been demonstrated to have both beneficial and negative effects on the environment. Issues have been raised about the environmental implications of utility-scale solar energy, emphasising the importance of understanding the interconnections between solar energy systems and the environment (Hernandez et al. 2014) and (Torres and Petrakopoulou 2022). Utility-scale sustainability in solar energy has received significant attention, highlighting the significance of considering ecological concepts to minimise adverse environmental effects. (Moore et al. 2017).

On the other hand, studies have been done on the potential for solar farms to require extensive land preparation and impact habitat conversion and loss of fauna and flora (Pouran et al. 2022). Apart from solar farm, numerous studies have identified potential negative impacts linked with wind energy production. The impact of wind farms on bird and bat populations, notably through mortality events caused by collisions (Barré et al. 2018). The growth of wind energy has aroused worries about its possible negative economic consequences, such as volatility and externalities (Dorrell and Lee 2020). The growth of wind energy projects has been linked to negative effects on vulnerable species and their habitats, raising worries about potential ecological disruption (Lemaître and Lamarre 2020). Aside from environmental issues, the visual impact of wind farms has been a source of disagreement, with public opposition frequently citing the perceived bad influence on the aesthetics of surrounding landscapes as a cause to oppose wind energy development (Hazboun et al. 2016). In respect to marine renewable energy, it is important to note that the installation of marine renewable energy projects may also have adverse impacts on the marine environment, including habitat loss, collision risks, noise, and electromagnetic fields (Siswandi 2017).

While renewable energy sources have considerable potential for environmental protection and mitigation of negative consequences, distinctive environmental concerns linked with different types of renewable energy projects must be considered to guarantee sustainable and responsible deployment. These energy sources provide significant environmental benefits, such as reduced greenhouse gas emissions and the potential displacement of non-renewable energy sources. However, it is critical to address the potential negative impacts associated with energy production to ensure sustainable and responsible deployment so that environmental justice will not be jeopardised by energy needs. Although studies show that renewable energy and alternative energy could better address both energy and environmental injustice, stringent regulations and monitoring systems are required to minimise any adverse impacts of their deployments. Despite the advantages of renewable energy, these technologies also have environmental justice implications, and the incorporation of energy justice principles in decision-making processes is crucial to mitigate negative environmental impacts such as biodiversity loss, natural resource degradation, and pollution. Apart from that, equal distribution of energy services and their costs for electricity access must also be addressed. The integration of social justice into the distribution of the burdens and benefits of renewable and alternative energy is essential, and procedural reforms to policy and planning are suggested to enable greater public expression of justice concerns.

3.5 Conclusion

In conclusion, this chapter delves into the relationship between energy justice and environmental justice, acknowledging significant overlap while also highlighting similarities in their objectives. Environmental justice's primary aims are to preserve the environment and promote clean energy sources, whereas energy justice's main objectives are ensuring people's right to access affordable, abundant, and reliable energy. While renewable energy sources offer significant potential for environmental protection and mitigating negative consequences, it is crucial to address distinct environmental concerns associated with different types of renewable energy projects to ensure their sustainable and responsible deployment. These energy sources offer notable environmental benefits, including reduced greenhouse gas emissions and the displacement of non-renewable energy sources. However, it is imperative to acknowledge potential negative impacts associated with energy production to avoid compromising environmental justice in meeting energy needs. Studies indicate that renewable and alternative energy could better address both energy and environmental injustice, but stringent regulations and monitoring systems are necessary to minimise adverse impacts. Despite the advantages of renewable energy, these technologies also have environmental justice implications, highlighting the importance of integrating energy justice principles into decision-making processes to mitigate negative environmental impacts such as biodiversity loss, natural resource degradation, and pollution. Overall, addressing these factors is essential for ensuring the sustainable usage of renewable energy while safeguarding environmental justice. By addressing these issues, the chapter contributes to the understanding of how energy justice and environmental justice intersect and can be integrated into energy system strategies for a sustainable future.

References

Addy M, Akunyumu S, Simons B (2019) Key risk factors affecting renewable energy independent power producer (IPP) set-up projects in developing countries. J Eng Des Technol 17(6):1340–1355. https://doi.org/10.1108/jedt-03-2019-0072

Ali H, Phoumin H, Suryadi B, Farooque AA, Yaqub R (2022) Assessing ASEAN's liberalised electricity markets: the case of Singapore and the Philippines. Sustain 14(18):11307

Anderson C, Kissel K, Field C, Mach K (2018) Climate change mitigation, air pollution, and environmental justice in California. Environ Sci Technol 52 (18):10829–10838. https://doi.org/10.1021/acs.est.8b00908

Arthur J (2020) Mitigating the environmental effects of oil and gas exploitation: issues of compliance, cost of production, and community awareness. J Power and Energy Eng 08(09):51–64. https://doi.org/10.4236/jpee.2020.89005

Barré K, Viol I, Bas Y, Julliard R, Kerbiriou C (2018) Estimating habitat loss due to wind turbine avoidance by bats: implications for European siting guidance. Biol Cons 226:205–214. https://doi.org/10.1016/j.biocon.2018.07.011

Bousnina R, Gabsi FB (2023) Energy poverty, government expenditure, and institution factors in Sub-Saharan Africa countries: evidence based on a panel threshold model. Environ Sci Pollut Res 30 (24):65512–65526. https://doi.org/10.1007/s11356-023-27005-1

Bouzarovski S (2013) Energy poverty in the European Union: landscapes of vulnerability. Wiley Interdisc Rev Energy Environ 3(3):276–289. https://doi.org/10.1002/wene.89

Burke MJ, Stephens JC (2017) Energy democracy: goals and policy instruments for sociotechnical transitions. Energy Res Soc Sci 33:35–48. https://doi.org/10.1016/j.erss.2017.09.024

Burlinson A, Giulietti M, Battisti G (2018) The elephant in the energy room: establishing the nexus between housing poverty and fuel poverty. Energy Econ 72:135–144. https://doi.org/10.1016/j.eneco.2018.03.036

Burnham A, Han J, Clark C, Wang M, Dunn J, Palou-Rivera I (2011) Life-cycle greenhouse gas emissions of shale gas, natural gas, coal, and petroleum. Environ Sci Technol 46(2):619–627. https://doi.org/10.1021/es201942m

Chen J, Li W (2022) Analysis on the current situation and trend of nuclear energy development. In: Proceedings of the 3rd international conference on green energy, environment and sustainable development (GEESD2022), vol 23. IOS Press, p 277

Chen J, Wu Y, Song M, Dong Y (2018) The residential coal consumption: disparity in urban–rural China. Resour Conserv Recycl 130:60–69. https://doi.org/10.1016/j.resconrec.2017.11.003

Chen XH, Tee K, Elnahass M, Ahmed R (2023) Assessing the environmental impacts of renewable energy sources: a case study on air pollution and carbon emissions in China. J Environ Manage 345:118525. https://doi.org/10.1016/j.jenvman.2023.118525

Córdoba P (2019) Emissions of inorganic trace pollutants from coal power generation. https://doi.org/10.5772/intechopen.79918

Demski C, Evensen D, Pidgeon N, Spence A (2017) Public prioritisation of energy affordability in the UK. Energy Policy 110:404–409. https://doi.org/10.1016/j.enpol.2017.08.044

Diaz-Rainey I, Tulloch DJ, Ahmed I, McCarten M, Taghizadeh-Hesary F (2021) An energy policy for ASEAN? Lessons from the EU experience on energy integration, security, and decarbonisation. In: ADBI working paper series, no 1217. Asian Development Bank Institute

Dorrell J, Lee K (2020) The cost of wind: negative economic effects of global wind energy development. Energies 13(14):3667. https://doi.org/10.3390/en13143667

Dupré la Tour MA (2023) Photovoltaic and wind energy potential in Europe—a systematic review. Renew Sustain Energy Rev 179:113189. https://doi.org/10.1016/j.rser.2023.113189

Enssle F, Kabisch N (2020) Urban green spaces for the social interaction, health and well-being of older people—an integrated view of urban ecosystem services and socio-environmental justice. Environ Sci Policy 109:36–44

Farghali M, Osman AI, Chen Z (2023) Social, environmental, and economic consequences of integrating renewable energies in the electricity sector: a review. Environ Chem Lett 21:1381–1418. https://doi.org/10.1007/s10311-023-01587-1

Feldman D, Bolinger M, Schwabe P (2020) Current and future costs of renewable energy project finance across technologies. Golden, CO: National Renewable Energy Laboratory. NREL/TP-6A20-76881. https://www.nrel.gov/docs/fy20osti/76881.pdf

Finer M, Jenkins C, Pimm S, Keane B, Ross C (2008) Oil and gas projects in the western amazon: threats to wilderness, biodiversity, and indigenous peoples. PLoS ONE 3(8):e2932. https://doi.org/10.1371/journal.pone.0002932

Finley-Brook M, Holloman EL (2016) Empowering energy justice. Int J Environ Res Public Health 13 (9):926

Fry M, Hilburn A (2020) The distributional justice of oil industry social development projects and oil field production activities. Extr Ind Soc 7(2):647–659. https://doi.org/10.1016/j.exis.2020.03.017

Ghose M, Banerjee S (2004) Assessment of air pollution indices for a coal washery project—an Indian case study. Int J Environ Stud 61(5):515–525. https://doi.org/10.1080/0020723042000212609

Hanke F, Guyet R, Feenstra M (2021) Do renewable energy communities deliver energy justice? Exploring insights from 71 European cases. Energy Res Soc Sci 80:102244

Hazboun S, Krannich R, Robertson P (2016) Public views on renewable energy in the rocky mountain region of the united states: distinct attitudes, exposure, and other key predictors of wind energy. Energy Res Soc Sci 21:167–179. https://doi.org/10.1016/j.erss.2016.07.002

Hazrati M, Heffron RJ (2021) Conceptualising restorative justice in the energy transition: changing the perspectives of fossil fuels. Energy Res Soc Sci 78:102115. https://doi.org/10.1016/j.erss.2021.102115

Heffron RJ, McCauley D (2017) The concept of energy justice across the disciplines. Energy Policy 105:658–667

Heffron RJ, Talus K (2016) The evolution of energy law and energy jurisprudence: insights for energy analysts and researchers. Energy Res Soc Sci 19:1–10

Heffron RJ, Rønne A, Tomain JP, Bradbrook A, Talus K (2018) A treatise for energy law. J World Energy Law Bus 11(1):34–48

Heffron RJ (2022) Energy law in crisis: an energy justice revolution needed. J World Energy Law Bus 15 (3):167–172

Hendryx M, Wolfe L, Luo J, Webb B (2011) Self-reported cancer rates in two rural areas of West Virginia with and without mountaintop coal mining. J Community Health 37(2):320–327. https://doi.org/10.1007/s10900-011-9448-5

Heo H, Kim J, Kim S, Kim J, Moon W (2019) Influence of heat treatment on the workability of modified 9cr-2w steel with higher b content. https://doi.org/10.20944/preprints201907.0193.v1

Hernandez R, Easter S, Murphy-Mariscal M, Maestre F, Tavassoli M, Allen E, Barrow C, Belnap J, Ochoa-Hueso R, Allen M (2014) Environmental impacts of utility-scale solar energy. Renew Sustain Energy Rev 29:766–779. https://doi.org/10.1016/j.rser.2013.08.041

Hess CEE, Ribeiro WC (2016) Energy and environmental justice: closing the gap. Environ Justice 9(5):153–158

Hess DJ, McKane RG, Pietzryk C (2022) End of the line: environmental justice, energy justice, and opposition to power lines. Environ Politics 31(4):663–683

Hoque M, Rashid F (2021) Gasification process using downdraft fixed-bed gasifier for different feedstock. https://doi.org/10.5772/intechopen.96227

Huseynov A, Huseynov E, Samusevych Y (2021) Innovative development of oil and gas industry: role of environmental taxation. Mark Manage Innov 5(4):79–91. https://doi.org/10.21272/mmi.2021.4-07

Institute for Democracy and Economic Affairs (IDEAS) (2023). Brief IDEAS No. 39: competitive neutrality in the Malaysian power sector: removing barriers for a greener and more innovative energy industry. https://www.ideas.org.my/ideas-to-achieve-net-zero-emissions-in-the-near-future-malaysias-energy-industry-players-must-compete-in-a-more-level-playing-field/

IRENA, ILO (2022) Renewable energy and jobs: annual review 2022. International Renewable Energy Agency. Abu Dhabi and International Labour Organization, Geneva

Jenkins K (2018) Setting energy justice apart from the crowd: lessons from environmental and climate justice. Energy Res Soc Sci 39:117–121

Jenkins K, McCauley D, Heffron R, Stephan H, Rehner R (2016) Energy justice: a conceptual review. Energy Res Soc Sci 11:174–182

Jiang H, Zhang Y, Qi Y, Zeng Y (2016) The life cycle assessment of the coal, oil, and LNG. https://doi.org/10.2991/aeecs-16.2016.8

Kabeyi MJB, Olanrewaju OA (2022) Sustainable energy transition for renewable and low carbon grid electricity generation and supply. Front Energy Res 9:1032

Keady W, Panikkar B, Nelson IL, Zia A (2021) Energy justice gaps in renewable energy transition policy initiatives in Vermont. Energy Policy 159:112608

Kim G, Lee H, Yoo S (2018) Willingness to pay for substituting coal with natural gas-based combined heat and power in South Korea: a view from air pollutants emissions mitigation. Sustainability 10(5):1554. https://doi.org/10.3390/su10051554

Kim S, Jacome V, Pellow D, Deshmukh R (2023) Evaluating equity and justice in low-carbon energy transitions. Environ Res Lett 18(12):123003. IOP Publishing. https://doi.org/10.1088/1748-9326/ad08f8

Klein F, Pieber S, Ni H, Stefenelli G, Bertrand A, Kilic D, Prévôt A (2018) Characterisation of gas-phase organics using proton transfer reaction time-of-flight mass spectrometry: residential coal combustion. Environ Sci Technol 52(5):2612–2617. https://doi.org/10.1021/acs.est.7b03960

Kumar M (2020) Social, economic, and environmental impacts of renewable energy resources. Wind Sol Hybrid Renew Energy Syst. IntechOpen. https://doi.org/10.5772/intechopen.89494

Lai HL (2022) Foregrounding the community: geo-historical entanglements of community energy, environmental justice, and place in Taihsi Village, Taiwan. Environ Plan E Nat Space 5(2):666–693

Lee J, Byrne J (2019) Expanding the conceptual and analytical basis of energy justice: beyond the three-tenet framework. Front Energy Res 7. https://doi.org/10.3389/fenrg.2019.00099

Lemaître J, Lamarre V (2020) Effects of wind energy production on a threatened species, the Bicknell's Thrush Catharus Bicknelli, with and without mitigation. Bird Conserv Int 30(2):194–209. https://doi.org/10.1017/s095927092000012x

Levenda AM, Behrsin I, Disano F (2021) Renewable energy for whom? A global systematic review of the

environmental justice implications of renewable energy technologies. Energy Res Soc Sci 71:101837
Liebe U, Dobers G (2020) Measurement of fairness perceptions in energy transition research: a factorial survey approach. Sustainability 12(19):8084. https://doi.org/10.3390/su12198084
Liu C, Zhang Q, Kang D, Wen X, Jiao Y (2020) Development and application of an intelligent test system for the model test on deep underground rock caverns. Energies 13(2):358. https://doi.org/10.3390/en13020358
Luderer G, Bosetti V, Jakob M, Leimbach M, Steckel J, Waisman H, Edenhofer O (2011) The economics of decarbonising the energy system—results and insights from the recipe model intercomparison. Clim Change 114(1):9–37. https://doi.org/10.1007/s10584-011-0105-x
Mangula M, Kuzilwa J, Msanjila S, Legonda I (2018) Indicators of energy access in rural areas of Tanzania: an application of confirmatory factor analysis approach. Independent J Manag Prod 9(4):1068. https://doi.org/10.14807/ijmp.v9i4.797
Marchand R, Genovese A, Koh SL, Brennan A (2019) Examining the relationship between energy poverty and measures of deprivation. Energy Policy 130:206–217. https://doi.org/10.1016/j.enpol.2019.03.026
Martin A, Armijos, MT, Coolsaet B, Dawson N, Edwards G, Few R, Gross-Camp N, Rodriguez I, Schroeder H, GL Tebbboth, White CS (2020) Environmental justice and transformations to sustainability. Environ Sci Policy Sustain Dev 62(6):19–30
Mas'ud A, Yusuf R, Udin N, Al-Dhamari R (2020) Enforce environmental tax compliance model for the oil and gas industry. Int J Energy Sect Manage 14 (6):1073–1088. https://doi.org/10.1108/ijesm-11-2019-0005
McCauley D, Brown A, Rehner R, Heffron R, Graaff S (2018) Energy justice and policy change: an historical political analysis of the German nuclear phase-out. Appl Energy 228:317–323. https://doi.org/10.1016/j.apenergy.2018.06.093
McCauley DA, Heffron RJ, Stephan H, Jenkins K (2013) Advancing energy justice: the triumvirate of tenets. Int Energy Law Rev 32(3):107–110
McGee JA, Greiner PT (2019) Renewable energy injustice: the socio-environmental implications of renewable energy consumption. Energy Res Soc Sci 56:101214. https://doi.org/10.1016/j.erss.2019.05.024
Milčiuvienė S, Kiršienė J, Doheijo E, Urbonas R, Milčius D (2019) The role of renewable energy prosumers in implementing energy justice theory. Sustain 11 (19):5286
Mittal V, Shah R, Huq N (2021) The future of coal as a transportation fuel. Energy Environ 33(7):1265–1278. https://doi.org/10.1177/0958305x211037973
Moore-O'Leary K, Hernandez R, Johnston D, Abella S, Tanner K, Swanson A, Lovich J (2017) Sustainability of utility-scale solar energy—critical ecological concepts. Front Ecol Environ 15(7):385–394. https://doi.org/10.1002/fee.1517
Muzanni A, Soesilo T, Martono D, Hamzah U, Berkademi W, Febraldo D (2022) Review of environmental management system implementation in the oil and gas company in Indonesia. IOP Conf Series Earth Environ Sci 1111(1):012018. https://doi.org/10.1088/1755-1315/1111/1/012018
Naderi R, Nikabadi M, Alem-Tabriz A, Pishvaee M (2020) Sustainable coal supply chain management using exergy analysis and genetic algorithm. Manage Syst Prod Eng 29(1):44–53. https://doi.org/10.2478/mspe-2021-0006
Najm S, Matsumoto KI (2020) Does renewable energy substitute LNG international trade in the energy transition? Energy Econ 92:104964
Neelakandan T, Sridharan S, Ramrao N (2021) Harnessing renewable energy at Kalasalingam Academy of Research and Education—a role model case. J Sustain Perspect 1(1). https://doi.org/10.14710/jsp.2021.11201
Nriagu J, Udofia E, Ekong I, Ebuk G (2016) Health risks associated with oil pollution in the Niger Delta, Nigeria. Int J Environ Res Public Health 13(3):346. https://doi.org/10.3390/ijerph13030346
Opal A, Nathwani J (2023) Global energy transition risks: evaluating the intergenerational equity of energy transition costs. In: Nadesan M, Pasqualetti MJ, Keahey J (eds) Energy democracies for sustainable futures. Academic Press, pp 301–310. https://doi.org/10.1016/B978-0-12-822796-1.00032-2
O'Rourke D, Connolly S (2003) Just oil? The distribution of environmental and social impacts of oil production and consumption. Annu Rev Environ Resour 28 (1):587–617. https://doi.org/10.1146/annurev.energy.28.050302.105617
Pata UK (2021) Linking renewable energy, globalisation, agriculture, CO_2 emissions and ecological footprint in BRIC countries: a sustainability perspective. Renew Energy 173:197–208
Pesch U, Correljé A, Cuppen E, Taebi B (2017) Energy justice and controversies: formal and informal assessment in energy projects. Energy Policy 109:825–834
Pidgeon N, Lorenzoni I, Poortinga W (2008) Climate change or nuclear power—no thanks! A quantitative study of public perceptions and risk framing in Britain. Glob Environ Chang 18(1):69–85. https://doi.org/10.1016/j.gloenvcha.2007.09.005
Pollit MG (2012) The role of policy in energy transitions: lessons from the energy liberalisation era. Energy Policy 50:128–137
Pouran H, Lopes M, Nogueira T, Branco D, Sheng Y (2022) Environmental and technical impacts of floating photovoltaic plants as an emerging clean energy technology. Iscience 25(11):105253. https://doi.org/10.1016/j.isci.2022.105253
Qin Y, Zhou M, Pan D, Klimont Z, Gingerich D, Mauzerall D, Bielicki J (2021) Environmental consequences of potential strategies for China to prepare for natural gas import disruptions. Environ Sci Technol 56(2):1183–1193. https://doi.org/10.1021/acs.est.1c03685

Rehner R, McCauley D (2016) Security, justice and the energy crossroads: assessing the implications of the nuclear phase-out in Germany. Energy Policy 88:289–298. https://doi.org/10.1016/j.enpol.2015.10.038

Ristic M, Radovanovic L, Perišić J, Vasović I, Đorđević L (2021) Analysis of the use of renewable energy sources in the republic of Serbia, pp 67–83. https://doi.org/10.1007/978-3-030-86009-7_4

Ristovska K, Ristovska A (2014) The impact of globalization on the business. Econ Analysis EA 47(3/4):83–89

Savić I (2019) Universal access to energy and sustainable development, pp 1–10. https://doi.org/10.1007/978-3-319-71057-0_54-1

Setyowati AB (2021) Mitigating inequality with emissions? Exploring energy justice and financing transitions to low carbon energy in Indonesia. Energy Res Soc Sci 71:101817

Seuret-Jiménez D, Robles-Bonilla T, Cedano K (2020) Measurement of energy access using fuzzy logic. Energies 13(12):3266. https://doi.org/10.3390/en13123266

Shahzad U (2020) Environmental taxes, energy consumption, and environmental quality: theoretical survey with policy implications. Environ Sci Pollut Res 27(20):24848–24862

Shao S, Chang W, Guo Y, Yang L, Chen S, Yan J, Guan D (2020) Enlarging regional disparities in energy intensity within China. Earth S Future 8(8). https://doi.org/10.1029/2020ef001572

Shi R, Meacham S, Davis G, You W, Sun Y, Goessl C (2019) Factors influencing high respiratory mortality in coal-mining counties: a repeated cross-sectional study. BMC Public Health 19(1). https://doi.org/10.1186/s12889-019-7858-y

Siswandi A (2017) Marine renewable energy, the law of the sea and the marine environment: an Indonesian perspective. Padjadjaran J Int Law 1(1):36–50. https://doi.org/10.23920/pjil.v1i1.274

Smith S (2021) Mobilising heterogeneity: understanding the dynamic qualities and diversity of energy access in Maputo. Urban Geogr 44(1):221–242. https://doi.org/10.1080/02723638.2021.1959769

Stojilovska A (2021) 'Energy poverty and the role of institutions: exploring procedural energy justice—ombudsman in focus. J Environ Planning Policy Manage 25(2):169–181. https://doi.org/10.1080/1523908x.2021.1940895

Svarstad H, Benjaminsen TA (2020) Reading radical environmental justice through a political ecology lens. Geoforum 108:1–11

Tang Q, Chen X, Yu L, Xue W, Xia X (2017) Estimation of the effectiveness of the current policies on coal-related air pollution control. Destech Trans Environ Energy Earth Sci (edep). https://doi.org/10.12783/dteees/edep2017/15558

Thomson H, Snell C, Bouzarovski S (2017) 'Health, well-being and energy poverty in Europe: a comparative study of 32 European countries. Int J Environ Res Public Health 14(6):584. https://doi.org/10.3390/ijerph14060584

Tomei J, Cronin J, Arias H, Machado S, Palacios M, Ortiz Y, Anandarajah G (2020) Forgotten spaces: how reliability, affordability and engagement shape the outcomes of last-mile electrification in Chocó, Colombia. Energy Res Soc Sci 59:101302. https://doi.org/10.1016/j.erss.2019.101302

Torres JF, Petrakopoulou F (2022) A closer look at the environmental impact of solar and wind energy. Global Chall 6(8):2200016. https://doi.org/10.1002/gch2.202200016

Tuğrul A (2017) Energy security and security relationship. Int RelatS Dipl 5(7). https://doi.org/10.17265/2328-2134/2017.07.005

Ullah S, Khan M, Yoon S (2021) Measuring energy poverty and its impact on economic growth in Pakistan. https://doi.org/10.20944/preprints202108.0314.v1

UNDP (2016) Delivering sustainable energy in a changing climate strategy note on sustainable energy 2017–2021. United Nations Development Programme. https://www.google.com/url?sa=t&rct=j&q=&esrc=s&source=web&cd=&ved=2ahUKEwjC2uex3ZqEAxVUR2cHHTR8B484ChAWegQIBhAB&url=https%3A%2F%2Fwww.un-energy.org%2Fwp-content%2Fuploads%2F2017%2F01%2FUNDP-Energy-Strategy-2017-2021.pdf&usg=AOvVaw0So_hPVN5gX-vHIG7D7scF&opi=89978449

UNEP (2022) The closing window: climate crisis calls for rapid transformation of societies. Emissions Gap Report 2022. United Nations Environmental Programme. https://www.unep.org/resources/emissions-gap-report-2022

Vainio A, Paloniemi R, Varho V (2016) Weighing the risks of nuclear energy and climate change: trust in different information sources, perceived risks, and willingness to pay for alternatives to nuclear power. Risk Anal 37(3):557–569. https://doi.org/10.1111/risa.12640

Vaish S, Sharma N, Kaur G (2022) Equivalent models of solar photovoltaic cells and discuss the effects of resistance on maximum output power curve. IOP Conf Series Mater Sci Eng 1228(1):012031. https://doi.org/10.1088/1757-899x/1228/1/012031

Van Uffelen N (2022) Revisiting recognition in energy justice. Energy Res Soc Sci 92:102764

Wallsgrove R (2022) Restorative energy justice. UCLA J Environ Law Policy 40(2). https://doi.org/10.5070/l540257928

Wang G, Guan J, Li L (2019) Energy justice and construction of community with a shared future for mankind. Energy Justice Across Borders, 217–235. https://doi.org/10.1007/978-3-030-24021-9_11

Wang X, Lo K (2023) Energy and environmental justice in China: literature review and research agenda. J Asian Energy Stud 7:91–106

Wood N, Roelich K (2019) Tensions, capabilities, and justice in climate change mitigation of fossil fuels. Energy Res Soc Sci 52:114–122

Wood N, Roelich K (2020) Substantiating energy justice: creating a space to understand energy dilemmas. Sustain 12(5):1917

Zhang Y, Chen Z, Chen H, Mao J (2023) Exploring the impact of "double carbon target" on environmental efficiency of coal cities in China. Front Environ Sci 11. https://doi.org/10.3389/fenvs.2023.1132251

Maizatun Mustafa is an Associate Professor with the Legal Practice Department, Ahmad Ibrahim Kulliyyah of Laws IIUM. She obtained her LL.B. in 1992, M.C.L. in 1993 from IIUM, and Ph.D. in 2001 from the School of Oriental and African Studies University of London. Her areas of expertise are environmental law and climate. She is the coordinator for the Climate Change Policy and Law Centre IIUM. In 2020, her book project on climate change litigation with the National University of Singapore and Yale University has been published by Cambridge University Press. In 2022, she published the fifth edition of her book entitled "Environmental Law Malaysia" with Wolters Kluwer. Dr. Maizatun involves as a trainer in various environmental programs including training Malaysia's judges on environmental law; training of legal and judicial officers on climate change; and training on ESG for the state government's listed companies. At the International level, she is a member of the International Union for Conservation of Nature (IUCN) Academy of Environmental Law, and the IUCN World Commission on Environmental Law.

Farahdilah Ghazali is a Senior Lecturer at School of Business and Economics, Universiti Putra Malaysia (UPM). Previously, she served at the Universiti Malaysia Terengganu (UMT) and Universiti Tenaga Nasional Malaysia (UNITEN). She obtained her PhD in Law from Ahmad Ibrahim Kulliyyah of Laws, International Islamic University Malaysia (IIUM). Her areas of interest are natural resources law and sustainability, particularly renewable energy law and policy. She has been appointed as a technical member for the enactment of the Malaysian Legal Framework on Seabed Mining in 2019. Being a passionate researcher and academic, she has widely published in peer-reviewed journals as well as book chapters and presented papers at several national and international conferences in her areas of interest.

Part II

Solutions and Strategies for Energy Justice

4 Patterns of Global Disparities in Access to Electricity and Clean Cooking Fuels

Rajesh Ramachandran

Abstract

Access to modern energy sources is a crucial aspect of energy justice. Utilizing data from 66 countries, covering over 1.67 million households from the Demographic and Health Survey, I provide insights into the percentage of households with access to modern energy sources. The findings underscore persistent challenges in accessing electricity and clean cooking fuels, particularly in Sub-Saharan Africa (SSA). Subsequently, I explore the association between socioeconomic status, rural residence, education and the gender of household heads with access to these two energy sources. Socioeconomic status and urban location are significant correlates of access to modern energy sources, even when comparing households within the same regions in the country. Additionally, households led by more educated individuals and women exhibit higher rates of using modern energy sources, highlighting the importance of education and gender in promoting energy access and equity.

Keywords

Modern energy sources · Energy justice · Energy access and equity · Sustainable development

4.1 Introduction

Energy justice is a multidimensional concept that comprises different aspects of fairness and equality. Essentially, it aims to ensure that everyone, regardless of their socioeconomic status or geographical location, has access to clean and affordable energy (Sovacool and Dworkin 2014).

Energy justice has three main principles: distributional justice, procedural justice and recognition justice. Distributional justice is concerned with sharing energy resources and advantages. The goal is to make sure no community or group carries a heavier burden from the downsides of energy production or misses out on its benefits. The other key aspect of distributional justice, which is the principal focus of the chapter, is concerned with the consumption perspective.[1] In particular, the agenda of fuel poverty which highlights the lack of affordable access to energy services (Jenkins et al. 2016). Having access to energy that is both affordable and sustainable is not just about fairness in society; it is also a necessary condition for attaining sustainable development and addressing the challenges of climate change.

R. Ramachandran (✉)
Monash University Malaysia,
Kuala Lumpur, Malaysia
e-mail: rajesh.ramachandran@monash.edu

A. Trakic et al. (eds.), *Energy Justice*, Sustainable Development Goals Series,
https://doi.org/10.1007/978-981-97-6059-6_4

Two crucial aspects of energy access are access to electricity and clean cooking fuels. Access to electricity is vital for various aspects of human well-being and societal development. It is a key driver of economic growth, and it directly correlates with improvements in income levels and living standards (Burke et al. 2018). Electricity can serve a crucial role in improving education, fostering the growth of businesses and enterprises, and supporting essential activities like lighting, refrigeration and the operation of household appliances. These functions are integral and not easily substituted by alternative forms of energy. The World Health Organization (WHO) emphasizes the importance of electricity for health care, stating that it is essential for the effective functioning of health systems and the delivery of medical services (WHO 2023).

Access to clean cooking fuels is of paramount importance for both public health and environmental sustainability. Traditional cooking methods, often reliant on solid fuels like wood and charcoal, contribute to indoor air pollution, leading to respiratory diseases and premature deaths (Hystad et al. 2019). According to the World Health Organization (WHO), over 3 billion people globally still rely on traditional cooking methods, and the resulting indoor air pollution is a major health risk, particularly for women and children. Clean cooking fuels, such as liquefied petroleum gas (LPG), bio-gas, electricity and natural gas offer a healthier alternative by producing fewer harmful emissions. Clean cooking solutions have been highlighted as essential in preventing respiratory diseases and reducing environmental impact (Anenberg et al. 2013). Transitioning to cleaner fuels not only improves air quality but also mitigates deforestation, as it decreases the demand for wood and reduces the associated environmental degradation (Jagger and Kittner 2017).

Today, however, a substantial number of people still lack access to electricity and clean cooking fuel. In 2019, approximately 791 million people still lacked access to electricity, with a significant majority residing in Sub-Saharan Africa (SSA). Regarding clean cooking fuel, the issue is even more extensive, as over 3 billion people lacked access as of 2016, affecting both Africa and Asia (Ritchie et al. 2019).

Any policy that wants to promote energy justice needs to address existing shortfalls in energy access. However, effectively addressing these gaps necessitates a comprehensive understanding of the present state of access. This chapter makes a contribution in this regard by presenting the latest estimates on challenges of access to energy sources drawing on the latest round of the demographic and health surveys from 66 countries from six regions across the globe (ICF 2005–2022). Based on data on more than 1.6 million households, I first present country-level estimates on the proportion of households having access to electricity and clean cooking fuel. Two, I explore how this access within countries, or by regions within a country, varies by socioeconomic status, rurality, education and gender of the household head. The findings indicate persistent challenges in access to electricity, particularly in Sub-Saharan Africa (SSA). The situation concerning clean cooking fuel is even more daunting, with considerable deficiencies observed across all regions covered by our data. The findings also highlight how socioeconomic status, geographical location and sex of household head continue to remain important correlates of access to modern energy sources even when comparing households living within the same regions in the country.

4.2 Data and Method

The primary data source to obtain a picture of the share of households who have access to electricity and clean cooking fuels is the demographic and health surveys (DHS) (ICF 2005–2022). I employ the latest round of the household surveys for each country conditional on the data being from 2005 or afterwards. The data set comprises 67 countries, encompassing a total of 1.67 million households. Table 4.1 shows the survey year along with the corresponding number of households in the sample for each country.

We consider two key proxies for capturing energy justice. The first is on access to electricity.

Table 4.1 Sample description

Country	No. of households	Survey year	Region
Afghanistan	24,395	2015	South Asia
Albania	15,823	2018	Europe and Central Asia
Angola	16,109	2016	Sub-Saharan Africa
Armenia	7893	2016	Europe and Central Asia
Azerbaijan	7180	2006	Europe and Central Asia
Bangladesh	19,457	2018	South Asia
Benin	14,156	2018	Sub-Saharan Africa
Bolivia	19,564	2008	Latin America and the Caribbean
Burkina Faso	13,251	2021	Sub-Saharan Africa
Burundi	15,977	2017	Sub-Saharan Africa
Cambodia	20,806	2022	East Asia and the Pacific
Cameroon	11,710	2019	Sub-Saharan Africa
Chad	17,233	2015	Sub-Saharan Africa
Colombia	44,614	2016	Latin America and the Caribbean
Comoros	4482	2012	Sub-Saharan Africa
Congo	11,632	2012	Sub-Saharan Africa
D.R. Congo	18,171	2014	Sub-Saharan Africa
Cote d'Ivoire	14,766	2021	Sub-Saharan Africa
Dominican Republic	14,033	2013	Latin America and the Caribbean
Egypt	28,175	2014	Sub-Saharan Africa
Ethiopia	8663	2011	Sub-Saharan Africa
Gabon	11,781	2021	Sub-Saharan Africa
Gambia	6549	2020	Sub-Saharan Africa
Ghana	11,835	2014	Sub-Saharan Africa
Guatemala	21,383	2015	Latin America and the Caribbean
Guinea	7912	2018	Sub-Saharan Africa
Guyana	5632	2009	Latin America and the Caribbean
Haiti	13,405	2017	Latin America and the Caribbean
Honduras	21,362	2012	Latin America and the Caribbean
India	636,699	2021	South Asia
Indonesia	47,963	2017	East Asia and the Pacific
Jordan	18,802	2018	Middle East and North Africa
Kenya	37,911	2022	Sub-Saharan Africa
Kyrgyz Republic	8040	2012	Europe and Central Asia
Lesotho	9402	2014	Sub-Saharan Africa
Liberia	9068	2020	Sub-Saharan Africa
Madagascar	20,510	2021	Sub-Saharan Africa
Malawi	26,361	2016	Sub-Saharan Africa
Maldives	6050	2017	South Asia

(continued)

Table 4.1 (continued)

Country	No. of households	Survey year	Region
Mali	9510	2018	Sub-Saharan Africa
Mauritania	11,658	2021	Sub-Saharan Africa
Moldova	11,095	2005	Europe and Central Asia
Mozambique	13,919	2011	Sub-Saharan Africa
Myanmar	12,500	2016	South Asia
Namibia	9849	2013	Sub-Saharan Africa
Nepal	13,786	2022	South Asia
Niger	10,750	2012	Sub-Saharan Africa
Nigeria	40,427	2018	Sub-Saharan Africa
Pakistan	14,540	2018	South Asia
Papua New Guinea	16,021	2018	East Asia and the Pacific
Peru	27,218	2012	Latin America and the Caribbean
Philippines	30,372	2022	East Asia and the Pacific
Rwanda	12,949	2020	Sub-Saharan Africa
Sao Tome and Principe	3536	2009	Sub-Saharan Africa
Senegal	4538	2019	Sub-Saharan Africa
Sierra Leone	13,399	2019	Sub-Saharan Africa
South Africa	11,083	2016	Sub-Saharan Africa
Tajikistan	7843	2017	Europe and Central Asia
Tanzania	15,705	2022	Sub-Saharan Africa
Timor-L'este	11,502	2016	East Asia and the Pacific
Togo	9549	2014	Sub-Saharan Africa
Uganda	19,588	2016	Sub-Saharan Africa
Ukraine	13,379	2007	Europe and Central Asia
Yemen	17,351	2013	Middle East and North Africa
Zambia	12,831	2019	Sub-Saharan Africa
Zimbabwe	10,534	2015	Sub-Saharan Africa
Total	1,665,243		

Notes The table shows the number of households and the year of survey for the list of countries included in our sample from the demographic and health survey (DHS)

The DHS asks each household whether it has electricity. We thus create a binary indicator which captures the share of households with access to electricity by country, along with the 95% confidence intervals. The second proxy for energy justice captures the dimension of the energy source employed for cooking purposes. The DHS asks households about the type of cooking fuel used. We classify households which report using electricity, liquefied petroleum gas, natural gas, or bio-gas as having access to clean cooking fuel. We again create a binary indicator which captures the share of households with access to electricity by country, along with the 95% confidence intervals.

The differences across countries provide a picture of the extent of disparities on two key proxies of energy justice across our sample of 67 countries. In the next step, I consider three *within* country features that affect access to electricity and clean cooking fuel.

The first is the wealth index of the household. According to the DHS, "The wealth index is a composite measure of a household's cumulative living standard. The wealth index is calculated using easy-to-collect data on a household's ownership of selected assets, such as televisions and bicycles; materials used for housing construction; and types of water access and sanitation facilities."[2] The DHS separates all interviewed households into five wealth quintiles and I explore how access for households varies by quintiles of wealth. The five wealth quintiles are labelled as follows: Quintile 1, that is, the bottom 20%-poorest; Quintile 2-poorer; Quintile 3-middle; Quintile 4-richer; Quintile 5, that is, the top 20%-richest.

Second, we explore how access varies by rural-urban residence. Infrastructure challenges along with the dispersion of population and often challenging geographical terrain can make it logistically difficult and economically less viable to extend power grids and fuel supply networks to rural areas. Limited financial resources in rural households can further impede their ability to afford and access modern energy services (Mainali et al. 2014; Puzzolo et al. 2019). On the other hand, urban areas generally benefit from more concentrated populations, better infrastructure and increased economic opportunities, making it comparatively easier to deploy and maintain electricity and clean cooking fuel infrastructure. We thus explore how access to electricity and clean cooking fuel varies by rural-urban residence.

The final facet we consider is the sex of the household. Gender dynamics often influence decision-making processes within households, impacting resource allocation and priorities. In many societies, female-headed households may face additional challenges, such as limited economic empowerment and societal constraints, which can hinder their ability to access and afford modern energy services (Oparaocha and Dutta 2011).

To analyse how these features—wealth, rural residence and sex of household head—moderate access to electricity and clean cooking fuel, we classify these 66 countries into six regions based on the regional classification of the World Bank. These are (i) East Asia and Pacific (EAP); (ii) Europe and Central Asia (ECA); (iii) Latin America and the Caribbean (LAC); (iv) Middle East and North Africa (MENA); (v) South Asia (SA); and (vi) Sub-Saharan Africa (SSA). The 66 countries classification to the six regions is given in table. We proceed by presenting regional-level averages, accompanied by 95% confidence intervals, for access to electricity and clean cooking fuel. These averages are stratified by wealth, rural-urban residence and the sex of the household, conditional on country-level dummies. In other words, the averages are derived by leveraging differences within each country.

In a final step, we estimate the following equation to explore the association between wealth, rural residence, gender and education of household head and outcomes of access to modern and clean sources of energy:

$$\begin{aligned} Y_{\text{hrc}} = {} & \alpha + \beta_j \text{Wealth\_Quintiles}_{\text{hrc}} + \gamma \text{Rural}_{\text{hrc}} \\ & + \zeta \text{Female\_Head}_{\text{hrc}} + \eta_{\text{rc}} + \delta_k \text{Educ\_Catg}_{\text{hrc}} \\ & + \pi \text{Age\_Head}_{\text{hrc}} + \varepsilon_{\text{hrc}} \end{aligned} \tag{4.1}$$

where Y_{hrc} is a dummy variable capturing whether household h in region r and country c has access to electricity or clean cooking fuel. The variable η_{rc} is a set of country dummies or administrative unit one fixed effects; in other words, Eq. 4.1 estimates are based on comparing households resident within the same country or within the same administrative division of the country.

The first coefficients of interest are β_j's, which captures the conditional effect of belonging to one of the four wealth quintiles on access to electricity or clean cooking fuel with the omitted category being the bottom 20% of the population on the wealth index. The second and third are γ and ζ which capture the effect of rural residence and having a female household head on the two outcomes capturing access to modern energy sources. The final coefficient of interest is δ_k's which captures the conditional effect of the household head belonging to one of the four education categories with the omitted category being no schooling. The standard errors are estimated, accounting for the correlation between error terms at the level of the primary sampling unit.

4.3 Access to Electricity

4.3.1 Country-Level Estimates

Figures 4.1 shows the proportion of households reporting access to electricity in East Asia and Pacific (EAP); Europe and Central Asia (ECA); Latin America and the Caribbean (LAC); Middle East and North Africa (MENA) and South Asia (SA) and the left-hand panel of Fig. 4.2 for Sub-Saharan Africa (SSA).

For the four countries in East Asia, the access varies from a low of 73% in Timor to a high of 97% in Indonesia. Papua New Guinea, in the Pacific, stands out with only 15% of households having access to electricity. Turning to Europe and Central Asia, the six countries covered are the the former Soviet Republic states and we observe access is almost universal in all the countries with the lowest being Tajikistan, where 99% of households have access to electricity. For the LAC region, we have seven countries in the

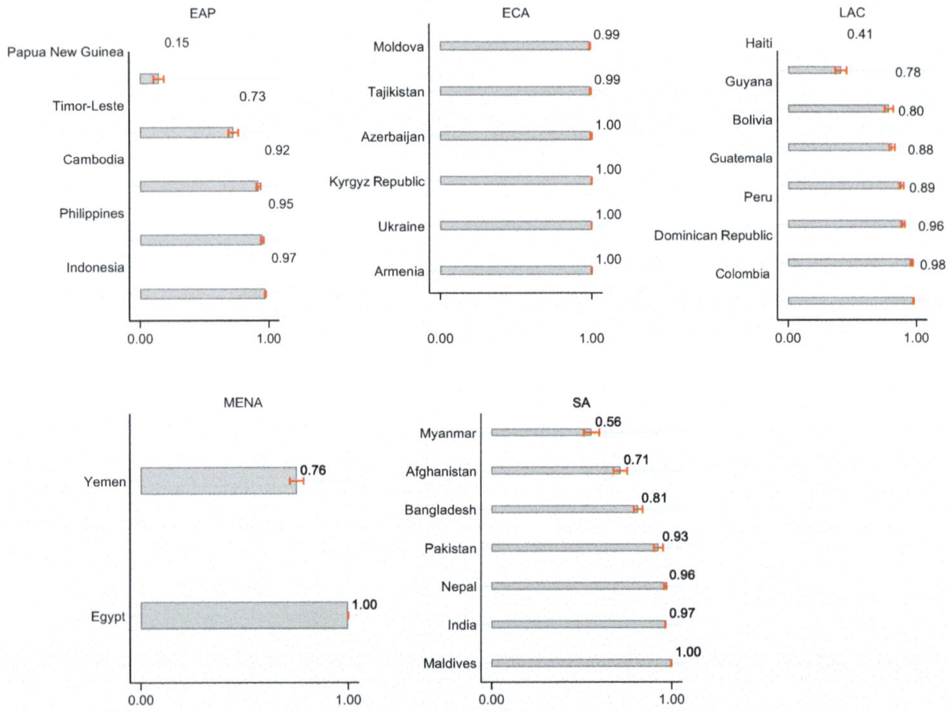

Fig. 4.1 Access to electricity for households—country-level estimates by region. *Notes* See Table 4.1 for year of survey and number of observations for each country. The 95% confidence intervals are calculated by accounting for correlation at the primary sampling unit level. EAP refers to East Asia and Pacific; ECA refers to Europe and Central Asia; LAC refers to Latin America and the Caribbean; MENA refers to Middle East and North Africa; SA refers to South Asia; SSA refers to Sub-Saharan Africa

Fig. 4.2 Access to electricity and clean cooking fuel for households in Sub-Saharan Africa. *Notes* See Table 4.1 for year of survey and number of observations for each country. The 95% confidence intervals are calculated by accounting for correlation at the primary sampling unit level

data set. In Haiti, one of the poorest countries in the Western.

Hemisphere, only 41% of households are seen to have access to electricity. The rest of the LAC countries are seen to have much better access with 78% of households in Guyana, 88 and 89% in Peru and Guatemala, respectively, and 98% in Colombia.

We only have two countries in the data set from the MENA region; we see that 76% of households in Yemen have access whereas in Egypt access is universal. There are seven countries from South Asia present in our data. Myanmar is seen to have the lowest access to electricity with 56% of households having access to electricity. In Afghanistan and Bangladesh, the commensurate figure is 71 and 81%, respectively. In India, Nepal and Pakistan the figure exceeds 90%, and in Maldives, the access is seen to be universal.

The broadest data coverage is for the continent of SSA, where we have data on 40 countries. It is seen to be the region where electricity access is still far from universal. The mean and median share of households having access to electricity is 41% and 0%, respectively. Looking at it by country, less than 10% of households in Chad, Burundi and Malawi have access to electricity. In Ethiopia, one of the fastest growing countries in SSA over the last two decades, still only 35% of households have access. In Nigeria, the most populous country in the continent, 59% of households have access. Senegal, Ivory Coast and Ghana all have between 70 and 80% of households having access to electricity and only in South Africa and Gabon does access exceed 90%.

4.3.2 Wealth, Rurality and Gender

Figures 4.3 and 4.4 show how access varies by wealth, rural-urban residence and sex of the household head. These present region-wide estimates based on averages of *within* country differences.[3]

The figures show that wealth or socioeconomic status plays an important role: in EAP 65% of the households in the bottom-most quintile of the wealth index have access to electricity and this increases to 96% for the top quintile. In ECA, access does not vary by wealth, which is to be expected as coverage is universal. In LAC, the proportion of households with access to electricity for the five quintiles of the wealth index is 57, 85, 93, 98 and 100%, respectively. Thus, more than 40% of the poorest households still lack access to electricity. In South Asia, wealth is seen to play a smaller role with 82% of households in the bottom-most quintile and 100% of households in the top quintile having access to electricity. Finally, in SSA for the five quintiles of the wealth index, 7, 16, 33, 59 and 88%, respectively, report having

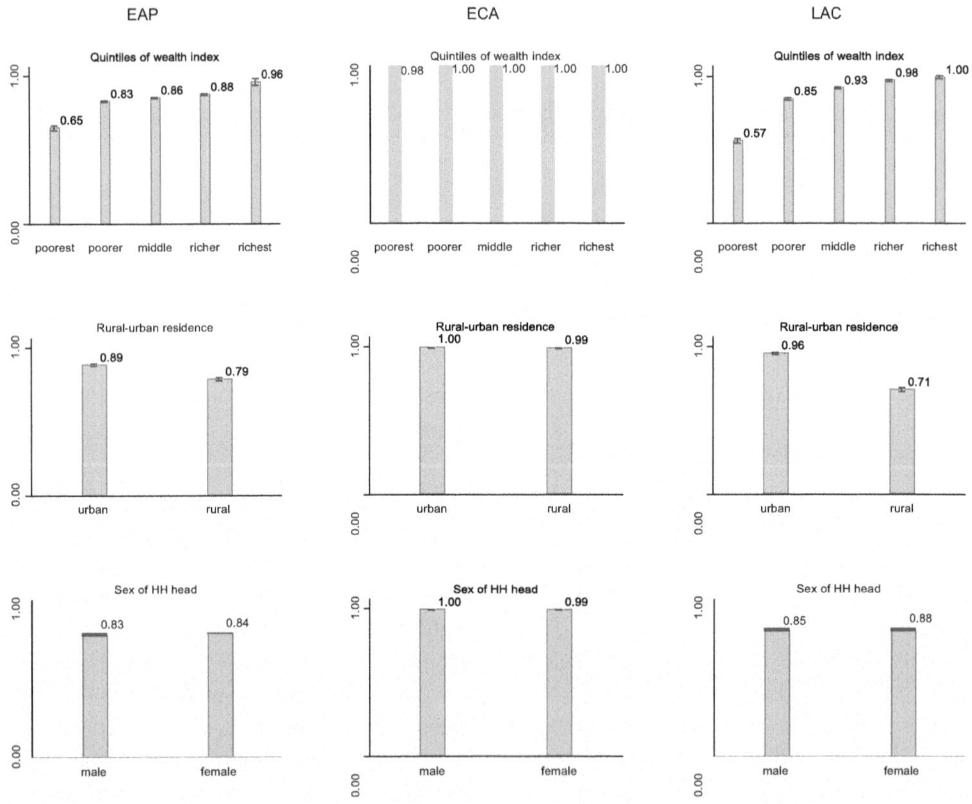

Fig. 4.3 Access to electricity for households by wealth, rurality and sex of household head-within country estimates by region-I. *Notes* See Table 4.1 for year of survey and number of observations for each country. The 95% confidence intervals are calculated by accounting for correlation at the primary sampling unit level. EAP refers to East Asia and Pacific; ECA refers to Europe and Central Asia; LAC refers to Latin America and the Caribbean

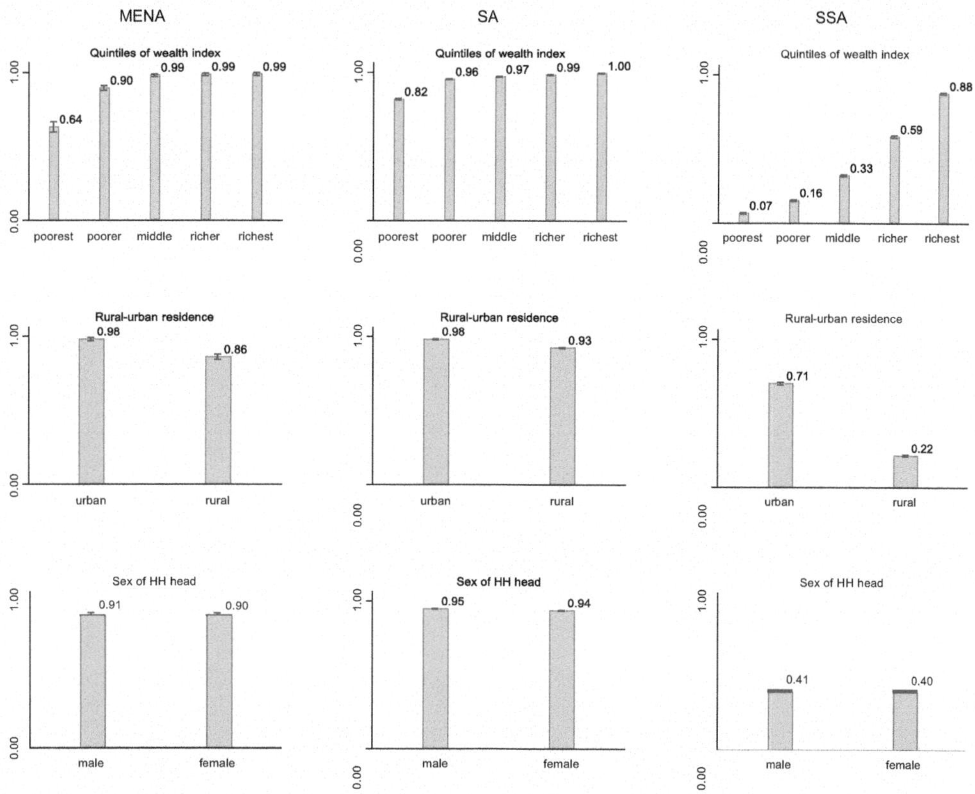

Fig. 4.4 Access to electricity for households by wealth, rurality and sex of household head-within country estimates by region-II. *Notes* See Table 4.1 for year of survey and number of observations for each country. The 95% confidence intervals are calculated by accounting for correlation at the primary sampling unit level. MENA refers to Middle East and North Africa; SA refers to South Asia; SSA refers to Sub-Saharan Africa

access to electricity. We can see that for the bottom 60% of the population, access to electricity remains a pressing concern.

Turning to rural-urban differences, there is a huge gap in SSA with 71% of urban residents and 22% of rural residents reporting access to electricity. In the LAC region, 96 and 71% of the urban and rural households have access and the commensurate figure for the EAP region is 89 and 79%, respectively. The sex of the household head is seen to be not such an important determinant of access to electricity with differences being small compared to the influence of wealth or rural residence.

4.4 Access to Clean Cooking Fuel

4.4.1 Country-Level Estimates

Figures 4.5 shows the proportion of households reporting access to clean cooking fuel in East Asia and Pacific (EAP); Europe and Central Asia (ECA); Latin America and the Caribbean (LAC); Middle East and North Africa (MENA) and South Asia (SA), and the right-hand panel of Fig. 4.2 for Sub-Saharan Africa (SSA). Access to clean cooking fuel is more limited than access to electricity, as evidenced by Figs. 4.5 and 4.2.

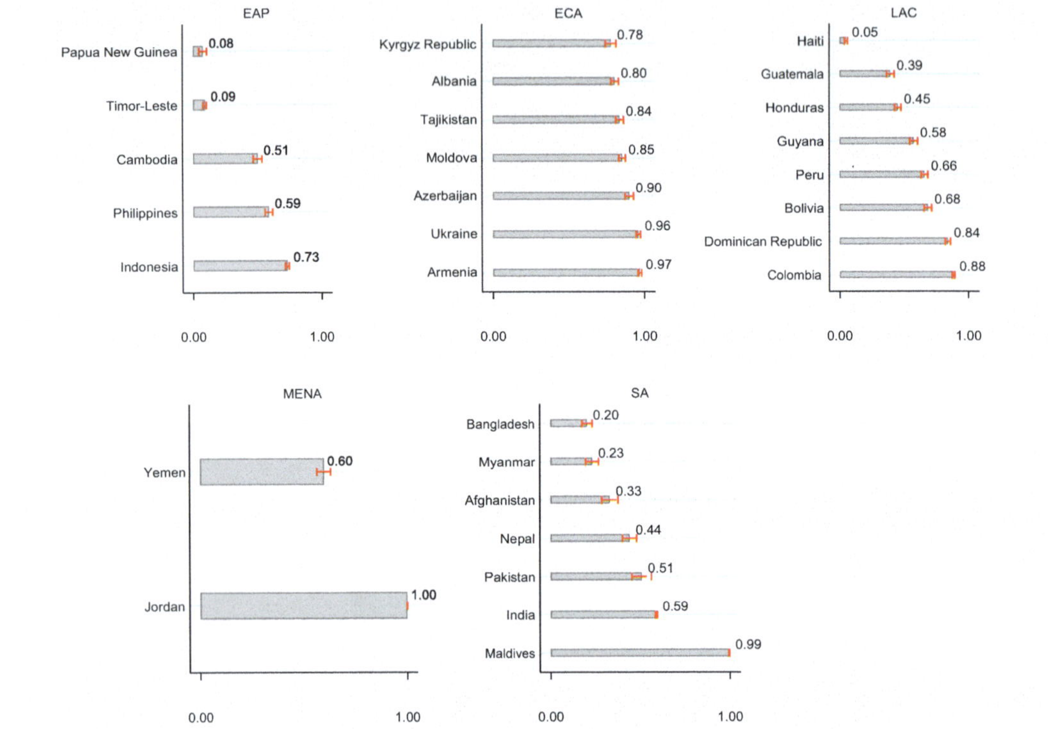

Fig. 4.5 Access to clean cooking fuel for households-country-level estimates by region. *Notes* See Table 4.1 for year of survey and number of observations for each country. The 95% confidence intervals are calculated by accounting for correlation at the primary sampling unit level. EAP refers to East Asia and Pacific; ECA refers to Europe and Central Asia; LAC refers to Latin America and the Caribbean; MENA refers to Middle East and North Africa; SA refers to South Asia; SSA refers to Sub-Saharan Africa

In East Asia and the Pacific, the access varies from a low of 8 and 9% in Papua New Guinea and East Timor to a high of 73% in Indonesia. The share of households having access to clean cooking fuel in Cambodia and the Philippines is 51 and 59%, respectively. Turning to Europe and Central Asia, the lowest access is in Kyrgyz Republic with 78% of households using clean cooking fuels to highs of 90, 96 and 97% in Azerbaijan, Ukraine and Armenia. For the LAC region, again Haiti exhibits a severe lack of access to modern energy sources with only 5% of households having access to clean cooking fuel. On the other end of the spectrum, 88 and 84% of households in Colombia and the Dominican Republic report having access. In Guatemala and Honduras, it is 39% and 45%, respectively, whereas the commensurate figures are 58, 66 and 68% for Guyana, Peru and Bolivia, respectively.

In the MENA region, data on clean cooking fuel is available for Yemen and Jordan and the share of households having access to clean cooking fuel is 60 and 100%, respectively. In South Asia, only Maldives has universal access, India is second with 59% of households having access to clean cooking fuel. On the other hand, for Bangladesh and Myanmar, this figure is as low as 20 and 23%, respectively. In Afghanistan, one-in-three and in Pakistan one-in-two households have access to clean cooking fuel.

The situation in SSA is seen to indicate an extreme situation where except for South Africa (81%) and Gabon (89%), no country has more than 50% of households having access to clean cooking fuel. In fact, in 21 of the 40 countries, the share is less than 10%.

4.4.2 Wealth, Rurality and Gender

Figures 4.6 and 4.7 show how access varies by wealth, rural-urban residence and sex of the household head. These, again, are region-wide estimates based on averages of *within* country differences.[4]

The figures show that wealth or socioeconomic status plays an important role: in EAP 11% of the households in the bottom, most quintile of the wealth index have access to clean cooking fuel and this increases to 86% for the top quintile. In ECA, the commensurate figures are 63 and 100%, respectively. In LAC, the proportion of households with access to electricity for the five quintiles of the wealth index is 15, 48, 68, 84 and 94%, respectively.

In South Asia, we see the same pattern of increasing access in wealth levels: 9, 34, 63, 83 and 96%, respectively, from the bottom to the top quintile. The commensurate figures for SSA are 3, 5, 10, 23 and 46%, respectively. Thus, in SSA, less than one-in-two households even in the top quintile have access to clean cooking fuel.

Turning to rural-urban differences, in SSA 34% of urban residents and 7% of rural residents report access to clean cooking fuel. In South Asia, the figures are 88 and 40%, respectively. In

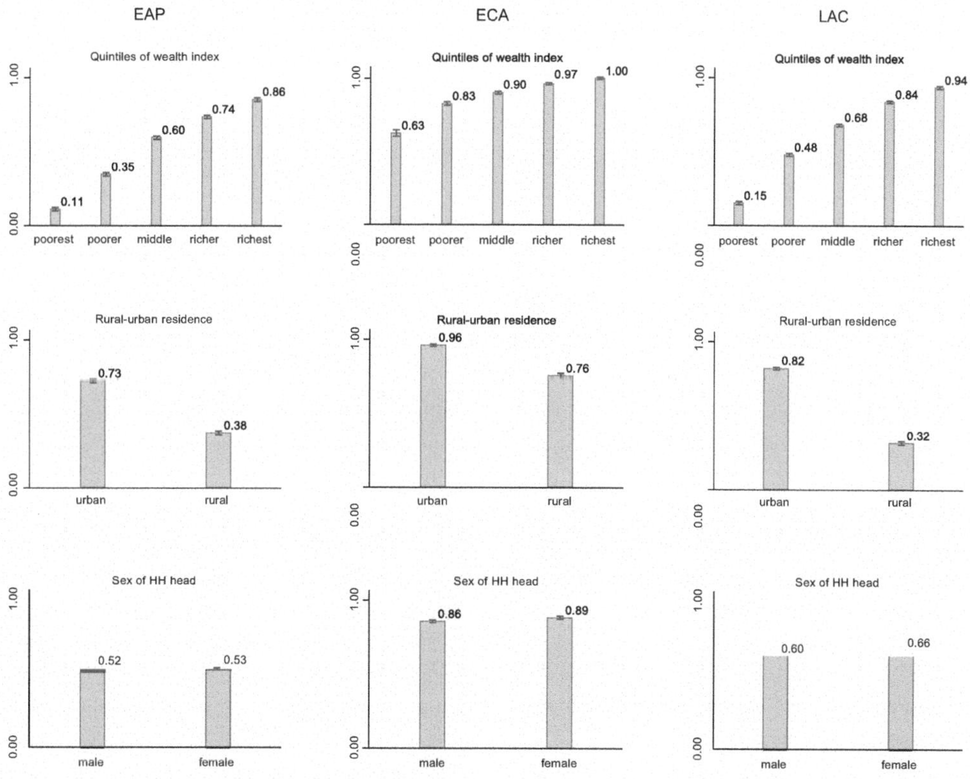

Fig. 4.6 Access to clean cooking fuel for households by wealth, rurality and sex of household head-within country estimates by region-I. *Notes* See Table 4.1 for year of survey and number of observations for each country. The 95% confidence intervals are calculated by accounting for correlation at the primary sampling unit level. EAP refers to East Asia and Pacific; ECA refers to Europe and Central Asia; LAC refers to Latin America and the Caribbean

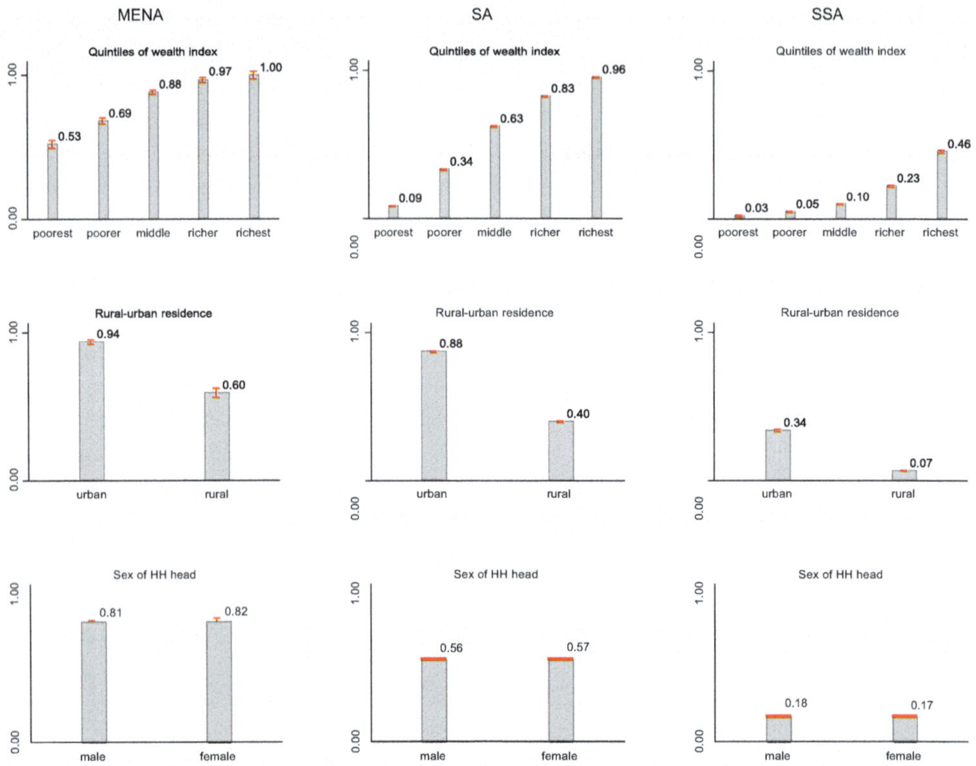

Fig. 4.7 Access to clean cooking fuel for households by wealth, rurality and sex of household head-within country estimates by region-II. *Notes* See Table 4.1 for year of survey and number of observations for each country. The 95% confidence intervals are calculated by accounting for correlation at the primary sampling unit level. MENA refers to Middle East and North Africa; SA refers to South Asia; SSA refers to Sub-Saharan Africa

the LAC region 82 and 32% and in the ECA region 96 and 76%, respectively. Finally, for EAP, 73% of urban residents and 38% of rural residents report access to clean cooking fuel. Finally, again, the sex of the household head is seen to be not such an important determinant of access to clean cooking fuel.

4.5 Correlates of Access to Modern Energy Sources

Table 4.2 gives the results of estimating Eq. 4.1. The dependent variables in columns (1) and (2) are a dummy for the household having access to electricity, and in columns (3) and (4), a dummy for the household having access to clean cooking fuel. Columns (1) and (3) include country fixed effects, whereas columns (2) and (4) include administrative unit one fixed effects. In other words, columns (1) and (3) are based on within country comparisons, whereas columns (2) and (4) restrict comparisons to households living within the same districts of the country. I below summarize results based on columns (2) and (4), which are able to account for differences in access stemming from differences in location or geography through the inclusion of district-fixed effects.

Column (2) shows that compared to the poorest 20% of households, the poorer, middle, richer and richest households are 12, 18, 26 and 36% points more likely to have access to electricity. Thus, we see that socioeconomic status is a very strong correlate of having access to electricity. On the other hand, rural households are

Table 4.2 Access to electricity and clean cooking fuel: association with wealth, rurality, sex and education of household head

	(1)	(2)	(3)	(4)
	Dummy for household has access to electricty		Dummy for household has access to clean cooking fuel	
	Electricity		Clean cooking fuel	
Poorer	0.12*** (0.0019)	0.13*** (0.0018)	0.17*** (0.0018)	0.15*** (0.0017)
Middle	0.18*** (0.0023)	0.19*** (0.0021)	0.33*** (0.0024)	0.30*** (0.0022)
Richer	0.26*** (0.0025)	0.26*** (0.0025)	0.45*** (0.0027)	0.42*** (0.0025)
Richest	0.36*** (0.0032)	0.34*** (0.0033)	0.56*** (0.0031)	0.53*** (0.0030)
Rural residence	− 0.057*** (0.0022)	− 0.036*** (0.0021)	− 0.14*** (0.0026)	− 0.13*** (0.0024)
Female HH head	0.011*** (0.00098)	0.0092*** (0.00088)	0.029*** (0.0010)	0.028*** (0.00092)
Primary	0.017*** (0.0013)	0.0096*** (0.0011)	0.0046*** (0.0014)	0.011*** (0.0012)
Secondary	0.022*** (0.0014)	0.015*** (0.0012)	0.036*** (0.0015)	0.043*** (0.0013)
Higher	0.018*** (0.0021)	0.012*** (0.0018)	0.073*** (0.0021)	0.082*** (0.0019)
Age HH head	− 0.00061*** (0.000030)	− 0.00040*** (0.000026)	− 0.00053*** (0.000033)	− 0.00062*** (0.000030)
Country fixed effects	Yes	No	Yes	No
Admin1 fixed effects	No	Yes	No	Yes
Observations	1,556,348	1,556,348	1,568,701	1,568,701
R-squared	0.560	0.607	0.546	0.597

Notes The table presents the results of estimating Eq. 4.1 using an linear probability model (LPM). The omitted category is the poorest, that is, quintile 1 or the bottom 20%; poorer refers to quintile 2; middle refers to quintile 3; richer refers to quintile 4; and richest refers to quintile 5, that is, the top 20%. The omitted education category is the set of households heads with no education. The list of countries and the sample size is shown in Table 4.1
SE clustered by PSU
$*p < 0.05$; $**p < 0.01$; $***p < 0.001$

3.6% points less likely to have access to electricity. Rurality is seen to matter less than wealth in determining access to electricity. Female-headed households are also more likely to have access to electricity. The effect though statistically significant is small in magnitude and suggests a 0.92% point increase in the likelihood of having access to electricity. Finally, education is also seen to be a statistically significant correlate of having access to electricity even after accounting for socioeconomic status or wealth. Compared to households heads with no education, heads with primary, secondary and higher education are 1.1, 2.2 and 1.8% points more likely to have access to electricity.

Column (4) displays the results for the dependent variable of having access to clean cooking fuel. The findings indicate that socioeconomic status or wealth has an even larger and more positive association compared to the outcome of having access to electricity. Households in the next four quintiles compared to the poorest 20% of households are 15, 30, 42 and 53% points more likely to be using clean cooking fuels. Rurality is also seen to have a stronger association with the outcome of having access to clean cooking fuels relative to electricity. Rural households conditional on wealth and education are a whole 13% points less likely to have access to clean cooking fuel. Again female-headed households are positively associated with using clean cooking fuels: conditional on wealth, rural residence, education and age, female households are 2.8% points more likely to be using clean cooking fuels. Finally, education also exhibits a stronger association with the outcome of clean cooking fuels relative to electricity. Compared to households heads with no education, those with primary, secondary and higher education are 1.1, 4.3 and 8.2% points more likely to be using clean cooking fuels.[5]

4.6 Conclusion

I use data from 66 countries encompassing 1.67 million households from the DHS to present a picture of global disparities in access to electricity and clean cooking fuels. The data presented in this chapter shows that the principle of distributive energy justice remains far from being met. Several countries, including the majority of Sub-Saharan Africa, lack access to electricity or clean cooking fuels. The data also highlights the significant associations between socioeconomic status and geographical location on access to modern energy sources. Households with higher socioeconomic status resident in the same region of the country are more likely to have access to electricity and clean cooking fuel. Additionally, urban areas within the same regions of the country generally have better access to modern energy services compared to rural areas.

An interesting aspect is the association between female-headed households and clean cooking fuels. Female-headed households are more likely to be using clean cooking fuels; for electricity, the effect is still positive and significant but not surprisingly much smaller. This suggests that where women have more decision-making power they make better choices for both health and the environment in terms of choice of energy source.

Furthermore, the findings emphasize the importance of addressing these disparities in access to modern energy sources. Efforts should be made to improve access in rural areas and among households with lower socioeconomic status. This can be achieved through the deployment and maintenance of electricity and clean cooking fuel infrastructure, as well as targeted interventions to support rural households in affording and accessing modern energy services. By addressing these issues, we can work towards achieving greater energy justice and ensuring that all households have equal access to modern energy sources.

Notes

1. See Sovacool and Dworkin (2014) for a discussion on the concepts of recognition and procedure as tenets of energy justice.
2. See the following link for more details.
3. Table 4.3 in the appendix shows how access to electricity varies by wealth, rural-urban residence and sex of household head for each country in the data set.
4. Table 4.4 in the appendix shows how access to clean cooking fuel varies by wealth, rural-urban residence and sex of household head for each country in the data set.
5. We also calculate the predicted values for the share of households with access to electricity and clean cooking fuel for the five wealth quintiles; by rural-urban residence; by sex of household head; and the education level of the household head based on estimating Eq. 4.1 and the results are shown in Figs. 4.8 and 4.9 in the appendix.

Table 4.3 Access to electricity: wealth, rurality and sex of household head

Country	Q1	Q2	Q3	Q4	Q5	Urban	Rural	Male	Female
Afghanistan	0.64	0.63	0.59	0.75	0.96	0.92	0.64	0.71	0.78
Angola	0.01	0.05	0.37	0.87	0.98	0.64	0.07	0.44	0.37
Armenia	1	1	1	1	1	1	1	1	1
Azerbaijan	1.01	1	1.01	1	1	1	1	1	1.01
Bangladesh	0.35	0.85	0.93	0.97	1	0.94	0.76	0.82	0.8
Benin	0.01	0.05	0.17	0.5	0.91	0.56	0.2	0.37	0.32
Bolivia	0.22	0.83	1.01	1.02	1.02	1.01	0.56	0.82	0.85
Burkina Faso	0	0.01	0.04	0.15	0.75	0.54	0.07	0.22	0.34
Burundi	0	0	0	0.02	0.46	0.6	0.02	0.1	0.07
Cambodia	0.64	0.98	1	1	1	0.99	0.88	0.91	0.94
Cameroon	0.01	0.19	0.74	0.97	0.99	0.9	0.27	0.6	0.67
Chad	0	0	0	0.02	0.39	0.32	0.01	0.08	0.06
Colombia	0.88	1	1	1	1	1	0.9	0.97	0.99
Comoros	0.16	0.5	0.85	0.97	1	0.86	0.62	0.71	0.69
Congo	0	0.04	0.24	0.89	0.98	0.6	0.12	0.44	0.34
D.R. Congo	0	0	0.01	0.05	0.7	0.42	0	0.14	0.12
Cote d'Ivoire	0.1	0.56	0.94	0.99	1	0.89	0.52	0.72	0.79
Dominican Republic	0.89	0.99	1	1	1	0.99	0.92	0.95	0.98
Egypt	0.99	1	1	1	1	1	1	1	1
Ethiopia	0.02	0.04	0.14	0.42	0.93	0.83	0.14	0.32	0.45
Gabon	0.7	0.98	0.99	0.99	1	0.98	0.54	0.93	0.92
Gambia	0.07	0.4	0.84	0.96	1	0.79	0.23	0.63	0.74
Ghana	0.26	0.58	0.85	0.98	1	0.91	0.63	0.78	0.8
Guatemala	0.44	0.89	0.98	1	1	0.97	0.8	0.87	0.9
Guinea	0.07	0.12	0.26	0.8	0.99	0.87	0.21	0.44	0.43
Guyana	0.34	0.79	0.96	1	1.03	1.02	0.75	0.8	0.87
Haiti	0	0.04	0.28	0.74	0.95	0.76	0.17	0.39	0.43
Honduras	0	0	0	0	0	0	0	0	0
India	0.86	0.98	1	1	1	0.99	0.95	0.97	0.95
Indonesia	0.89	0.99	0.99	1	1	0.99	0.96	0.97	0.96
Kenya	0.03	0.14	0.51	0.93	0.99	0.9	0.36	0.59	0.55
Kyrgyz Republic	1	1	1	1	1	1	1	1	1
Lesotho	0	0.03	0.07	0.31	0.9	0.62	0.12	0.29	0.25
Liberia	0	0.02	0.11	0.35	0.74	0.39	0.04	0.24	0.25
Madagascar	0	0.04	0.17	0.46	0.93	0.73	0.23	0.34	0.32
Malawi	0	0	0.01	0.02	0.53	0.49	0.04	0.12	0.08
Maldives	1	1	1	1	1	1	1	1	1
Mali	0.12	0.24	0.43	0.67	0.97	0.85	0.38	0.5	0.43
Mauritania	0	0.04	0.36	0.84	0.98	0.88	0.07	0.49	0.41

(continued)

Table 4.3 (continued)

Country	Q1	Q2	Q3	Q4	Q5	Urban	Rural	Male	Female
Moldova	0.96	1	1	1	1.01	1.01	0.98	0.99	1
Mozambique	0	0	0.02	0.21	0.89	0.54	0.05	0.2	0.2
Myanmar	0.19	0.36	0.52	0.78	0.98	0.92	0.42	0.55	0.58
Namibia	0.02	0.07	0.27	0.88	1	0.72	0.21	0.5	0.44
Nepal	0.84	0.99	1	1	1	0.98	0.94	0.97	0.96
Niger	0	0.01	0.02	0.08	0.67	0.62	0.06	0.16	0.13
Nigeria	0.06	0.25	0.64	0.88	0.98	0.83	0.39	0.58	0.66
Pakistan	0.68	0.96	0.99	1	1	0.99	0.88	0.92	0.95
Papua New Guinea	0	0	0.01	0.07	0.74	0.55	0.1	0.14	0.18
Peru	0.53	0.95	0.99	1	1	0.98	0.7	0.88	0.91
Philippines	0.75	1	1	1	1	0.98	0.91	0.94	0.96
Rwanda	0.06	0.2	0.41	0.73	0.95	0.86	0.37	0.48	0.42
Sao Tome and Principe	0.04	0.3	0.78	0.95	1	0.7	0.44	0.55	0.59
Senegal	0.04	0.47	0.93	0.98	1	0.94	0.46	0.67	0.78
Sierra Leone	0	0	0	0.19	0.9	0.51	0.02	0.22	0.25
South Africa	0.51	0.98	1	1	1	0.92	0.87	0.89	0.91
Tajikistan	0.98	0.99	1	1	0.99	0.99	0.99	0.99	0.99
Tanzania	0	0.01	0.09	0.6	0.99	0.75	0.18	0.37	0.36
Timor-Leste	0.36	0.66	0.83	0.97	1	0.98	0.65	0.74	0.7
Togo	0.02	0.05	0.2	0.72	0.99	0.83	0.16	0.45	0.47
Turkey	0	0	0	0	0	0	0	0	0
Uganda	0.01	0.04	0.12	0.31	0.85	0.59	0.18	0.3	0.27
Ukraine	1.01	1	1	1	1	1	1	1	1
Yemen	0.17	0.74	0.97	1	1	0.98	0.65	0.76	0.74
Zambia	0	0.02	0.06	0.62	0.98	0.69	0.08	0.35	0.32
Zimbabwe	0.01	0.02	0.06	0.57	0.93	0.81	0.1	0.35	0.32
Total	0.30	0.43	0.55	0.70	0.88	0.77	0.45	0.58	0.58

Table 4.4 Access to clean cooking fuel: wealth, rurality and sex of household head

Country	Q1	Q2	Q3	Q4	Q5	Urban	Rural	Male	Female
Afghanistan	0.02	0.06	0.12	0.48	0.94	0.84	0.15	0.33	0.5
Albania	0.37	0.78	0.89	0.95	0.99	0.92	0.62	0.79	0.85
Angola	0	0.07	0.67	0.98	1	0.77	0.08	0.53	0.45
Armenia	0.89	0.97	0.99	1	1	1	0.92	0.97	0.97
Azerbaijan	0.61	0.87	0.98	1	1	0.99	0.77	0.9	0.91
Bangladesh	0	0.01	0.05	0.3	0.68	0.56	0.06	0.21	0.17
Benin	0	0	0	0.01	0.21	0.1	0.01	0.05	0.05
Bolivia	0.02	0.43	0.95	1	1	0.96	0.29	0.67	0.73
Burkina Faso	0	0	0	0.09	0.59	0.42	0.04	0.16	0.22
Burundi	0	0	0	0	0	0	0	0	0
Cambodia	0.03	0.1	0.56	0.88	0.98	0.78	0.33	0.51	0.49
Cameroon	0	0	0.03	0.34	0.83	0.46	0.04	0.27	0.27
Chad	0	0	0	0	0.17	0.14	0	0.03	0.02
Colombia	0.4	0.98	1	1	1	0.99	0.5	0.86	0.92
Comoros	0	0	0	0.03	0.2	0.1	0.02	0.05	0.04
Congo	0	0	0.04	0.16	0.67	0.25	0.03	0.18	0.13
D.R. Congo	0	0	0	0	0.11	0.06	0	0.02	0.01
Cote d'Ivoire	0	0.06	0.17	0.77	0.93	0.64	0.1	0.4	0.45
Dominican Republic	0.49	0.95	0.99	0.99	1	0.95	0.66	0.81	0.87
Egypt	0	0	0	0	0	0	0	0	0
Ethiopia	0	0	0	0.01	0.29	0.21	0.01	0.06	0.11
Gabon	0.55	0.97	0.99	1	1	0.95	0.45	0.91	0.87
Gambia	0	0.01	0.05	0.07	0.2	0.09	0	0.07	0.07
Ghana	0	0.01	0.08	0.3	0.79	0.41	0.1	0.29	0.23
Guatemala	0	0.01	0.12	0.66	0.97	0.68	0.16	0.38	0.44
Guinea	0	0	0	0.04	0.08	0.06	0	0.02	0.03
Guyana	0.1	0.35	0.62	0.83	0.97	0.76	0.5	0.55	0.63
Haiti	0	0	0	0.01	0.22	0.1	0.01	0.05	0.05
Honduras	0	0.05	0.34	0.77	0.97	0.72	0.17	0.43	0.52
India	0.09	0.37	0.68	0.87	0.97	0.9	0.43	0.59	0.59
Indonesia	0.25	0.68	0.88	0.96	0.98	0.88	0.59	0.75	0.67
Jordan	1	1	1	1	1	1	1	1	1
Kenya	0	0	0.07	0.4	0.88	0.64	0.1	0.33	0.29
Kyrgyz Republic	0.63	0.58	0.63	0.89	1	0.95	0.67	0.75	0.85
Lesotho	0	0.03	0.15	0.6	0.94	0.79	0.16	0.38	0.33
Liberia	0	0	0.01	0.01	0.03	0.02	0	0.01	0.01
Madagascar	0	0	0	0	0.07	0.06	0.01	0.02	0.02
Malawi	0	0	0	0	0.11	0.12	0	0.02	0.02
Maldives	0.97	1	1	1	1	1	0.99	0.99	0.99

(continued)

Table 4.4 (continued)

Country	Q1	Q2	Q3	Q4	Q5	Urban	Rural	Male	Female
Mali	0	0	0.01	0.01	0.03	0.02	0.01	0.01	0.01
Mauritania	0	0.12	0.42	0.77	0.97	0.76	0.2	0.48	0.44
Moldova	0.49	0.83	0.98	1	1	0.99	0.76	0.85	0.85
Mozambique	0	0	0	0	0.19	0.11	0	0.04	0.03
Myanmar	0	0.02	0.1	0.33	0.74	0.6	0.1	0.22	0.26
Namibia	0	0.06	0.27	0.76	0.99	0.75	0.11	0.47	0.4
Nepal	0.01	0.08	0.37	0.76	0.97	0.56	0.2	0.43	0.45
Niger	0	0	0	0	0.05	0.06	0	0.01	0.01
Nigeria	0	0	0.01	0.11	0.54	0.27	0.04	0.15	0.16
Pakistan	0.02	0.18	0.52	0.82	0.97	0.89	0.27	0.5	0.52
Papua New Guinea	0	0	0	0	0.43	0.38	0.04	0.07	0.11
Peru	0.02	0.43	0.89	0.99	1	0.89	0.16	0.64	0.7
Philippines	0.04	0.31	0.72	0.91	0.98	0.79	0.37	0.56	0.68
Rwanda	0	0	0	0.01	0.22	0.19	0.01	0.05	0.04
Sao Tome and Principe	0	0	0	0	0.02	0.01	0	0	0
Senegal	0	0.03	0.14	0.44	0.82	0.55	0.09	0.27	0.39
Sierra Leone	0	0	0	0	0.03	0.02	0	0.01	0
South Africa	0.27	0.83	0.97	0.99	1	0.9	0.62	0.84	0.78
Tajikistan	0.68	0.74	0.8	0.91	0.99	0.98	0.77	0.83	0.87
Tanzania	0	0	0	0.08	0.34	0.23	0.03	0.09	0.09
Timor-Leste	0	0.01	0.04	0.13	0.34	0.25	0.04	0.09	0.09
Togo	0	0	0	0.02	0.29	0.15	0.01	0.07	0.06
Turkey	0	0	0	0	0	0	0	0	0
Uganda	0	0	0	0	0.04	0.03	0	0.01	0.01
Ukraine	0.85	0.98	0.99	1	1	0.98	0.91	0.96	0.96
Yemen	0.04	0.34	0.76	0.94	1	0.95	0.44	0.6	0.63
Zambia	0	0	0	0.03	0.43	0.19	0.02	0.1	0.08
Zimbabwe	0	0	0.01	0.45	0.91	0.74	0.06	0.3	0.28
Total	0.13	0.23	0.33	0.46	0.63	0.52	0.23	0.36	0.37

Appendix

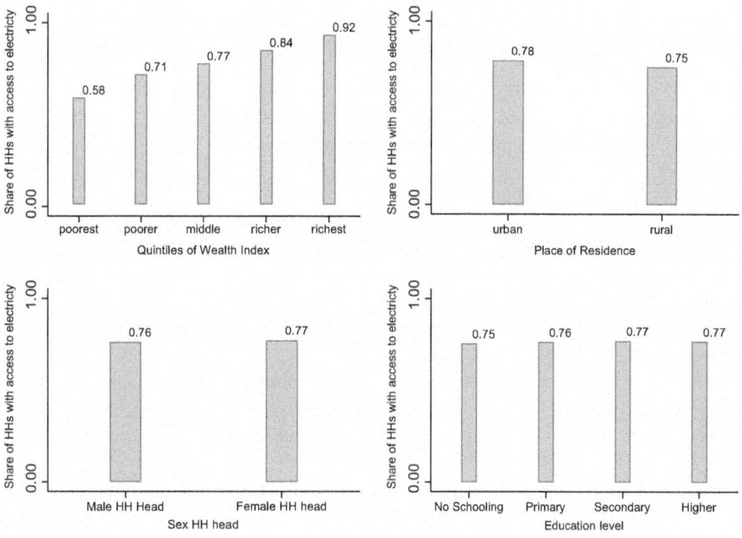

Fig. 4.8 Access to electricity for households by wealth, rural residence, education sex of house-hold head-within country estimates. *Notes* The predicted values arise from estimating Eq. 4.1 using an linear probability model (LPM). The list of countries and the sample size is given in Table 4.1

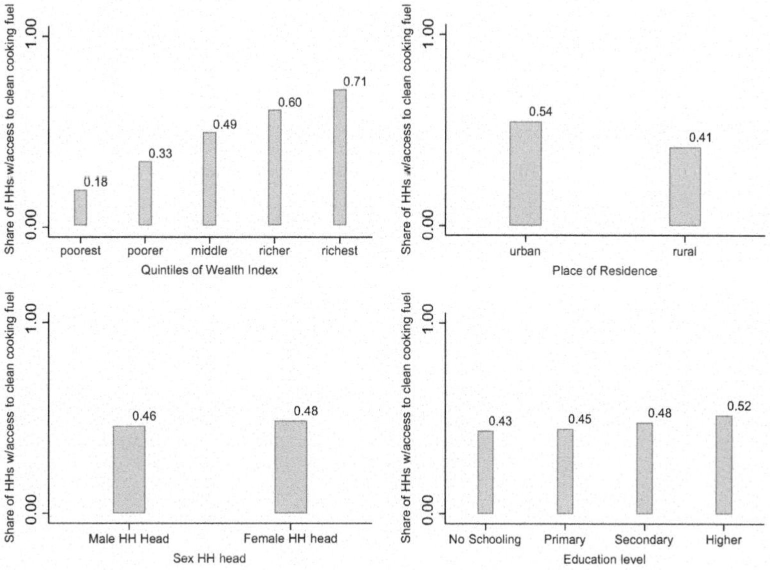

Fig. 4.9 Access to clean cooking fuel for households by wealth, rural residence, education sex of household head-within country estimates. *Notes* The predicted values arise from estimating Eq. 4.1 using an linear probability model (LPM). The list of countries and the sample size is given in Table 4.1

References

Anenberg SC, Balakrishnan K, Jetter J, Masera O, Mehta S, Moss J, Ramanathan V (2013) Cleaner cooking solutions to achieve health, climate, and economic cobenefits. Environ Sci Technol 47 (9):3944–3952

Burke PJ, Stern DI, Bruns SB (2018) The impact of electricity on economic development: a macroeconomic perspective. Int Rev Environ Resour Econ 12 (1):85–127

Hystad P, Duong M, Brauer M, Larkin A, Arku R, Kurmi OP, Fan WQ, Avezum A, Azam I, Chifamba J et al (2019) Health effects of household solid fuel use: findings from 11 countries within the prospective urban and rural epidemiology study. Environ Health Perspect 127(5):057003

ICF (2005–2022) Demographic and health surveys (various) [Datasets]. Funded by USAID. ICF [Distributor], Rockville, MD

Jagger P, Kittner N (2017) Deforestation and biomass fuel dynamics in Uganda. Biomass Bioenerg 105:1–9

Jenkins K, McCauley D, Heffron R, Stephan H, Rehner R (2016) Energy justice: a conceptual review. Energy Res Soc Sci 11:174–182

Mainali B, Pachauri S, Rao ND, Silveira S (2014) Assessing rural energy sustainability in developing countries. Energy Sustain Dev 19:15–28

Oparaocha S, Dutta S (2011) Gender and energy for sustainable development. Curr Opin Environ Sustain 3 (4):265–271

Puzzolo E, Zerriffi H, Carter E, Clemens H, Stokes H, Jagger P, Rosenthal J, Petach H (2019) Supply considerations for scaling up clean cooking fuels for household energy in low-and middle-income countries. GeoHealth 3(12):370–390

Ritchie H, Rosado P, Roser M (2019) Access to energy. Our world in data. https://ourworldindata.org/energy-access

Sovacool BK, Dworkin MH (2014) Global energy justice. Cambridge University Press

WHO (2023) Energizing health: accelerating electricity access in health-care facilities. World Health Organization

Rajesh Ramachandran is a Senior Lecturer at Monash University Malaysia. He obtained his Ph.D. at Universitat Autonoma de Barcelona and previously held positions at the Goethe University and Heidelberg University. His principal areas of research lie at the intersection of development economics and its political economy. His primary research interests include political linguistics, economics of discrimination, economics of education, institutional economics and social identity.

5 Exploring the Dynamics of Access to Electricity and Educational Outcomes in Asia

Pauline Khoo

Abstract

This chapter explores the impact of electricity access on socio-economic development, with a particular focus on educational outcomes in India, Nepal, and Pakistan. Access to energy is a significant determinant of education outcomes, and there has been a substantial increase in electricity access within the region from the 1980s to 2023. Utilising data from the Integrated Public Use Microdata Series (IPUMS) Demographic and Health Surveys, the study examines how increased electrification has influenced educational attainment across different socio-economic groups. Findings indicate improvements in both electricity access and educational achievements, particularly among lower-income households and women. However, the analysis highlights the need for further research to establish causal relationships and consider the broader socio-economic and policy contexts. This work underscores the importance of equitable energy distribution in fostering sustainable development and addressing educational disparities.

Keywords

Energy justice · Electricity access · Education · Asia

5.1 Introduction

In this chapter, I delve into the transformative influence of electricity on the socio-economic landscapes of the Asian region. The focus here is to understand how increased access to electricity correlates with educational advancements and overall societal well-being, which thus contributes to the narrative of sustainable development. There is a wide consensus that electricity plays an integral role in economic activities. It not only powers industries and businesses but is also essential for technological progress. This is in alignment with the United Nations' Sustainable Development Goal (SDG) 7, which emphasises the need for universal access to affordable, reliable, sustainable, and modern energy. Shifting from economic impacts to social dimensions, electricity also plays a role in enhancing human development. Its impact on crucial areas such as health, education, and gender equality is supported by examples from various global regions.

Central to the chapter is the detailed exploration of the dynamics between electricity access and education outcomes. Given the electrification efforts in Asia, there has been a substantial

P. Khoo (✉)
Monash University Malaysia, Kuala Lumpur, Malaysia
e-mail: pauline.khoo@monash.edu

A. Trakic et al. (eds.), *Energy Justice*, Sustainable Development Goals Series,
https://doi.org/10.1007/978-981-97-6059-6_5

increase in electricity access within the region. Through a descriptive analysis of data from demographic health surveys, the chapter investigates the correlation between the expansion of electricity access and educational attainment across various socio-economic groups. While those with higher socio-economic status have traditionally enjoyed consistent access to electricity, I examine whether this expansion has extended to bridge the gap for the lower-income households. By tracing the trajectory of electricity access from near non-existence to substantial levels in these income groups, the chapter seeks to understand whether this improved infrastructure is mirrored by a corresponding improvement in educational attainment. I also consider the other factors that influence the interplay of electricity access and education, such as gender roles.

By highlighting the nuances and subtleties of this relationship, this chapter contributes to the discourse on sustainable development. It underscores the importance of inclusive and equitable growth, where advancements in infrastructure like electricity do not benefit only the affluent but reach every stratum of society. This exploration is particularly salient in the context of the SDG, highlighting the need for an integrated approach that considers the interdependencies of economic growth, social equity, and environmental sustainability. In essence, this chapter not only provides a descriptive analysis of the trends in electricity access and its social implications but also reflects on the broader issues of sustainable development and energy justice. It offers insights into how equitable distribution of energy resources can be a transformative force for societal change, aligning with global goals and addressing longstanding disparities in education and gender roles.

5.2 Access to Electricity and Growth

Electricity is a cornerstone of modern civilisation, universally acknowledged as a key enabler of economic growth and a critical component in the fight against poverty. As highlighted by the SDG 7 to ensure access to affordable, reliable, sustainable and modern energy for all, electricity access is at the core of economic development. Access to electricity as a measure of economic development and growth has many compelling reasons. As a fundamental driver of economic activities, it powers industries, businesses, and manufacturing processes, ensuring stability and efficiency. Additionally, electricity acts as a catalyst for technological advancements. In a world increasingly dependent on technology, a continuous and reliable power source is indispensable for fostering innovation and research, and for the adoption of new technologies—elements vital for sustained economic growth (IEA 2020).

Figure 5.1 demonstrates the relationship between primary energy consumption per capita (measured in kilowatt-hours per person) and GDP per capita (measured in 2017 international dollars) for various countries in the year 2022. The figure illustrates a positive correlation between the two variables: as GDP per capita increases, the primary energy consumption per capita tends to increase as well. In the upper right quadrant of the plot, countries like Singapore and the United Arab Emirates are positioned, indicating both high GDP per capita and high energy consumption per capita. In contrast, countries such as India, Pakistan, and Sri Lanka are positioned in the lower left quadrant, reflecting lower values in both GDP and energy consumption per capita. This suggests that wealthier countries (with higher GDP per capita) also consume more energy per person, which might be due to a higher standard of living that includes more energy-intensive goods and services. It also suggests that economic prosperity, as measured by GDP per capita, is often accompanied by increased energy consumption, which may reflect broader access to electricity and higher usage levels. This motivates the discussion on the potential effects of energy consumption on other social and economic factors.

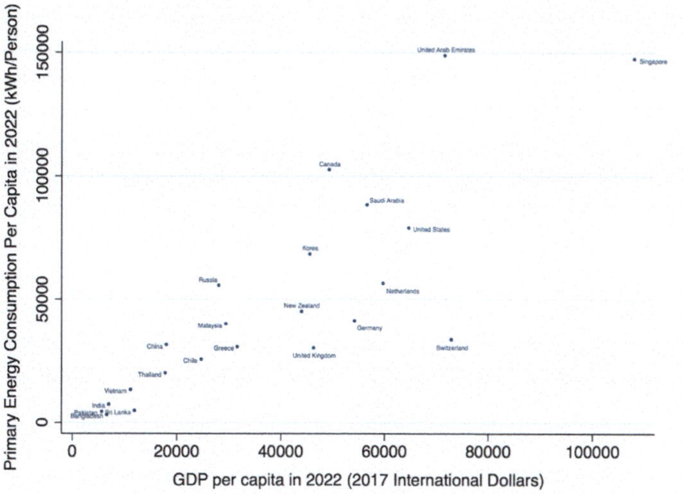

Fig. 5.1 Primary energy consumption per capita and GDP per capita in 2022. *Note* The data on primary energy consumption per capita is from the IEA's WEO, while the data on GDP per capita is from the IMF's World Economic Database. GDP per capita is converted to international dollar using purchasing power parity rates. For example, an international dollar has the same purchasing power over GDP as the Malaysian Ringgit has in Malaysia. Data is in constant 2017 international dollars. *Source* International Energy Agency's *World Energy Outlook* (WEO), 2023 and International Monetary Fund, World Economic Outlook Database, October 2023

5.3 Access to Electricity and Social Determinants

Continuing from the insights provided by the relationship between energy consumption and economic output, there is a growing interest in understanding how access to electricity influences various social determinants. Beyond the already acknowledged influence of electricity on employment and agriculture (Gaye 2007; IEA 2020; UNDP 2018; WHO 2018; DESA 2020), further research from diverse regions has consistently demonstrated the pivotal role of electricity and clean energy in enhancing human development. This encompasses improvements in health, education, and gender equality, as evidenced by various studies (da Silveira Bezerra et al. 2017; Njiru and Letema 2018; Oum 2019; Sambodo and Novandra 2019; Acheampong et al. 2021). For instance, schools and hospitals rely on a reliable power supply to enhance the quality of services offered. In health care, the significance of a dependable energy source cannot be overstated, often making the difference between life and death. A study conducted in Ghana, a country that experiences chronic power outages, shows that there is a high correlation between the frequency of power outages and in-facility mortality, indicating that the risk of mortality increases by 43% for each day the power remains out for over 2 h (Apenteng et al. 2018). This situation is also not unique to Ghana. The World Health Organization in 2023 estimates around 12 to 15% of healthcare facilities in low- and lower-middle-income countries in South Asia and sub-Saharan Africa still have no access to electricity (WHO 2023).

At the household level, access to electricity as well as modern cooking fuel, substantially improves the overall quality of life. In the context of modern cooking fuel, the shift away from traditional biomass sources, such as wood or charcoal, not only reduces indoor air pollution—a known health hazard—but also improves the overall cleanliness and safety of the cooking environment. The reduced exposure to harmful fumes consequently increases better health outcomes, especially for women and children who traditionally engage in cooking activities. A study by Pachauri et al. (2013) shows that total rural electrification and universal access to cleancombusting cooking fuels and stoves have the potential to avert between 0.6 and 1.8 million premature deaths annually by 2030. In addition to clean cooking fuel, reliable access to electricity facilitates efficient food preparation methods, which enables households to diversify their diets and include a wider range of nutrient-rich foods. Furthermore, modern cooking appliances powered by electricity offer precise temperature control, reducing the likelihood of nutrient loss during cooking and enabling the preparation of healthier meals. The availability of electricity also supports the refrigeration and preservation of perishable food items, ensuring a more stable and varied food supply. This, in turn, contributes to improved dietary diversity, addressing nutritional deficiencies that may arise from a limited range of available foodstuffs. In fact, recent studies have demonstrated the positive impact of electricity access and modern cooking fuel on increasing caloric intake and reducing malnutrition (Njiru and Letema 2018; Sambodo and Novandra 2019).

The linkage between electricity access and well-being underscores the integral role of reliable energy in shaping the welfare of individuals and communities. Figure 5.2 further illustrates this notion by providing the relationship between the Human Development Index (HDI) and the primary energy consumption per capita across nations in 2022. Developed by the United Nations Development Programme (UNDP), the Human Development Index (HDI) serves as a measure of average achievement in three dimensions of human development, encompassing health, standard of living and education. These dimensions are assessed through life expectancy at birth, gross national income per capita, and a composite of mean years of schooling for adults aged 25 years and above, along with the expected years of schooling for children entering school. The positive relationship between primary energy consumption per capita and HDI indicates that countries that have broader access to electricity and higher usage levels are often associated with a higher HDI.

Additionally, a reliable energy source is not only important for economic growth but also a fundamental determinant in shaping various social outcomes, including education. In regions where electricity is unreliable or inaccessible, educational activities face disruptions, hindering the learning environment. Furthermore, the absence of electricity in schools limits the use of

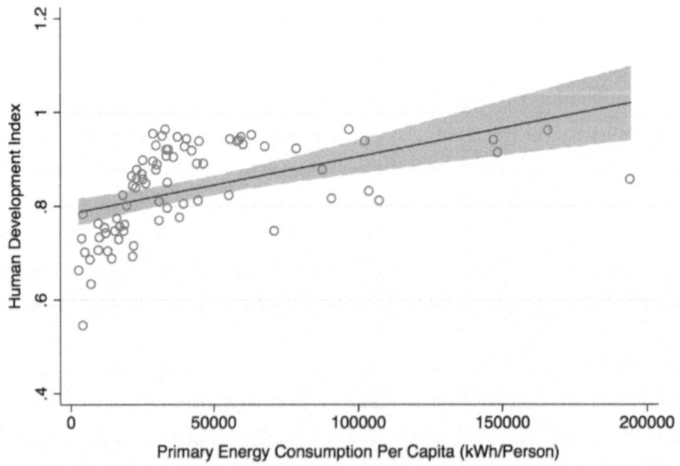

Fig. 5.2 Human development index and primary energy consumption in 2022. *Note* The data on the Human Development Index (HDI) is from the UNDP's Human Development Reports, while the data on primary energy consumption per capita is from the IEA's WEO. The HDI is a geometric mean of normalised indices for the three key dimensions of human development: a long and healthy life, having a decent standard of living, and being knowledgeable. *Source* United Nations Development Programme (2022) and Human International Energy Agency's *World Energy Outlook* (WEO) (2023)

modern teaching tools and technologies, impeding the adoption of innovative educational methods. This challenges the development of essential skills among students, in particular, digital literacy skills. As governments and organisations strive to bridge the electricity access gap, the aim is not only to promote economic growth but also to create an enabling environment that fosters educational development.

As this chapter delves into exploring the link between access to electricity and education outcomes, it specifically isolates the education measure from the HDI and demonstrates its correlation with energy consumption per capita across nations in 2022 in Fig. 5.3. Notably, Fig. 5.3 mirrors the pattern observed in Fig. 5.2, illustrating that countries with higher per capita electricity consumption exhibit a higher rank in the education index. Access to electricity allows households to access various amenities, such as lighting, refrigeration, and electronic devices. However, these amenities not only serve their functions but also free up considerable time from time-intensive household tasks for more productive or enjoyable activities, especially for the female population, who mostly bear the burden of home production. With the reduction of time spent on helping with household chores, children potentially will have more opportunities to attend school and other educational activities. With access to electricity, children who previously did not have sufficient luminescence for studying during the dark and access to Information and Communication Technologies are able to spend more time at home to study. Access to electricity has been linked to higher literacy levels and an increase in average school years (Njiru and Letema 2018; Oum 2019).

5.4 Electrification Efforts in Asia

The historical trajectory of electricity access paints a clear picture of disparity: the affluent has long enjoyed the fruits of technological advancements, including reliable electricity, while the less privileged, often in remote or underserved regions, grapple with considerable barriers to access. The link between wealth and electricity access manifests on a global scale, with the top 10% of the world's richest consuming approximately 20 times more energy than the bottom 10%, a statistic that starkly highlights the imbalance in energy consumption

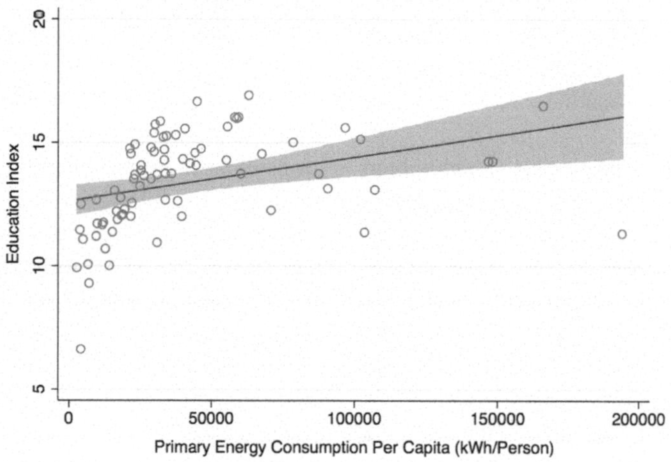

Fig. 5.3 Education index and primary energy consumption in 2022. *Note* The data on the education index is from the UNDP's Human Development Reports, while the data on primary energy consumption per capita is from the IEA's WEO. The education index is measured by the mean of years of schooling for adults aged 25 years and more and expected years of schooling for children of school entering age, and it is used to form the HDI. *Source* United Nations Development Programme (2022) and Human International Energy Agency's *World Energy Outlook* (WEO) (2023)

(Oswald et al. 2020). The inequality is also suggested by Figs. 5.1 and 5.2, which shows a comparison of energy consumption across countries and is anticipated to be mirrored within individual countries as well.

Such disparities do more than outline consumption patterns; they expose the inequities faced by lower-income households and underscore the fact that the benefits of energy transitions are not being enjoyed equitably. Sovacool (2012) examined the issue of 'energy poverty' and found that within nations, rural and remote communities, as well as marginalised urban populations, often lag behind in terms of access to reliable and affordable electricity. In India, e.g. Kanagawa and Nakata (2008) demonstrated significant variability in electricity access between urban and rural areas, attributing this to both infrastructural challenges and policy shortcomings. Similar patterns were observed by Grogan and Sadanand (2013) in sub-Saharan African countries where access to electricity in urban centres often contrasts sharply with inadequate infrastructure in rural areas.

There are strong implications of disparities within countries. Kooijman-van Dijk and Clancy (2010) have shown how unequal electricity access can deepen income inequalities by limiting the economic engagement of the poor in electrification-dependent activities, a situation that can entrench intergenerational poverty through educational disparities. Gender dimensions of this inequality are particularly pronounced, with Winther et al. (2017) highlighting that women, often the primary managers of household energy, are disproportionately disadvantaged by inadequate electricity, affecting their educational and economic prospects. While past government interventions, as studied by Urpelainen (2011) and Aklin et al. (2016), frequently succeed in broadening electricity coverage but commonly overlook the nuanced inequalities that hinder the most marginalised populations, national governments are increasingly aware that the push for modern energy services must be inclusive.

This expanding recognition translates into policies and initiatives aimed at democratising energy access, which the UNDP has advocated for over a decade (UNDP 2010). For example, India launched the Saubhagya Scheme, which is a concerted effort by the government aimed to provide electricity access to all households, even those in remote and marginalised communities have access to electricity, in the country (Government of India Ministry of Power 2023). Another notable illustration is Nepal's Rural Energy Development Programme (REDP), initiated by UNDP and the Nepalese government in 1996, which began as a small-scale pilot in five districts and has expanded to over 75 districts that serve approximately 1 million people in remote areas (United Nations Office for South-South Cooperation 2011). With all these initiatives, progress has been made on this front. Just a decade ago, 17% of the global population lacked access to electricity (IEA 2014). Substantial improvements have occurred over time, and currently, around 90% of the world's population is expected to have access to electricity (IEA 2023a, b).

The aim of this chapter is to dissect the interplay between the surge in electricity access and educational progress within the Asian region. The concerted push towards universal electrification to broader segments of the population in Asia has catalysed significant progress. Figure 5.4 depicts the progress in electricity access across various global regions from 2000 to 2020. While Latin America and the Caribbean, Europe and Central Asia, and the Middle East and North Africa show improvements or consistently high levels of electricity access, Asia demonstrates the most considerable growth, starting from just above 70% in the year 2000 with the percentage approaching the 90% threshold by 2020. Despite the dampening effect of the pandemic on advancements in Asia, focused policies like the one in India and Nepal, as well as easier access to financing, have contributed to enhanced electricity access.

Building on the positive trajectory of electrification in Asia, it is imperative to explore how this has influenced educational progress, especially across socio-economic status. While those with higher socio-economic status have traditionally enjoyed consistent access to electricity, I

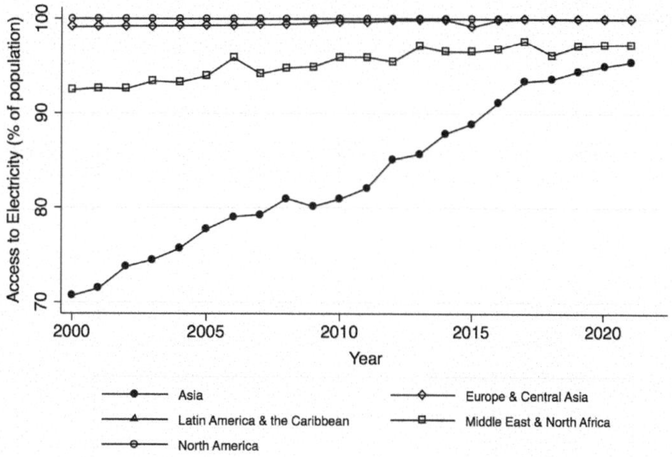

Fig. 5.4 Access to electricity by region. *Source* The World Bank (2023)

investigate whether this expansion has extended to bridge the gap for lower-income households. Utilising data from the Integrated Public Use Microdata Series (IPUMS) Demographic and Health Surveys (DHS), this chapter provides a case study on three Asian countries that it has availability for—India, Nepal, and Pakistan—to compare access to electricity and the status of education attainment across different socio-economic groups. The data is compiled from the India Demographic Health Survey, conducted between 1992 and 2015; the Nepal Demographic Health Survey, conducted between 1996 and 2016; and the Pakistan Demographic Health Survey, conducted between 1991 and 2017. In all three respective countries, members of randomly selected households are surveyed on household characteristics, which include access to electricity at their dwelling and the year of schooling completed.

Before delving into the disparities in educational attainment across socio-economic groups within India, Nepal, and Pakistan, it is critical to first ascertain the trajectory of electricity access within these countries. As illustrated in Fig. 5.5, all three nations have experienced substantial increases in electrification rates over two decades, setting a foundational context for subsequent analyses. India's growth in electricity access has been steady, with rates nearing 80% by 2020. Nepal's rise has also been remarkable, soaring from about 20% to 90% in the same timeframe, and Pakistan's access increased from just under 60% around 2000 to approaching 80%

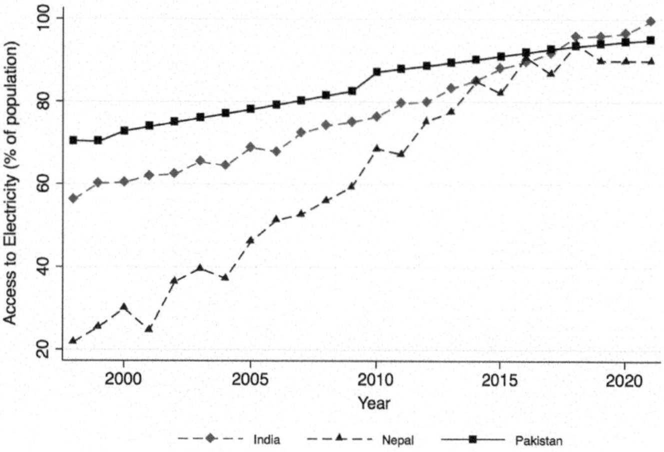

Fig. 5.5 Access to electricity for India, Nepal, and Pakistan. *Source* The World Bank (2023)

by 2020. This upward trend in electricity access across the three countries provides a backdrop for investigating whether such improvements in infrastructure have translated into enhanced educational outcomes, especially for those in lower-income brackets, and whether the patterns of progress in electrification correlate with strides made in educational equity.

5.4.1 Dynamics of Access to Electricity and Education Outcomes in India

Since significant strides have been made in electrification across India, Nepal, and Pakistan, I can start examining whether the improved access to electricity has been able to bridge the educational gaps across varying socio-economic statuses. I consider the case of India first, where I investigate the interconnections between improved electricity provisions and educational outcomes from 1992 to 2015. The analysis will scrutinise the changes in years of education completed, providing insight into whether the increase in electricity access has corresponded with the equalisation of educational opportunities, particularly for the lower-income households.

To do so, I show the progression of electricity access across different wealth quintiles in India for the sampled years. Simultaneously, I also show the progression of years of schooling completed across the same index. I can also then compare how years of schooling completed change across the different wealth quintiles. These are illustrated in Table 5.1. The quintiles range from the poorest (1st quintile) to the wealthiest (5th quintile) households. Panel (a) of the table tracks the progress of electricity access across five wealth quintiles in India for the years 1992, 1998, 2005, and 2015. It shows that access to electricity has increased over time for all income groups, with the most substantial gains seen in the lower quintiles. The poorest quintile witnessed an increase from a mere 0.4 to 57.1%, reflecting significant strides in electrification efforts targeting low-income households. The second quintile improved from 10.1 to 91%, and the middle-income third quintile rose from 52.8 to 98.3%, indicating progress towards universal access. The upper-middle fourth quintile, starting from 93.2%, achieved near-complete electrification at 99.7%. Lastly, the wealthiest quintile maintained near-universal access throughout, increasing marginally from 99.7 to 99.9%. As discussed, those with higher socio-economic status have always had access to electricity, but

Table 5.1 Access to electricity and years of education in India

Year of sample	Household wealth index in quintiles				
	1	2	3	4	5
Panel (a): access to electricity					
1992	0.004	0.101	0.528	0.932	0.997
1998	0.009	0.283	0.71	0.956	0.996
2005	0.213	0.527	0.836	0.973	0.999
2015	0.571	0.91	0.983	0.997	0.999
Panel (b): years of education					
1992	4.954	5.373	5.785	6.653	8.828
1998	4.115	4.881	5.711	6.687	9.09
2005	4.415	5.337	6.221	7.353	9.823
2015	5.471	6.524	7.414	8.353	10.079

Note Sample is restricted to adults aged 25 to 60
Source IPUMS-Demographic and Health Surveys: Version 9

by 2015, even the poorest quintile had over half the households with electricity, a dramatic increase from virtually none in 1992.

In Panel (b), there is also a notable upward trend for all quintiles for the years of education completed. In 1992, the access levels ranged from 4.954 in the poorest quintile to 8.828 in the wealthiest. Over the years, every quintile saw an increase in access. This shows a differential percentage gains across the socio-economic spectrum. The first quintile experienced a gain of 10.44% in educational attainment, while the second quintile saw a more pronounced increase of 21.42%. The third quintile marked the most significant improvement, with a 28.16% gain, reflecting substantial advancements in access to education for the middle-income group. The fourth quintile also exhibited a notable rise of 25.55%, suggesting that education has become more attainable across the board. However, the wealthiest quintile (5th quintile) showed the smallest increase at 14.17%, potentially indicating a saturation point in educational levels for the higher-income bracket. These disparities in percentage gains highlight a trend of increasing educational equity, with the most marked progress in the middle-income strata hinting at a closing gap between the different socio-economic groups over the 23-year span.

5.4.2 Dynamics of Access to Electricity and Education Outcomes in Nepal

Table 5.2 also gives the progression of electricity access and years of completed schooling across different wealth quintiles in Nepal for the sampled years of 1996 to 2016. Panel (a) here reveals an encouraging trend towards equitable electrification. In 1996, the poorest 20% of households (1st quintile) had virtually no access to electricity, which remarkably increased to 66.8% by 2016, indicating a focused effort to reach the most underserved populations. The second quintile experienced a similar upsurge, from no access to 90.6% in the same period, suggesting widening electrification efforts. The third quintile, initially at 0.6%, saw its access rise to 97.6%, reflecting the middle-income group's improved situation. The fourth quintile began with relatively higher access of 17.6% in 1996, which escalated to nearly universal access at

Table 5.2 Access to electricity and years of education in Nepal

Year of sample	Household wealth index in quintiles				
	1	2	3	4	5
Panel (a): access to electricity					
1996	0	0	0.006	0.176	0.753
2001	0	0	0.031	0.32	0.871
2006	0.026	0.236	0.524	0.813	0.975
2011	0.313	0.726	0.915	0.979	0.999
2016	0.668	0.906	0.976	0.997	1
Panel (b): years of education					
1996	3.132	3.455	4.147	4.61	6.399
2001	3.001	3.438	3.853	4.448	6.275
2006	3.18	3.725	4.186	5.289	6.743
2011	3.758	4.573	5.281	6.031	7.257
2016	4.988	5.917	6.002	6.854	8.128

Note Sample is restricted to adults aged 25 to 60
Source IPUMS-Demographic and Health Surveys: Version 9

99.7% by 2016. Finally, the richest quintile had the smallest relative increase, starting from a high 75.3% and reaching full access, indicative of their already advantaged position. The data illustrates significant strides towards reducing electricity access disparity, with even the lowest income households benefitting from the drive towards universal electrification.

Panel (b) of the table, focusing on years of completed schooling across different wealth quintiles in Nepal from 1996 to 2016, showcases significant educational improvements. In the first quintile, the average years of schooling increased from 3.132 to 4.988, a significant gain of 59.26%. The second quintile saw an even greater increase, from 3.455 to 5.917 years, translating to a 71.26% gain. This indicates substantial strides in educational access and quality for the lower-income groups. The third and fourth quintiles, representing middle-income groups, also experienced notable improvements of 44.73% and 48.68%, respectively. The fifth quintile, starting from a higher baseline, had a smaller relative gain of 27.02%, increasing from 6.399 to 8.128 years of schooling. This data suggests a nationwide effort to enhance educational opportunities, particularly benefiting lower-income groups. The substantial percentage gains in the lower quintiles indicate a narrowing educational disparity, with the most significant strides in the lower and lower-middle-income brackets. However, the consistently lower percentage gain in the wealthiest quintile suggests a possible ceiling effect in educational attainment for this group. This trend of increasing educational equity aligns with the broader development goals of reducing socio-economic disparities.

5.4.3 Dynamics of Access to Electricity and Education Outcomes in Pakistan

Table 5.3 also gives similar information as Table 5.1 and Table 5.2 but for Pakistan for the years 1991 to 2017. In Panel (a), I analyse electricity access across wealth quintiles in Pakistan over the years, and the data displays a dramatic shift, particularly in the lower quintiles. The poorest quintile experienced a meteoric rise of approximately 3567% in electricity access, growing from a mere 2.1 to 77%, indicating focused policy interventions to electrify the most underprivileged households. The second quintile also witnessed significant progress, with a 70.27% increase, highlighting the expansion of access to more broadly cover lower-middle-income groups. The middle quintile, starting from a higher baseline of 82.1%, saw a relatively

Table 5.3 Access to electricity and years of education in Pakistan

Year of sample	Household wealth index in quintiles				
	1	2	3	4	5
Panel (a): access to electricity					
1991	0.021	0.565	0.821	0.956	0.999
2006	0.61	0.932	0.985	0.998	1
2012	0.83	0.976	0.995	0.999	1
2017	0.77	0.962	0.991	0.998	0.999
Panel (b): years of education					
1991	3.702	4.092	4.579	5.324	7.24
2006	3.735	4.488	5.145	6.114	7.922
2012	5.285	6.002	6.804	7.471	9.636
2017	5.287	6.295	6.998	8.008	9.855

Note Sample is restricted to adults aged 25 to 60
Source IPUMS-Demographic and Health Surveys: Version 9

modest gain of 20.71%, reaching near-universal access. The fourth quintile had a smaller increment of 4.39%, indicating that this group already enjoyed relatively high access from the start. Remarkably, the wealthiest quintile maintained nearly complete access throughout the period, starting and ending at 99.9%. This pattern reflects a concerted effort in the country to bridge the access gap between the richest and poorest, with the most striking improvements seen in the lower-income brackets, signifying a commitment to reducing energy inequality and enhancing socio-economic development through electrification.

5.5 Comparative Analysis

Across the three countries examined, the data reveals a clear trend of increasing electricity access over the years, particularly among the poorest quintiles. In each country, the poorest segments began with extremely low access to electricity—often near zero—but saw significant improvements over the study period. This substantial increase in electricity access among the poorest quintiles represents a major stride towards achieving energy equity. Notably, however, while the gap between the poorest and the wealthiest quintiles has reduced considerably, it has not completely closed. The wealthiest quintiles, which started with higher access levels, have mostly maintained their lead, achieving near or actual universal access. In terms of educational attainment, all three countries display a consistent improvement across all wealth quintiles. The lower-income groups, particularly the first and second quintiles, recorded significant percentage gains in the average years of completed schooling. This suggests enhanced educational opportunities and possibly better educational outcomes for these groups. Despite these gains, the data also indicates that the wealthiest quintiles, beginning with a higher baseline of educational attainment, continue to be ahead. The gap in education between the lowest and highest income groups, while narrowed, remains evident.

In this exploration of educational trends within these three countries, I have also plotted the highest level of education attainment among adult males and females of different wealth quintiles over time separately in Figs. 5.6, 5.7, and 5.8. In Fig. 5.6, I show these trends for India. The data reveals a striking trend: a universal decline in the proportion of both males and females attaining only primary education across all wealth quintiles. The trends highlighted in both panels (a) and (b) suggest a significant shift in educational trajectories, with a notable increase in adults pursuing and attaining higher levels of education beyond primary school. This indicates a shift in educational priorities or access and secondary as well as tertiary educational achievements are becoming more prevalent and accessible across various socio-economic strata, as observed in panels (c), (d), (e), and (f). Both secondary education and tertiary education show a trajectory of growth, albeit more stable growth for tertiary education.

The most dynamic changes are observed within secondary education, where the data reveals how educational outcomes intertwine with gender dynamics. In panel (c), the figure illustrates the progression of secondary school attainment as the highest level of education for males in India, categorised by wealth quintiles from the poorest (Q1) to the wealthiest (Q5) over approximately two decades. In 1992, the lowest quintile (Q1) showed minimal attainment, barely above zero, indicating very few males in this group completed secondary education as their highest level. Each subsequent quintile shows a higher level of attainment, with the wealthiest quintile (Q5) demonstrating the highest level, although still under 0.5 on the scale provided. Over time, the data reveals an upward trend for all quintiles. By 1998, there was a noticeable increase in secondary school attainment across all wealth groups, with Q1 showing a more pronounced rise compared to its 1992 level. This upward trajectory continues into 2005, with all quintiles, especially the lower and middle quintiles (Q1, Q2, and Q3), showing significant gains. By 2015, the graph indicates that secondary school attainment has continued to improve for all quintiles, with the most substantial progress observed in the lower wealth

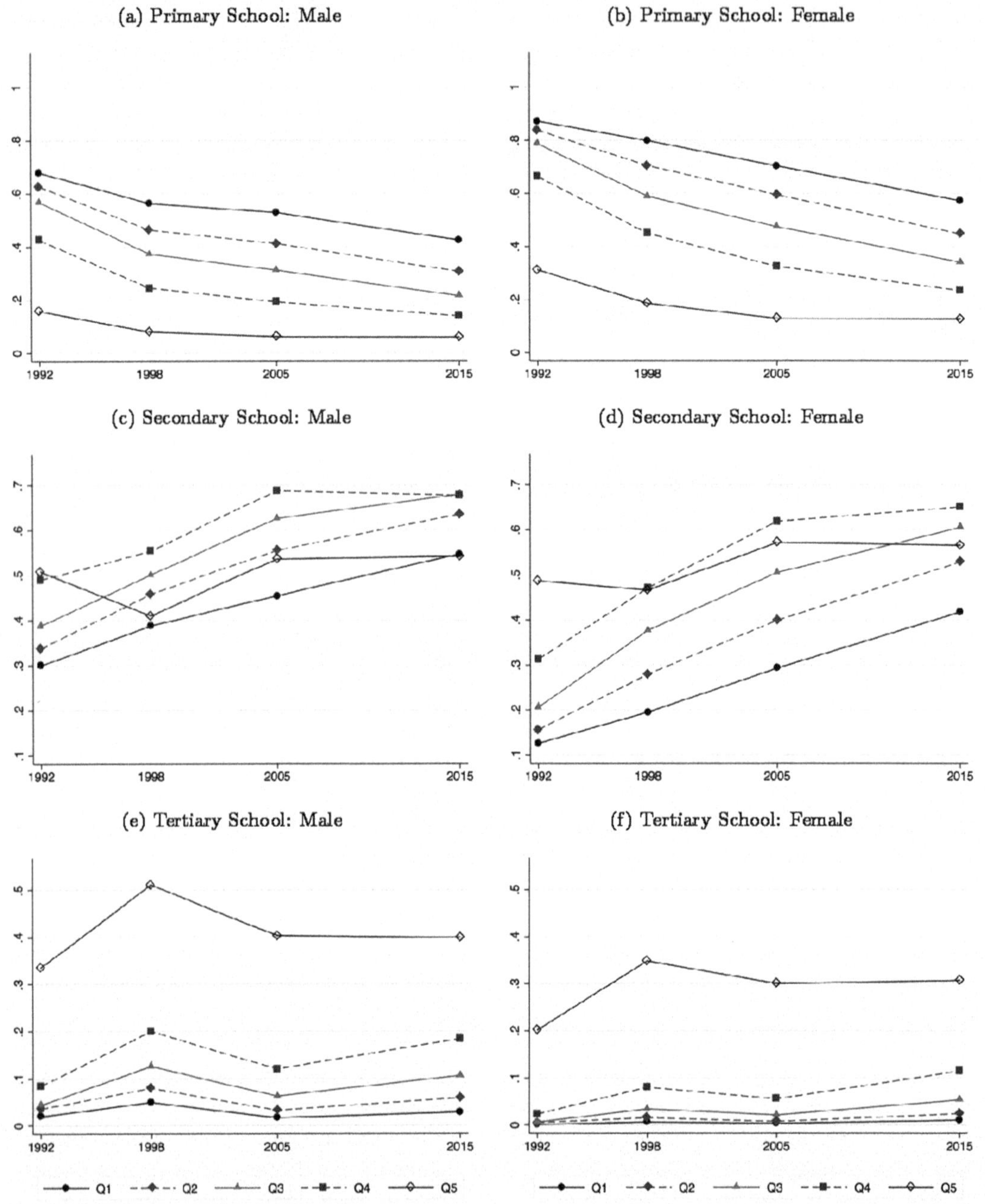

Fig. 5.6 Highest education level attained for adults by wealth quintile and gender over time in India. *Source* IPUMS-Demographic and Health Surveys

quintiles, suggesting an overall increase in educational attainment among males in India across all socio-economic groups.

This pattern is even more pronounced for females. Panel (d) presents the evolution of secondary school attainment as the highest education level for females in India, sorted by wealth quintiles. Similar to the male graph, the poorest quintile (Q1) starts with the lowest level of attainment in 1992. By 1998, there is a noticeable

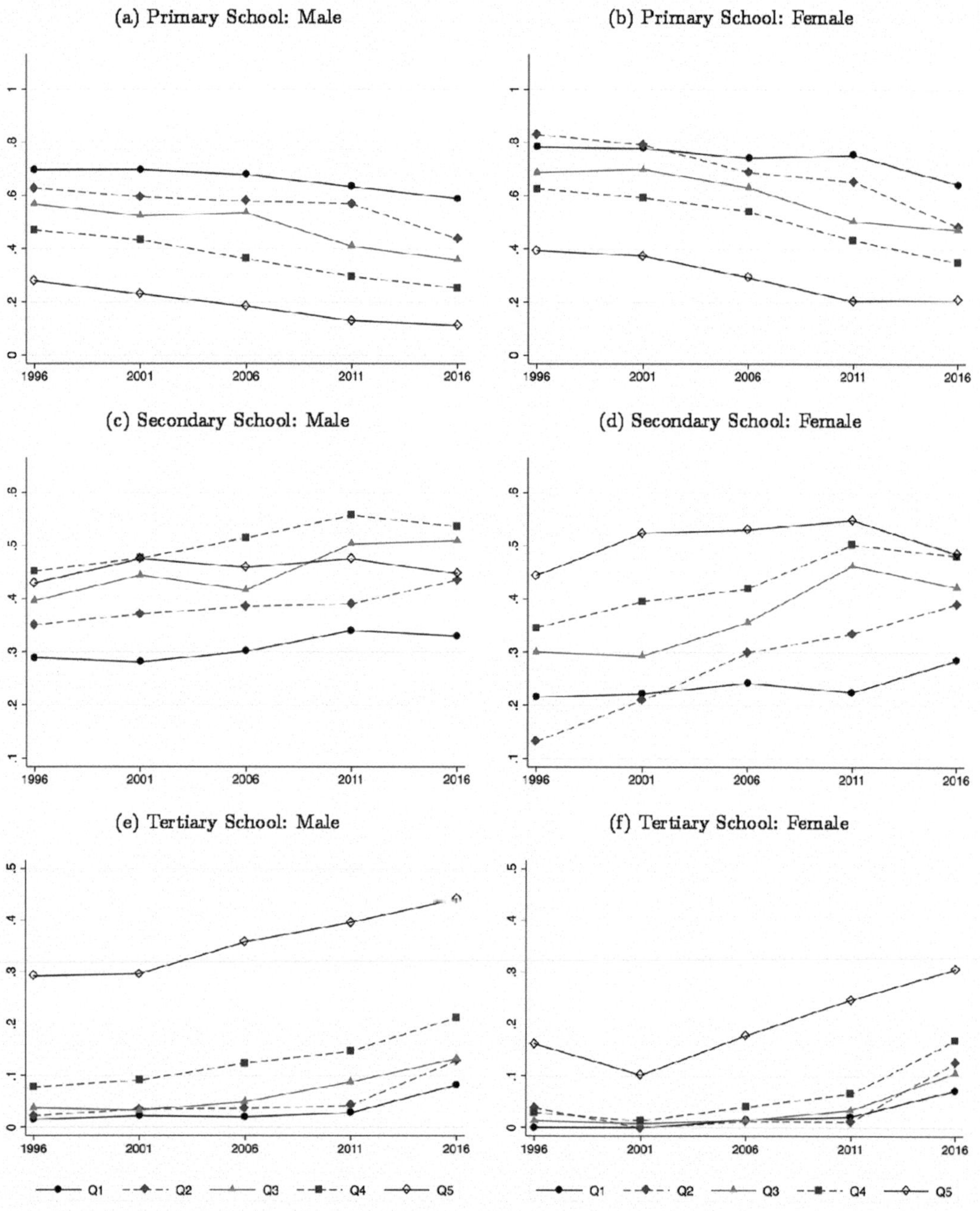

Fig. 5.7 Highest education level attained for adults by wealth quintile and gender over time in Nepal. *Source* IPUMS-Demographic and Health Surveys

improvement for all quintiles, with a particularly sharp rise for the second quintile (Q2), indicating an increase in female educational attainment. As the years progress to 2005, all quintiles continue to ascend, showcasing a significant enhancement in educational achievement for females, especially in the lower and middle quintiles. By 2015, the graph underscores a continual upward trend

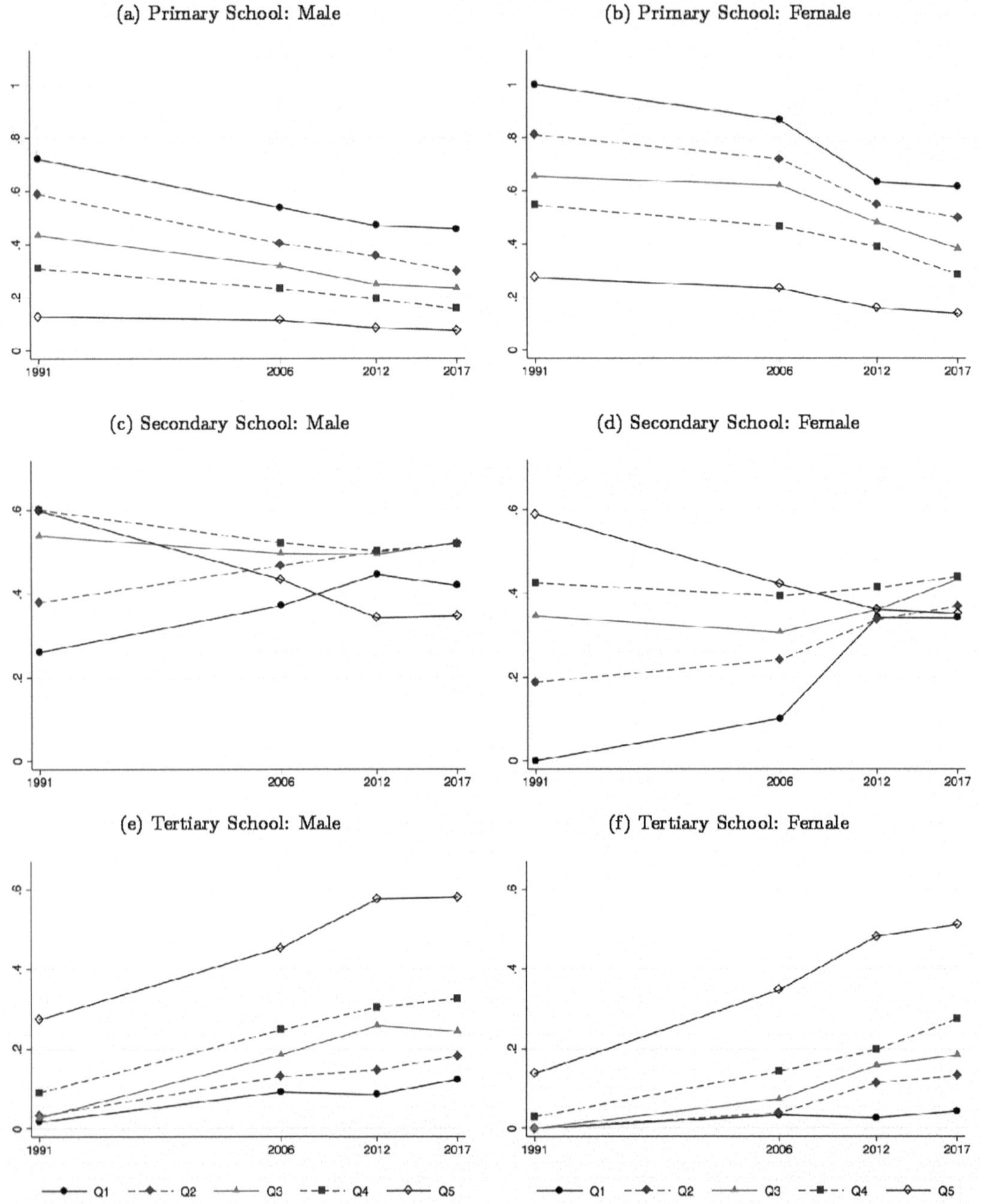

Fig. 5.8 Highest education level attained for adults by wealth quintile and gender over time in Pakistan. *Source* IPUMS-Demographic and Health Surveys

with the lower quintiles again demonstrating a more marked increase, reducing the gap with the higher quintiles.

In the context of Nepal, as depicted in Fig. 5.7, the relationship for primary education attainment follows a similar trajectory for both genders, with decreasing rates of primary school as the highest level of attainment. The secondary education attainment, shown in panels (c) and (d), initially reveals an upward trend across all wealth quintiles, with females experiencing a particularly sharp increase. However, data from the most recent survey in 2016 indicates, for both genders, a notable decrease in secondary education as the highest level of attainment for the middle to upper wealth quintiles (Q3, Q4, and Q5). This decline for the wealthier quintiles appears to coincide with a transition towards tertiary education, suggesting a reallocation of educational focus to more advanced levels of education, as evidenced in panels (e) and (f).

A similar trend is observed in Fig. 5.8 for Pakistan, where the wealthier quintiles have already reached a saturation point in secondary education and are making even more pronounced strides in tertiary education. In both Nepal and Pakistan, the data indicates that while the lower quintiles are making significant progress in catching up in terms of secondary education, the higher quintiles are moving towards acquiring more advanced educational qualifications. Throughout these patterns, females show the most significant rates of increase, highlighting a noteworthy shift towards higher educational achievements among women in these regions. This dynamic suggests a broader transformation in educational priorities and access, particularly for female students who historically have had lower levels of educational attainment.

The comparison suggests that while males and females from wealthier quintiles have consistently high levels of education attainment, females, particularly from poorer quintiles, are closing the educational gap over time. The correlation with increased access to electricity could be one of the factors contributing to this trend. Access to electricity, a proxy for modern infrastructure and development, has likely played a pivotal role in this educational evolution. By providing the amenities necessary for a conducive learning environment—such as lighting for study after dusk, refrigeration for preserving food, and power for electronic devices—electricity has potentially alleviated some of the time-intensive domestic burdens from women. This liberation of time may have afforded young girls and women the opportunity to pursue education more vigorously. Furthermore, the availability of electricity can transform household dynamics, where women, traditionally bound to domestic roles, can engage in more productive or educational activities. This shift not only enhances the educational prospects for women but also aligns with the global narrative of empowerment and gender equality. As women gain access to education, they are able to contribute more significantly to the workforce, drive economic growth, and challenge the status quo of gender roles.

5.6 Conclusion

As I conclude this chapter, the observed interdependence between increased access to electricity and the rise in educational attainment across India, Nepal, and Pakistan underscores a wider narrative of socio-economic empowerment and equity. The strides made in electrification, notably in impoverished regions, resonate deeply with the tenets of distributive justice, advocating that the benefits of technological progress should extend beyond the affluent. The data vividly illustrates this evolution, showing that even the most marginalised households have witnessed substantial enhancements in access to electricity, closely followed by educational improvements. Particularly, the upward educational trends among women and lower-income groups highlight the transformative potential of equitable energy distribution. This progress aligns with the ambitions of SDG 7, which promotes sustainable energy for all as a pathway to a more educated, empowered, and equitable society.

This discourse on the influence of electricity on socio-economic development is a testament to its essential role in modern civilisation. The chapter has drawn a correlation between increased electricity access and economic growth, showcasing higher GDP per capita as a reflection of broader access to modern amenities and a higher standard of living. Specifically, the data from India, Nepal, and Pakistan reveal a significant trend: As electricity access has expanded, educational attainment, especially in lower-income households, has concurrently improved. This pattern is not just about economic measures; it embodies the capacity for social change. Electricity has emerged as a catalyst for educational advancement, a facilitator of gender equality, and a driver of improved health outcomes. The significant expansion of electricity access in Asia, as documented in the data, has been accompanied by enhanced educational outcomes, suggesting that increased access to electricity may contribute to a more equitable distribution of educational opportunities.

However, this chapter's descriptive analysis, while thorough in detailing trends in electricity access and educational attainment, does not establish causality. It highlights a correlation but stops short of dissecting the complex interplay of improved electricity access with socio-economic changes and policy developments in these Asian countries. This gap underscores the need for comprehensive studies that delve deeper into causal relationships and explore the long-term impacts of electrification on education and socio-economic status. Further research should employ a multifaceted approach, incorporating demographic factors, longitudinal data, and policy analysis to understand the full extent of electricity's role in educational and social advancement. Such studies are crucial in shaping energy policy as a tool for promoting energy justice and achieving global development goals. They must recognise that access to electricity is a means to facilitate fundamental human rights, such as education and gender equality. As the data suggests, increased access to electricity enables women and girls to shift their focus from domestic chores to educational pursuits, challenging gender norms and contributing to procedural and recognition justice. Yet, despite these promising developments, the journey towards energy justice is far from complete. Policies must be inclusive, participatory, and responsive, especially to the needs of the vulnerable, to realise the aspirations of SDG 7 and the broader objectives of energy justice outlined in this book.

References

Acheampong AO, Boateng E, Amponsah M, Dzator J (2021) Revisiting the economic growth–energy consumption nexus: does globalization matter?. Energ Econ 102:105472

Aklin M, Cheng CY, Urpelainen J, Ganesan K, Jain A (2016) Factors affecting household satisfaction with electricity supply in rural India. Nat Energy 1(11):1–6

Apenteng BA, Opoku ST, Ansong D, Akowuah EA, Afriyie-Gyawu E (2018) The effect of power outages on in-facility mortality in healthcare facilities: evidence from Ghana. Glob Public Health 13(5):545–555

da Silveira Bezerra PB, Callegari CL, Ribas A, Lucena AF, Portugal-Pereira J, Koberle A, Szklo A, Schaeffer R (2017) The power of light: socio-economic and environmental implications of a rural electrification program in Brazil. Environ Res Lett 12(9):095004

Gaye A (2007) Access to energy and human development. Human development report, 2008, p 2007

Government of India Ministry of Power (2023). Saubhagya. Retrieved from https://powermin.gov.in/en/content/saubhagya

Grogan L, Sadanand A (2013) Rural electrification and employment in poor countries: evidence from Nicaragua. World Dev 43:252–265

International Energy Agency (2014) World energy outlook 2014. Retrieved from https://www.iea.org/reports/world-energy-outlook-2014

International Energy Agency (2020) Defining energy access: 2020 methodology. Retrieved from https://www.iea.org/articles/defining-energy-access-2020-methodology

International Energy Agency (2023a) SDG7 database. Retrieved from https://www.iea.org/data-and-statistics/data-product/sdg7-database#energy-intensity

International Energy Agency (2023b) SDG7: data and projections access to electricity. Retrieved from https://www.iea.org/reports/sdg7-data-and-projections/access-to-electricity

International Monetary Fund (2023) World economic outlook database. Retrieved from https://www.imf.org/en/Publications/WEO/weo-database/2023/October

Kanagawa M, Nakata T (2008) Assessment of access to electricity and the socio-economic impacts in rural areas of developing countries. Energy Policy 36(6): 2016–2029

Kooijman-van Dijk AL, Clancy J (2010) Impacts of electricity access to rural enterprises in Bolivia, Tanzania and Vietnam. Energy Sustain Dev 14(1): 14–21

Njiru CW, Letema SC (2018) Energy poverty and its implication on standard of living in Kirinyaga, Kenya. J Energy 2018:1–12

Oum S (2019) Energy poverty in the Lao PDR and its impacts on education and health. Energy Policy 132:247–253

Oswald Y, Owen A, Steinberger JK (2020) Large inequality in international and intranational energy footprints between income groups and across consumption categories. Nat Energy 5(3):231–239

Pachauri S, van Ruijven BJ, Nagai Y, Riahi K, van Vuuren DP, Brew-Hammond A, Nakicenovic N (2013) Pathways to achieve universal household access to modern energy by 2030. Environ Res Lett 8(2):024015

Sambodo MT, Novandra R (2019) The state of energy poverty in Indonesia and its impact on welfare. Energy Policy 132:113–121

Sovacool BK (2012) The political economy of energy poverty: a review of key challenges. Energy Sustain Dev 16(3):272–282

The World Bank (2023) Access to electricity. Retrieved from https://data.worldbank.org/indicator/EG.ELC.ACCS.ZS

Winther T, Matinga MN, Ulsrud K, Standal K (2017) Women's empowerment through electricity access: scoping study and proposal for a framework of analysis. J Dev Eff 9(3):389–417

World Health Organization (2018) Policy brief for health and energy linkages—maximising health benefits from the sustainable energy transition. Retrieved from https://sustainabledevelopment.un.org/content/documents/17486PB_10_Draft.pdf

World Health Organization (2023) Database on electrification of healthcare facilities. Retrieved from https://www.who.int/data/gho/data/themes/database-on-electrification-of-health-care-facilities

United Nations Department of Economic and Social Affairs (2020) World social report 2020 inequality in a rapidly changing world. Retrieved from https://www.un.org/development/desa/dspd/wp-content/uploads/sites/22/2020/02/World-Social-Report2020-FullReport.pdf

United Nations Development Programme (2010). *UNDP and Energy Access for the Poor: Energising the Millennium Development Goals.* Retrieved from https://www.undp.org/sites/g/files/zskgke326/files/publications/2593.EnergyAccess_Booklet_Revision02.pdf

United Nations Development Programme (2018). *Policy brief on the interlinkages among energy, poverty and inequality.* Retrieved from https://sustainabledevelopment.un.org/content/documents/17480PB8.pdf

United Nations Development Programme (2022) Human development reports. Retrieved from https://hdr.undp.org/

United Nations Office for South-South Cooperation (2011) Rural energy development programme. Retrieved from https://www.southsouth-galaxy.org/wp-content/uploads/2020/08/Nepal_REDP_web.pdf

Urpelainen J (2011) Export orientation and domestic electricity generation: effects on energy efficiency innovation in select sectors. Energy Policy 39(9): 5638–5646

Pauline Khoo is a Lecturer in Economics at Monash University Malaysia. She holds a Ph.D. and M.A. in Economics from the University of Illinois at Chicago, as well as a B.Sc. in Mathematics from Hollins University. Dr. Khoo's principal areas of interest lie in the economics of education, labour economics, and applied microeconomics. She aims to enhance understanding of the mechanisms that drive education and labour markets, contributing her expertise towards the development of these critical sectors.

Hydrogen Exports and Energy Justice: the Case for Australia

6

Diane Kraal and Richard Taylor

Abstract

Significant research, funding and advocacy by government and industry in favour of hydrogen technologies saw investment in Australia almost double between 2021 and 2022 to A $230–303 billion. However, in late 2023, Australia still had no hydrogen producing facilities, while renewable energy contributed 35.9% of electricity generation in 2022. Contrast this support for hydrogen with Bruce Pascoe's book Dark Emu (2014) which presents material on Australia's First Nations people, whose sustainable agriculture and aquaculture activities were evident before British colonisation. Pascoe's book proffers a need for reflection on the use of natural resources. The need for respect raises the energy justice concept as the push for hydrogen gains traction. This chapter uses a targeted desktop review of the literature and the frame of energy justice theory, including its tenets of recognition, restorative, distributive and cosmopolitan justice, to query whether large-scale hydrogen production is a reasonable path for a just transition.

Keywords

Hydrogen · Taxation · Energy justice

6.1 Introduction

Significant investment in research and development of hydrogen technologies has occurred in Australia and internationally. For instance, the EU has established a funding bank of €3 billion for hydrogen initiatives; in addition, there is separate funding within member states (Kneebone 2023). This recent investment in hydrogen by the EU and the US (Osborne 2023) has put pressure on Australia for similar initiatives.

Since 2021, investments in Australia for hydrogen projects by government, industry and investors have almost doubled from an estimated range of A$133–185 billion in 2021 to A$230–303 billion in 2022—corresponding to some 48 projects, or 32 more than in the previous year. This did not include four regional 'hubs', which have been classified as hydrogen-related infrastructure (DISER 2022: 11). In early 2022, Australia's Clean Energy Council (2023: 50) reported on the successful recipients of federal funding for hydrogen projects comprising A$525 million for hubs in Western Australia, New South Wales, Northern Territory, Queensland, South Australia and Tasmania.

D. Kraal (✉) · R. Taylor
Monash University, Melbourne, VIC, Australia
e-mail: diane.kraal@monash.edu

R. Taylor
e-mail: richard.taylor1@monash.edu

A. Trakic et al. (eds.), *Energy Justice*, Sustainable Development Goals Series,
https://doi.org/10.1007/978-981-97-6059-6_6

The new Australian federal government, elected in May 2022, confirmed the above-mentioned development of its renewable energy hydrogen industry by announcing A$2 billion of funding in the 2023 Budget (Chalmers et al. 2023). However, in late 2023, Australia still had no hydrogen facilities in production (Clean Energy Council 2023) despite releasing a national hydrogen strategy back in 2019 (COAG Energy Council 2019). Contrast this gap with Australia's growing renewable energy industry, which contributed 35.9% of electricity generation in 2022, which is up from 32.5% in 2021.

Renewable energy for electricity in Australia comprises hydro 7.1%, wind 12.8%, small-scale solar photovoltaic (PV) 9.3%, bioenergy 1.4%, medium-scale solar PV 0.4% and large-scale solar PV 5%. To address the underdevelopment of the hydrogen industry in Australia, the Energy and Climate Change Ministerial Council (DCCEEW 2023a) agreed to review the national hydrogen strategy (COAG Energy Council 2019) during 2023 to ensure it positions Australia on the path of global hydrogen leadership by 2030 for both export profits and decarbonisation of its industries.

The hydrogen review website (DCCEEW 2023b) claims that hydrogen in Australia has the potential to be an important contributor to the transition to net zero through its usage as energy for transport, grid firming[1] and the production of chemicals and metals. It posits that Australia has upwards of AU$300 billion of potential hydrogen investments, including domestic projects as well as large export projects.

The hydrogen review site notes that while Australia was the third nation to publish a hydrogen strategy in 2019, hydrogen opportunities have emerged and grown to generate international interest, with now more than 30 nations posting hydrogen strategies (DCCEEW 2023b). The Federal Government initiated the review of the national hydrogen strategy in collaboration with the States and Territories. Between July and August 2023, public submissions were called, and further opportunities were forthcoming for community consultation. (DCCEEW 2023b).

Contrast this advocacy for Australia as a frontier nation for hydrogen production and export—with Bruce Pascoe's book *Dark Emu* (2014). Pascoe presents material in support of his argument that over the 65,000 years of continuous presence by Australia's First Nations people, their sustainable agriculture and aquaculture activities were evident before British colonisation and settlement from 1788. This chapter invokes the *Dark Emu* inference for modern Australia to respectfully utilise the knowledge of its First Nations people. Pascoe's book proffers a need for reflection in relation to the use of natural resources. In the context of hydrogen, the heavy-handed approach of production for export is arguably not a respectful use of land resources. The need for respect raises the energy justice concept of recognition justice, which is due to First Nations people, as the push for hydrogen energy projects gains traction.

Energy justice is about ensuring access to affordable, reliable, sustainable and modern energy for all (Heffron and McCauley 2017). It is imperative for a just transition to a low-carbon economy (Heffron 2021) that First Nations peoples' land knowledge be recognised in formulating Australia's hydrogen policy.

This chapter uses a targeted desktop review of the literature and the frame of energy justice theory to answer the main research question about whether Australian government's aspirations for large-scale production of hydrogen for export is a reasonable path for a just transition. The subsidiary research question considers the distributional justice issue of proposing hydrogen production without adequate tax considerations. It is noted that income tax applies to all commercial businesses, and that special tax treatment is applied to natural resource ventures, such as gas extraction in Australia—for they are also subject to a secondary resource tax. The tax system is pivotal to transferring the benefits from one part of the economy to the wider community—and especially for natural resurce projects in the form of an additional production or profits tax.

[1] Grid firming is defined as the addition of another energy resource to renewables—to balance the intermittent variation of renewables for the purpose of electricity generation.

There is a multi-faceted debate about whether to position Australia as the next frontier for hydrogen production—using energy sources of wind and solar installations or even gas and coal. The contribution of this chapter to the debate is that it marshals the thoughts from the literature, such as Pascoe's *Dark Emu*, and draws upon the energy justice tenets of recognition, restorative, distributive and cosmopolitan justice to query the veracity of large-scale production of hydrogen for export as a path to a just transition.

6.1.1 Chapter Roadmap

This chapter begins with an overview of hydrogen for a general readership. This is followed by recent literature on hydrogen examined in the context of an alternative energy source to address high carbon emissions sectors. Then, the methodology is explained, followed by the findings and analysis that applies the energy justice framework. The chapter ends with the conclusion.

6.2 Hydrogen: Overview

There are three primary forms of hydrogen production. First is electrolysis, which extracts hydrogen from water using electricity for energy. Then, there are two thermochemical reactions, using natural gas for energy in a process known as steam methane reforming; and coal for energy in a process known as gasification (COAG Energy Council 2019: 3).

Only the first production method of electrolysis, whereby an electrical charge is used, does not produce carbon emissions if the electricity is generated from renewable energy. The electrolysis process splits water molecules into hydrogen and oxygen. Thus, hydrogen produced using electrolysis from energy comprising 100% renewable electricity (such as wind and solar) is known as green hydrogen.

The latter two thermochemical processes are the most common; however, and the production processes result in the fossil fuels yielding hydrogen but with residual emissions, such as carbon dioxide. Given the overall aim is to reduce greenhouse gases, these emissions require carbon capture and storage. Thus, hydrogen extracted from natural gas via the process of steam methane reforming (SMR) and utilising carbon capture and storage (CCS) is known as blue hydrogen. Under this process, methane from natural gas is heated with steam to produce hydrogen but with a mixture of carbon monoxide and carbon dioxide.

Further, the use of either black or brown coal via the process of gasification and using CCS is known as black or brown hydrogen. Through gasification, coal turns into a very hot synthesis gas (known as syngas), to produce hydrogen, along with other trace greenhouse gases and particulate matter. Grey hydrogen is the name given to the use of natural gas via the SMR process with no CCS.

Currently, in Australia, there is no operating facility to produce hydrogen for energy via electrolysis using grid electricity or 100% renewable electricity. Neither is there an operative facility for hydrogen via coal gasification with CCS, nor coal gasification with no CCS, nor gas coupled with SMR processes (COAG Energy Council 2019: xiv). However, by 2019, around 350,000 tonnes of grey hydrogen has been produced annually for making ammonia, composed of hydrogen and nitrogen, for fertiliser and other chemicals (COAG Energy Council 2019: 40).

Green hydrogen (produced from renewable energy) could, over time, become cheaper than blue hydrogen (produced from natural gas but with carbon capture) and with less emissions than grey hydrogen (produced from natural gas without carbon capture).

Scientists have adopted standard colour designators to indicate the different sources of energy for hydrogen production (Bartlett 2019: 51; De Vos and Graham 2022), which enables the wider community to understand the basics of hydrogen production. The most common colour codes are listed in Table 6.1.

In terms of potential for hydrogen production, the REN21 global renewables resources report found that 'Australia had the largest number of announced renewable hydrogen plants worldwide as of 2022; due to its abundant solar and

Table 6.1 Colour codes: sources of energy for hydrogen production

Green for renewable energy used as the energy feedstock. Wind and solar energy are examples. The electricity generated from these renewable sources split water molecules into hydrogen and oxygen. Carbon emissions are minor
Blue for gas used as the energy feedstock. Hydrogen is produced via steam methane reforming (SMR). The production process uses carbon capture and storage (CCS) technology for carbon emissions
Black or brown for coal used as the energy feedstock. The black and brown colours refer to bituminous (black) and lignite (brown) coal, respectively. Hydrogen is produced via gasification. This type of hydrogen production uses CCS technology for carbon emissions
Grey for gas used as the energy feedstock. Hydrogen is produced via SMR, but without CCS

Source Created by the authors in 2024

wind resources, the country is expected to see some of the lowest costs for producing renewable hydrogen by 2050' (REN21 2023: 50). But, as noted in this chapter's introduction, in 2023, there are no commercially operating hydrogen plants in Australia. This compares to almost 1400 MW of capacity in the European Union and 300 MW of capacity in the United States, where hydrogen projects have reached final investment decisions (DCCEEW 2023c: ii). With those facts in mind, the chapter now turns to a wider literature review.

6.3 Recent Literature

At a 2023 conference roundtable in France on the role of hydrogen in the energy transition (Energy Justice Project 2023), the keynote speakers gave European science-oriented perspectives—both of which were even-handed, about the promise of hydrogen to lower carbon emissions (Burlet 2023; Rigollet 2023). Another roundtable participant warned of the negatives of mass hydrogen production from accumulations of toxic brine from production that uses desalinated water (Merad 2023). One participant gave an uncomfortable insight about the EU focus on hydrogen, which is the expectation of the use of land and associated infrastructure in former colonies (such as Australia) to produce the bulk of hydrogen for export to the northern hemisphere countries (Kneebone 2023). The information imparted by the last roundtable speaker, in particular, was a motivation to write this chapter.

A selected overview of recent international literature on pro- and anti-hydrogen positions for domestic and export was undertaken. Search terms included hydrogen transportation, hydrogen production, hydrogen storage, hydrogen export and hydrogen in Asia, Europe, Africa, the Middle East and South America.

6.3.1 Pro-Hydrogen Views

6.3.1.1 Environmental

The positive case for hydrogen as a potential source of greenhouse gas emission reduction is well represented in the literature. When hydrogen is burned, it produces only water vapour and heat (Hassan et al. 2023: 5), unlike fossil fuels, which produce hazardous gases, including carbon dioxide, carbon monoxide and methane (Olabi et al. 2023: 672). As part of the energy transition from fossil fuels to renewables, hydrogen created from excess renewable energy and stored locally can be used to produce energy to stabilise energy grids when renewables are insufficient (Kolb et al. 2022). Hydrogen can provide energy security for countries reliant on energy imports, such as Japan, when global fossil fuel production transitions in supplier countries, such as Australia (Hancock and Ralph 2021). An effective hydrogen industry is seen as an imperative to address climate change (Garlet et al. 2023: 2).

The environmental advantages of hydrogen are particularly prevalent when considering green hydrogen produced from water electrolysis from renewable energy sources (Panchenko et al. 2023: 4552). However, Incer-Valverde et al. (2023: 8) point out that as recently as 2021, only 0.04% of the 94 million tonnes of hydrogen produced was made from renewable energy sources. The research notes that the majority of production is for hydrogen derived from natural gas.

Hauglustaine et al. (2022) consider scenarios for a future hydrogen transition. A green hydrogen economy is more beneficial as there are less $C0_2$ emissions and lower leakage rates compared to blue hydrogen that is referred to as the 'climate penalty'. The transition to cleaner energy sources must include renewables. Hjeij et al. (2023) aim to quantify the hydrogen export competitiveness of a range of countries, with the best ranked being the United States, Australia, Canada, United Kingdom, China, Norway, India, Russia, Netherlands and Germany.

The benefits from green hydrogen are seen as greatest where production is for domestic use rather than long-distance transport for export. Storage increases both the cost and the need to use fossil fuels in the process. Borsboom-Hanson et al. (2022: 31,549) find that by keeping distances short and throughput high, transport might be inexpensive and with reduced impact. Squadrito et al. (2023: 10) find that hydrogen production can be water-use neutral if used locally, but that is not the case when large volumes are exported vast distances.

6.3.1.2 Economic Opportunities

Beyond the direct environmental benefits, hydrogen-producing or potentially producing countries see benefits from new industrial developments and training opportunities for a young workforce, particularly in Africa (Agyekum 2023: 5). Works include new infrastructure, such as ports, pipelines and transport hubs (Agyekum 2023: 5; Cardinale 2023: 113,623). Jurisdictions with abundant solar or wind resources have advantages in the use of excess renewables, such as the case of Brazil (Garlet et al. 2023: 2).

6.3.1.3 Differences by Country and Region

The benefits of hydrogen in the green economy have led to more than 30 countries and the European Union developing hydrogen strategies (Dillman and Heinonen 2022: 1). The nature of pro-hydrogen views differed somewhat by country and by region. Countries with surplus renewable energy have been outspoken proponents of hydrogen's potential. Brazil and parts of South America have excess renewable energy that could be energy sources to produce hydrogen for export (Garlet et al. 2023: 2; Galván et al. 2022). North Africa, particularly Egypt, Algeria and Morocco, have high solar energy potential and port facilities, and in some cases, natural gas pipelines with spare capacity that could arguably be repurposed to transmit hydrogen (Cardinale 2023: 113,623; Kolb et al. 2022: 133,289). Van der Zwann et al. (2021) claim that producing solar energy in North Africa and sending it to Europe is better than production in Europe.

The varying countries and political regions have increased interest in the geopolitical implications of large-scale hydrogen exports (Squadrito et al 2023: 5). Nunez-Jimenez and De Blasio (2022) state that the roles and views will be determined by a country's renewable potential, its water endowment and infrastructure capabilities.

Australia, with its abundant natural resources, large land mass and high solar potential, is also in the category of potentially large hydrogen exporters (Panchenko et al. 2023: 4553). Burke et al. (2022) investigate three pathways by which Australia operates as a 'global superpower in energy' by the export of zero-carbon energy and products: direct exports of renewable electricity via sub-sea cables, exports of zero-carbon fuels such as green hydrogen and the export of 'green' metals processed from Australian ores using renewable energy. Other research finds that Australia has the land and renewable energy resources to become a key exporter of these commodities, but the article did not include an environmental impact investigation (Walsh et al. 2021).

Large importers looking for replacement of fossil fuels have also been active in hydrogen strategies and developing hydrogen plans, particularly to support their industrial and manufacturing segments of the economy. In the case of the potentially large importers such as Germany, Brauner et al. (2023: 5) connect German support for hydrogen to economic viability, its profitability and its ability to be very beneficial to the its economy in terms of diversifying the country's energy sources and reducing energy costs in the future. Similarly, in the Asia–Pacific, energy importers like Korea, Japan, China and Singapore could become centres for high hydrogen demand (Lebrouhi et al. 2022: 7041).

Lebrouhi et al. (2022: 7038) usefully highlight the differences by dividing hydrogen-potential countries into (1) export champions (Australia, the United States, Morocco, Norway); (2) water-constrained producers (China and Saudi Arabia); (3) major importers (Japan, Korea and parts of Europe); (4) self-sufficient producers (Turkey, Spain, Thailand); and (5) infrastructure-constrained producers (India, Russia and South America).

The pro-hydrogen case becomes more complicated when looking beyond the countries with large renewable energy potential or with major needs for replacement of fossil fuels. Middle Eastern countries have high renewable energy potential (Gado and Hassan 2023; Alsaba, Al-Sobhi and Qyyum, 2023) but have water and infrastructure challenges. The Middle East remains the world's leading region for the production of fossil fuels, with significant economic incentives to adopt blue rather than green hydrogen production. For instance, Turkey has a large hydropower potential (Karayel et al. 2023), but increasing the number of hydropower dams will have environmental consequences. In India, high population density and a large agricultural economy create a trade-off with farmland (Pawar et al. 2021: 27,249).

In Niger, just supplying domestic energy needs and the transport sector alone is thought to require 5% of the total land area (Bhandari 2022: 808). In terms of Sub-Saharan Africa, the continent is seen as having great potential for renewables and would benefit from trade (Agyekum 2023: 3). However, some governments appear more focussed on addressing energy poverty and improving energy access than on developing large-scale energy exports (Brauner et al. 2023: 1).

6.3.2 Anti-Hydrogen Views

6.3.2.1 Cost of Production

Green hydrogen has the major barrier of high operational costs from using renewable energy, and seen as a serious impediment to its expanded use (Hassan et al. 2023: 5; Guan et al. 2023: 17). Contributing to the higher operational costs is the current energy inefficiency of existing water electrolysis technologies and energy losses incurred in the production and transportation process of hydrogen (Hassan et al. 2023: 8) The cost differential in regard to the type of energy input is particularly important, noting the status quo of cheaper fossil fuels favours blue hydrogen production (Durakovic et al. 2023; Agyekum 2023: 1).

Higher costs extend to the investment capital required for renewable energy sources, ports and pipelines (Hinkley et al. 2022). The costs vary significantly depending on the availability of renewables and their price of delivery in different areas (Guan et al. 2023: 17).

6.3.2.2 Land Use

Land for solar farms and wind turbines must compete with other land uses by populous nations. Agyekum (2023: 3) highlights Africa's 1.34 billion population and its growth rates as factors needing consideration. If Niger's case, where 5% of land is required to meet domestic and transport energy needs—were to be replicated across other countries (Bhandari 2022: 808), the potential for conflict over land use becomes obvious.

6.3.2.3 Access to Water

Access to water resources for hydrogen production is a major issue in some areas, even though there may be abundant solar energy. Large exports of hydrogen will require significant water resources, whether desalinated water or fresh water (Lebrouhi et al. 2022: 7021). Water resource requirements for hydrogen may be in direct conflict with hydropower reservoirs that supplement solar and wind-based renewables (Lebrouhi et al. 2022: 7021; Karayel et al. 2023). Desalination from seawater is energy-intensive, costly to undertake and creates environmental impacts. However, it may present an opportunity for island nations with abundant wind or other renewable forms (Santos Ayllón 2022).

6.3.2.4 Competition for Infrastructure

The infrastructure requirements for hydrogen export are considerable, particularly around transport and port facilities (Garlet et al. 2023: 9). Countries that take quick advantage of a hydrogen export sector are likely to have existing infrastructure, possibly repurposed for hydrocarbons (Brauner et al. 2023: 9). These factors may reinforce past patterns of investment and post-colonial dependencies (Brauner et al. 2023: 10; Nuñez-Jimenez and De Blasio, 2022: 35,553).

Finally, for Africa, corruption in the energy sector, political instability and the lack of a regulatory framework are also identified as barriers to the establishment of a hydrogen industry (Agyekum 2023: 1).

6.3.2.5 Justice Issues

The literature suggests that the perceived benefits of a hydrogen industry need to be balanced with the potential impacts on regions and countries with high and various natural resource endowments; and that the consideration of hydrogen should not be purely technocratic (Hanusch and Schad 2021: 85). Müller et al. (2022) highlight the risk of hydrogen industry injustices if large hydrogen exports were to emerge that mirror historical, one-sided energy trading patterns that disadvantage source countries. Dillman and Heinonen (2022: 10) see inherent social risks that, if not managed, will impede a transition to hydrogen.

Le et al. (2023: 15) have found that there is a widespread perception that hydrogen is costly to produce, lacks dependability and is not safe. Further, Cremonese et al. (2023: 19,431) see good governance, social inclusion and community engagement as important to a just hydrogen transition. Likewise, Bharadwadj et al. (2023: 7) highlight the importance of transparency, particularly when communities will be impacted. Brauner et al. (2023: 243) notes that African countries prioritise energy access over exports.

The gap in the literature is that no research advocated for sharing the load of the production of hydrogen equitably across countries and regions. This chapter seeks to address that omission as well as taxation necessary to raise funds to implement a hydrogen economy.

6.4 Methodology

A targeted desktop review of the literature is conducted to answer the main question of whether the Australian government's aspirations for large-scale production of hydrogen for export is a reasonable path for a just transition. This question raises recognition and restorative justice issues. The subsidiary question asks about the distributional justice issue of proposing hydrogen production without adequate tax considerations. Cosmopolitan justice, explained below, arises as an issue as well. Income tax applies to activities that include for-profit ventures and is a generally accepted (albeit unpopular) impost. Natural resources extracted in Australia are also subject to a secondary resource tax and can be more contentious, as will be discussed.

This chapter's focus is limited to the latest Australian reports and reviews on hydrogen, current government funding, initiatives for commercial hydrogen projects, and domestic test sites for hydrogen production and carbon capture and storage (CCS). Databases such as Science Direct, AGIS and other sources, such as current conference materials, and government and institutional websites, have been accessed for relevant data.

6.4.1 Framework

The framework draws on energy justice theory, being a decision-support tool for energy policymakers. Energy policy formulation, such as for hydrogen, is 'dominated by economists and industry, where economic costing is the prime tool for decision-making' (Heffron and McCauley 2017: 664). Energy justice emerged from, inter alia, the principle of distributive justice (Heffron 2021). Energy justice has been defined as 'a global energy system that fairly disseminates both the benefits and costs of energy' and includes procedural fairness, such as due process in energy decision-making as well as transparency and accountability (Sovacool et al. 2016: 4).

In this chapter, the energy justice focus is primarily on the justice tenets of recognition, restoration, cosmopolitanism and distribution (Heffron 2023). Energy justice is specifically used for analysis with a focus on recognition justice, which directs efforts towards recognising the rights of indigenous communities generally, and in particular for this work, Australia's First Nations people (Altman et al. 2023). It follows how restorative justice covers the belief that any injustice caused by the energy sector should be rectified. Energy justice is also concerned with distribution justice, whereby the benefits and the negatives of the energy sector are addressed. Taxation is prominent as a key tool for the redistribution of revenues and for levying imposts on negative outcomes. Cosmopolitan justice underpins the belief that we are all citizens of the world and must consider the effects beyond our borders and from a global context.

Energy justice theory has already been used for analysis in the area of taxation and energy extraction, for example, to provide insights on Australia's petroleum and coal problematic policy issues (Kraal 2024); in comparing the taxation legislation for gas in Australia and the Netherlands (Kraal et al. 2020; Kraal and Heffron 2022); and in decision-making on the mineral industry's tax incentives (Kraal 2019; Kraal and Odhuno 2019). Energy justice theory has been used in inquiries into energy policy implications for the comparator countries of Australia, Indonesia, Malaysia and PNG; and in highlighting Australian government concerns about low tax revenues on gas from Chevron's Gorgon LNG project (Kraal 2017).

The energy justice framework relates to the United Nations' sustainable development goal (SDG) 7 that is about ensuring access to affordable, reliable, sustainable and modern energy for all. The concept of a 'right to energy' is fundamental to addressing issues such as energy poverty and energy injustice (Shyu 2021).

6.5 Findings and Analysis

The Council of Australian Governments' (COAG) Energy Council set up a Hydrogen Working Group in 2018, chaired by Australia's Chief Scientist, Dr Alan Finkel, and resulted in a National Hydrogen Strategy report. The report's chapters cover Australia's hydrogen potential and summarise the ways in which States and Territories could take up the opportunity of producing 'clean' hydrogen, (COAG Energy Council 2019: 9). Fossil fuels are included energy sources for hydrogen production. The strategy requires 'the journey' of building hydrogen hubs. It advocates the need to grow a hydrogen production industry for local and export uses that requires a common vision across the national and sub-national governments. The strategy also covers the benefits of hydrogen to the community and progress to date and beyond 2030.

The National Hydrogen Strategy report's executive summary states the nascent hydrogen industry will be premised on Australia's abundant resources (COAG Energy Council 2019: viii), but these resources are not specified. The Australian experience in hydrogen production is promoted (COAG Energy Council 2019: viii); however, by 2019, it was limited to producing hydrogen to make small amounts of ammonia for fertilisers and other minor products. The report places emphasis on the creation of new jobs, economic growth, increased energy security and lower emissions. Production scale and cost competitiveness are seen as necessary. A key feature is the idea of the creation of hydrogen hubs around Australia, whether at ports, cities, regional or remote areas.

The National Hydrogen Strategy claims hubs are necessary to promote the use of hydrogen in transport, industry and gas distribution networks. Hubs require a clear regulatory framework, and the strategy expects that they would have a positive influence on energy prices. It states that the growth of hydrogen production will not compromise the safety, cost of living, water availability or the environment (COAG Energy Council 2019: viii).

Since the Australian government's dissemination of its 2019 National Hydrogen Strategy, many other countries have released their hydrogen strategies, reflecting many shifts in thinking. As a result of competitive positioning in the hydrogen sector, in February 2023, the Australian government directed the Energy and Climate Change Ministerial Council (ECMC) to review the National Hydrogen Strategy.

The aim of the ECMC review is to ensure the next iteration of the hydrogen strategy 'positions Australia on the path of global hydrogen leadership by 2030 for both export profits and decarbonisation of its industries' (DCCEEW 2023b: 1). In July 2023, the ECMC released a public consultation paper with a call for public submissions to respond to a range of questions and issues on the development of a hydrogen industry in Australia by August 2023. The main ECMC hydrogen review questions and issues are listed in Table 6.2 (DCCEEW 2023c: 21–23).

Table 6.2 items A to H are further discussed below.

A. How can Australia enable decarbonisation through the development of a hydrogen industry? The review asks sub-questions on decarbonisation and the hydrogen industry. They include prioritising ammonia production (as ammonia can serve as a hydrogen and energy carrier), economic sectors to be identified as early adopters of hydrogen, barriers to those sectors and associated remedial action.
B. How could Australia further activate its hydrogen and related industries? The review asks for ideas about a sector-driven approach to hydrogen.
C. Targets and mandates. Whether hydrogen production targets are a possibility, and if so, which sectors and posits transport. Descriptions of regulatory mandates and voluntary mandates are requested.
D. Supply chain risks. Insights are requested on significant supply chain barriers and how the government could help. Asks how local capabilities could be developed to secure the hydrogen supply chain, for instance, electrolyser manufacturing. Wants information on supply chain barriers, including workforce issues. Data requested on domestic manufacturing opportunities. For example, hydrogen buses.
E. How can we ensure our hydrogen industry attracts the necessary investment? Financial questions are posed about the cost issue with Australian hydrogen projects. Insights are called for to attract overseas and local institutional investment. A query is raised on whether a leaner technical approach is required. Extra research on the regulations could be necessary, including legal barriers to hydrogen development.

Table 6.2 2023 Review of national hydrogen strategy: questions and issues

A	How can Australia enable decarbonisation through the development of a hydrogen industry?
B	How could Australia further activate its hydrogen and related industries?
C	Targets and mandates
D	Addressing Supply Chain Risks
E	How can we ensure our hydrogen industry attracts the necessary investment?
F	How can we ensure our hydrogen industry develops in a way that benefits all Australians?
G	How should we develop the necessary infrastructure needed to support the development of our hydrogen industry?
H	How can we enable a hydrogen export industry (& export of goods manufactured with H_2)?

Source Created by the Authors in 2024

F. How can we ensure our hydrogen industry develops in a way that benefits all Australians? Sub-questions refer to the hydrogen benefits to the wider community, and specifically First Nations peoples. Water security is raised, including opinions on the type of hydrogen-related information that should be disseminated. Opinions on the adequacy of revenues to the government from a nascent hydrogen sector are requested, particularly as the sector matures.
G. How should we develop the necessary infrastructure needed to support the development of our hydrogen industry? Insights are requested on infrastructure assessments—whether new or existing, or used in common, such as hubs. Opinions are asked about the balance of infrastructure and synergies between gas, hydrogen energy and electricity generators.
H. How can we enable hydrogen (including goods manufactured) export? Submissions are needed on trade agreements and ideas for inward investment for export projects that might benefit local communities. This section furnishes an open question on revising the current hydrogen strategy.

6.5.1 Hydrogen Review: 'Item F'

'Item F' (from Table 6.2) asks 'How can we ensure our hydrogen industry develops in a way that benefits all Australians?' It raises the most relevant issues for this chapter's research questions on whether hydrogen production for export is a reasonable path and likely tax revenues to the government.

The ECMC hydrogen review consultation paper mentions that hydrogen developments will lead to increased demand on land, transport infrastructure, water resources and social services. For example, high on the list of community concerns is the point that hydrogen production may impact existing water uses and availability, especially in resource-constrained areas (DCCEEW 2023c: 14). A recent technical report to the government provides basic data on Australia's water resources for hydrogen, but does not offer solutions on the issue of water availability (Arup Engineering Australia 2022).

The ECMC hydrogen review consultation document states that, 'Much of the anticipated hydrogen industry growth will occur in remote and regional areas [and] will create new economic opportunities …for First Nations communities' (DCCEEW 2023c: 14). This statement provides a segue to the book *Dark Emu* (Pascoe 2014) to facilitate a response to the 'Item F' reference to the development of the hydrogen community and opportunities for First Nations people.

6.5.2 Dark Emu: The Need for a Light Footprint

Pascoe's *Dark Emu* (2014) claims that Australia's First Nations society was agricultural, or at least on the cusp of agriculture when the British colonists arrived in 1788. The book challenges the accepted orthodoxy that First Nations people were merely primitive hunter-gatherers, as was disseminated in earlier Australian school texts. Many critics of Pascoe's work emerged. The most well-known critics are Sutton and Walshe (2021), who decried *Dark Emu* for its lack of 'methodological and theoretical sophistication' of written accounts produced by formally trained social scientists and archaeologists (as cited in Davis 2020: 59).

In the years since the 2014 publication of *Dark Emu*, Australia has embraced the book's many core concepts to the extent that it has become almost mainstream reading (Langton 2023), and sparked academic interest in First Nations peoples' knowledge systems (e.g. Barber and Jackson 2017; Weir et al. 2020; Altman et al. 2023). A version of the book for primary school children is now available (Barlow and Horyniak 2019).

The central claims in *Dark Emu* were overviewed by a television documentary (ABC TV 2022), which used a 'talking heads' approach to elicit comments from academics as well as First Nations people. The following quotes from the television documentary are of interest in the context of this chapter's inquiry into the value of scaled-up hydrogen production and the potential impact on land and water resources:

> The greatest achievement of the Australian Indigenous peoples is the light footprint that they have had on the landscape. The footprint has been so light over the past 65,000 years...that some people might be misled into thinking that they [the First Nations people] hadn't managed this country so carefully.
> Pascoe's book has provoked a conversation. It has moved Australian understanding along in the way that it needs to...
> It's our job to watch over land and protect it.
> [Pascoe's book] 'has raised an argument about the future. It has tried to make the case for a better respect of country.'
> [First Nations people] 'have lived here for over 60,000 years and pretty successfully, before we got colonised. To use our knowledge of the land and to respectfully work with the land, is a key to moving on in the future...' (ABC TV 2022: at 67–118 mins)

Dark Emu cites sources from eighteenth-century journals of colonial explorers in Australia, a few of whom were 'in Australia to marvel at a new civilisation... but most were simply describing a landscape from which settlers could profit'(Pascoe 2014: 5). This view of Australia's vast resources as ready for exploitation persists through to today, and the results of poor land management can be seen in many degraded landscapes. Pascoe's book describes the First Nations peoples' light footprint as the best way to manage natural resources, as was the case until the arrival of the European colonists.

6.5.3 The Push for Hydrogen Production

Dark Emu's light footprint approach is at odds with those who push for hydrogen production by suggesting that Australia could lead the world in the domain of clean energy. For instance, Australian economist Ross Garnaut calls for investment in clean technologies, including batteries, hydrogen, solar, wind and hydro-electricity. He sees these technologies as facilitating Australia's rise as a 'global superpower in energy, low carbon industry and absorption of carbon in the landscape' (Garnaut 2019: 8). Garnaut argues that Australia is a natural place for Japan, Korea and other countries to base their heavy industries such as steel, aluminium and energy-intensive chemicals.

Other advocates to push the 'urgent' case to support green hydrogen production research includes a report by Deloitte Access Economics (2023), originally commissioned by the West Australia-based energy and minerals company Fortescue. The report covers utilising natural gas in Western Australia to produce ammonia as the chemical 'hydrogen carrier' for mass export. Ammonia is more stable and easier to transport but is not easily transformed back into hydrogen at the customer site. However, the technical issues around hydrogen transport for export purposes remain unresolved (DISER 2022: 12; Ferris 2023; IEEFA 2023). Indeed, future market demand seems underpinned by a belief that investment and government action on climate change would also promote the advancement in technologies, such as hydrogen export, with transport that becomes competitively priced (De Vos and Graham 2022).

As noted in the introduction, planning for hydrogen production and debate on associated laws (Bruton 2018; Bartlett 2019) are underway in Australia. For example, in the lead time to South Australia's Hydrogen Bill through the parliament, its government saw it as a world first in hydrogen laws (SA Dept of Energy and Mining 2023). The passing of South Australia's *Hydrogen and Renewable Energy Act 2023 is* designed to facilitate and regulate the generation of hydrogen and renewable energy in certain areas of the state (SA Parliament 2023).

Under South Australia's *Hydrogen and Renewable Energy Act 2023*, state government-owned land and waters where renewable energy projects might be hosted will require companies to compete for licences to deliver hydrogen projects. First Nations people's rights and interests will be considered early and throughout the regulatory processes. Multiple land use provisions will be sought to deliver fair outcomes for landowners, communities and other pre-existing land rights holders.

It will be of interest to follow the considerations of the First Nations people's rights and interests when commercial bids for land and water for hydrogen get underway. Recognition justice requires that the rights of First Nations people are addressed. They must be represented in discussions about scaled-up hydrogen projects. As mentioned above (Sect. 6.5.1) the Hydrogen Strategy Review consultation paper 'Item F' states that many proposed hydrogen projects will be located in remote locations on land managed by First Nations people.

There is legitimate consternation about the rush and push to produce green hydrogen for export without prior thoughts on sharing the proceeds with the wider community. Two immediate problems concern water and land impacts. Water required for producing green hydrogen via electrolysis can be 11 litres per one kilogram of hydrogen (Arup Engineering Australia 2022: 32). Given that Australia is a dry continent, hydrogen production can only draw water from limited sources, such as:

- Variable surface water, e.g. lakes, dams, rivers and creeks
- Groundwater, e.g. well water, aquifers and bore water
- Recycled water, e.g. treated wastewater effluent
- Brackish water sources, e.g. saline surface water and groundwater
- High salinity water sources e.g. seawater, estuary water (Arup Engineering Australia 2022: 74).

The most likely Australian water source for hydrogen is seawater, which would require the extensive construction of desalination plants. Desalination accumulates toxic brine from hydrogen production that is typically disposed of back into coastal waters. A UN environmental report notes:

> Desalination is the process of removing salts from water and a by-product of the process is toxic brine, which can degrade coastal and marine ecosystems unless treated.
> In most desalination processes, for every litre of potable water produced, about 1.5 litres of liquid polluted with chlorine and copper are created. When pumped back into the ocean, the toxic brine depletes oxygen and impacts organisms along the food chain. (UN Environmental Program 2019: 1)

Whether the community can trust businesses to dispose of toxic brine in an environmentally friendly way is a social licence issue that is open for discussion. Restorative justice concerns the belief that any injustice caused by the energy sector should be rectified, for instance, that a robust policy on the disposal of toxic brine from desalination should be underpinned by adequate laws.

An example of a business transgression of the community's 'social licence to operate' is Chevron's Gorgon LNG Project, which is one of the largest integrated natural gas-to-liquid operations off the northwest coast of Western Australia (DISER 2022). Its large production capacity is 15.6 million tonnes per annum. Since commencing operations in 2016, the Gorgon LNG project has disclosed its high CO_2 emissions. The project has large CO_2 carbon capture and storage (CCS) facilities; however, by 2023, it still has not met the regulated CCS obligations to lower emissions (Milne 2023). Chevron has been penalised, but the same levels of extraction of gas for export continue (Chevron Australia 2021).

6.5.4 Secondary Taxation

The chapter's subsidiary question concerns the distributional justice issue of proposing hydrogen production without adequate tax considerations. This omission in regard to taxation occurs at both state (e.g. South Australia) and federal levels. A resource tax is normally applied to mineral resources. For instance, the value of coal production in Australia is subject to a state royalty as well as the federal income tax. A second tax is generally justified on the premise that mineral resources are finite and their extraction can only occur once (Kraal 2021: 162).

Hydrogen is not explicitly considered a fossil fuel in mineral resource tax regimes. For instance, in the *Petroleum Resource Rent Tax*

Assessment Act 1987 (Australia), section 2E states which natural gas components are subject to the tax. Natural gas is the feedstock for the hydrogen produced from the steam methane reforming (SMR) process. Hydrogen is not included under section 2E and thus is not subject to this tax.

Australia's National Hydrogen Strategy report states:

> As hydrogen production and use grows, appropriate taxation, excises, fees or levies could help ensure that the community shares in the economic benefits from developing a hydrogen industry. Australian governments recognise the importance of the Australian public receiving a share of future benefits from a hydrogen industry and for investors to have certainty about future revenue arrangements. Governments will continue with the revenue arrangements that now apply to hydrogen, but may review them in the future. Governments will consult industry and the community before making any changes to current revenue arrangements that are specific to hydrogen. (COAG Energy Council 2019: 53)

Bruton (2018: 8) called for 'taxation treatment of transformed fuels: the taxation treatment of transformed fuels like hydrogen (both at the point of production and the point of sale) is another consideration that needs to be addressed…' However, from 2019 until now, a secondary tax on hydrogen is yet to be discussed.

Interestingly, some call for tax credits rather than tax imposts. For instance, Rowell et al. (2022: 144) write that 'in addition to technological developments to decrease the cost of hydrogen production, economic incentives (such as subsidies, tax credits and carbon markets) can also be used to lower the cost for producers'.

Limited consideration has been given in the literature to the taxation of the nascent hydrogen industry, with the main focus being on subsidies needed to promote hydrogen markets. This is anomalous taxpayer funding required for investment to expand infrastructure and system capacity (Dillman and Heinonen 2022: 6).

The harm to the environment from large-scale hydrogen production requires tax funds to remedy injustices. Toxic brine from desalination has already been discussed. Another problem is that hydrogen usage has leakage issues. When leaked into the atmosphere, it is 40 times more damaging than CO_2. Thus, hydrogen leakages add substantially to climate impact (Hauglustaine et al. 2022; Jones 2023).

Distribution justice concerns the distribution of the benefits and the costs of the negatives from the energy sector. Energy infrastructure for green hydrogen—wind turbines, solar farms and desalination plants—at a scale necessary for export has an impact on the environment that comprises physical space (e.g. land) and flow resources (e.g. coastal waters). While the concept of a company tax or income tax is not normally contested for commercial operations, a secondary resource tax of say 10% on the resource market value is necessary. The proceeds from the secondary tax collections would pay for the negative externalities of hydrogen production as well as some allocation to the wider community.

Distribution justice also calls for a fair global spread of green hydrogen infrastructure. The rise of not-in-my-back-yard or NIMBYism forces governments to look to outsourcing hydrogen production to other jurisdictions in order to meet nationally determined contributions (NDCs), which are a core part of the Paris Agreement to meet its long-term climate goals (UNFCCC 2023).

For Australia, which purports to be a potential superpower in the production of green hydrogen, and for those former colonial countries that are 'selected' by their former colonisers to be producers of hydrogen for export, the concentration of hydrogen activities in limited areas and subsequent environmental impacts—transgresses cosmopolitan justice. Cosmopolitan justice stems from the belief that we are all citizens of the world, and we must look to the effects of a low-carbon energy transition beyond our borders. In cases where precious materials and flow resources are concentrated in specific areas for intensive hydrogen production—the global community loses out. The land resources used will not recover, and waterways will become degraded.

6.6 Conclusion

This chapter has used a targeted desktop review of the literature and the frame of energy justice theory to answer research questions about the Australian government aspirations for a hydrogen industry and its associated taxation. This conclusion section is a summary of outcomes, the research contribution and future options.

The literature shows dominant support for a hydrogen industry in the chapter's selected regions and/or countries in Asia, Western Europe, Africa, the Middle East, South America and Australia. The gap in the literature is that no research was uncovered that clearly advocates for sharing the load of the production of hydrogen equitably across countries and regions.

Hydrogen is touted as a miracle energy for a low-carbon future. Cynics might even suggest that it is a 'stalking horse' for the fossil fuels lobby. Nonetheless, in terms of the research question about whether the Australian government's aspirations for large-scale production of hydrogen for export is a reasonable path for a just transition—the findings indicate that the answer is in the negative. The research looked at the Australian government's review of its 2019 National Hydrogen Strategy, which is about ensuring a hydrogen industry develops in a way that benefits all Australians. The review encompasses references to the benefits of hydrogen production and export to the wider community, and specifically First Nations peoples. Water security, which could be impacted by hydrogen production, is a review issue, mainly in relation to the large-scale desalination and toxic brine outputs.

It is advanced that a hydrogen review requires a multi-disciplinary response. Thus, the chapter drew upon literature, such as *Dark Emu*, that covers Australia's First Nations peoples' knowledge of land management and culture. This source suggests that a light footprint is the best way to manage natural resources. The findings infer, if and when hydrogen is required at an industrial level, that Australia should be one of many global jurisdictions that contribute to hydrogen production, but mainly for domestic use. It should not be seen as an opportunity for large-scale production, export and profit.

A just transition to a low-carbon economy looks to a range of justice outcomes. It was acknowledged that 'recognition justice' requires that the rights of First Nations people be addressed. They must be represented in discussions about large-scale hydrogen projects. Further, 'restorative justice' concerns the belief that any injustice caused by the energy sector should be rectified. For instance, a policy on the disposal of toxic brine from desalination should be underpinned by robust and adequate laws. Whether the community, including First Nations people, can trust businesses to dispose of toxic brine environmentally is an example of the priority of the social contract between businesses and the ordinary public.

The associated question about proposing hydrogen production and export without prior consideration on taxation collection and application towards hydrogen infrastructure, and as a mechanism to share the benefits of a hydrogen industry—calls on 'distribution justice'. This type of justice concerns the distribution of the benefits and the costs of the negatives from the energy sector.

And for those former colonial possessions, such as in Africa, that are selected by their former colonisers to be producers of hydrogen for export—the concentration of hydrogen activities in limited areas and subsequent environmental impacts transgresses 'cosmopolitan justice'.

This chapter contributes to the debate about the growth of the hydrogen industry in Australia beyond domestic needs. It is argued that hydrogen decisions must be driven by greater public interests and not narrow corporate profits that benefit only a certain segment of society. This chapter has lessons for other jurisdictions to draw upon.

The review of Australia's 2019 National Hydrogen Strategy is underway at the time of writing. Future research will cover the recommendations that might arise for changes to the national plan. It will be timely then to compare how a revised hydrogen strategy might align with other jurisdictions that are former colonies, including those in Asia, Africa, the Middle East and South America.

References

Agyekum EB (2023) Is Africa ready for green hydrogen energy takeoff? a multi-criteria analysis approach to the opportunities and barriers of hydrogen production on the continent. Int J Hydrog Energy 49:219–233

Alsaba W, Al-Sobhi SA, Qyyum MA (2023) Recent advancements in the hydrogen value chain: opportunities, challenges, and the way forward-middle east perspectives. Int J Hydrogen Energy 48(68):26408–26435

Altman J, Kneebone J, Nolan K (2023) First nations perspectives on decarbonisation. In Natgzaam G, Beaufoy M, O'Bryan K (eds), Legal pathways to deep decarbonisation in Australia. Lexis Nexis, Chatswood, pp 121–141

Arup Engineering Australia (2022) Water for hydrogen: technical paper department of climate change, energy, the environment and water and the Australian Hydrogen Council

Australian Broadcasting Commission (2022) The Dark Emu story. https://iview.abc.net.au/show/dark-emu-story

Barber M, Jackson S (2017) Identifying and categorizing co-benefits in state-supported Australian indigenous environmental management programs: international research implications. Ecol Soc 22(2):1–17

Barlow S, Horyniak A (2019) Dark Emu in the classroom. Magabala Books

Bartlett E (2019) The developing hydrogen sector in australia: taking stock: why hydrogen and why now? Aust Energy Resour Law Bull 3(4):50–52

Bhandari R (2022) Green hydrogen production potential in West Africa—case of Niger. Renew Energy 196:800–811

Bharadwaj B, Weder F, Ashworth P (2023) More support for hydrogen export than its domestic application in Australia. HumIties Soc Sci Commun 10(1):7–8

Borsboom-Hanson T, Patlolla S, Herrera O, Merida W (2022) Point-to-point transportation: the economics of hydrogen export. Int J Hydrogen Energy 47 (74):31541–31550

Brauner S, Lahnaoui A, Agbo S, Böschen S, Kuckshinrichs W (2023) Towards green hydrogen? a comparison of German and African visions and expectations in the context of the H2 Atlas-Africa project. Energy Strat Rev 50:101204

Bruton A (2018) Hydrogen in Australia: designing an effective regulatory framework for the use of hydrogen in a clean energy future. Aust Energy Resour Law Bull 3(1):6–9

Burke PJ, Beck FJ, Aisbett E, Baldwin KGH, Stocks M, Pye J, Venkataraman M, Hunt J, Bai X (2022) Contributing to regional decarbonization: Australia's potential to supply zero-carbon commodities to the Asia-Pacific. Energy 248:123563

Burlet H (2023) Roundtable: discussion on hydrogen's impact on energy justice. World Energy Justice Congress, Anglett, France

Cardinale R (2023) From natural gas to green hydrogen: developing and repurposing transnational energy infrastructure connecting North Africa to Europe. Energy Policy 181:113623

Chalmers J, Bowen C, McAllister J (2023) Joint media release. Hydrogen headstart to power new jobs & industry. https://ministers.treasury.gov.au/ministers/jim-chalmers-2022/media-releases/hydrogen-headstart-power-new-jobs-industry

Chevron Australia (2021) Chevron announces $40 million Western Australian lower carbon investment. https://australia.chevron.com/news/2021/chevron-announces-aud40-million-western-australian-lower-carbon-investment

Clean Energy Council (2023) Clean energy Australia report 2023. https://assets.cleanenergycouncil.org.au/documents/Clean-Energy-Australia-Report-2023.pdf

Cremonese L, Mbungu GK, Quitzow R (2023) The sustainability of green hydrogen: an uncertain proposition. Int J Hydrogen Energy 48(51):19422–19436

COAG Energy Council (2019) Australia's national hydrogen strategy. Commonwealth of Australia, Canberra

Davis R (2020) Black agriculture, white anger: arguments over aboriginal land use in Bruce Pascoe's Dark Emu. Borderlands 1(1):57–70

De Vos R, Graham T (2022) Anticipating Australia's hydrogen future. Aust Environ Rev 36(6):136–142

Deloitte Access Economics (2023) Australia's hydrogen tipping point: the urgent case to support renewable hydrogen production Deloitte Access Economics

Department of Climate Change Energy the Environment and Water (DCCEEW) (2023a) Energy and climate change ministerial council. https://www.energy.gov.au/government-priorities/energy-and-climate-change-ministerial-council

Department of Climate Change Energy the Environment and Water (2023b) National hydrogen strategy review. Commonwealth of Australia, Canberra. https://consult.dcceew.gov.au/review-of-the-national-hydrogen-strategy

Department of Climate Change Energy the Environment and Water (2023c) National Hydrogen strategy review July 2023: consultation paper. Commonwealth of Australia, Canberra

Department of Industry Science and Resources (DISER) (2022) Resources and energy major projects (Report 2022). Australian Government, Canberra

Dillman KJ, Heinonen J (2022) A "just" hydrogen economy: a normative energy justice assessment of the hydrogen economy. Renew Sustain Energy Rev 167:112648

Durakovic G, del Granado PC, Tomasgard A (2023) Are green and blue hydrogen competitive or complementary? Insights from a decarbonized European power system analysis. Energy 282:128282

Energy Justice & the Social Contract Project (2023) World Energy Justice Congress, Anglett, France. https://ejsc-worldcongress.sciencesconf.org/

Ferris N (2023) Why LNG terminals will not be transporting hydrogen any time soon. Energy Monitor. https://www.energymonitor.ai/hydrogen/why-lng-terminals-will-not-be-transporting-hydrogen-any-time-soon/?cf-view
Gado MG, Hassan H (2023) Potential of prospective plans in MENA countries for green hydrogen generation driven by solar and wind power sources. Solar Energy 263:111942
Galván A, Hass J, Moreino-Leiver S, Osorio-Aravena J, Nowak W, Palma-Benke R, Breyer C (2022) Exporting sunshine: planning South America's electricity transition with green hydrogen. Appl Energy 325:119569
Garlet T, de Souza Savian F, Duarte R, Mairesse S (2023) Unlocking Brazil's green hydrogen potential: overcoming barriers and formulating strategies to this promising sector. Int J Hydrog Energy
Garnaut R (2019) Superpower—Australia's low-carbon opportunity. Black Inc. with La Trobe University Press, Melbourne
Guan D, Wang B, Zhang J, Shi R, Jiao K, Li L, Wang Y, Xie B, Zhang Q, Yu J, Zhu Y, Shao Z, Ni M (2023) Hydrogen society: from present to future. Energy Environ Sci 16(11):4926–4943
Hancock L, Ralph N (2021) A framework for assessing fossil fuel "retrofit" hydrogen exports: security-justice implications of Australia's coal-generated hydrogen exports to Japan. Energy 223:119938
Hanusch F, Schad M (2021) Hydrogen research: technology first, society second? Gaia 30(2):82–86
Hassan Q, Sameen A, Salman H, Jaszczur M, Al-Jiboory A (2023) 'Hydrogen energy future: advancements in storage technologies and implications for sustainability. J Energy Storage 72:108404
Hauglustaine D, Paulot F, Collins W, Derwent R, Sand M, Boucher O (2022) Climate benefit of a future hydrogen economy. Commun Earth Environ 3(1):295
Heffron RJ (2021) Achieving a just transition to a low-carbon economy. Springer International Publishing AG, Cham
Heffron RJ (2023) Energy justice—the triumvirate of tenets revisited and revised. J Energy Nat Resour Law, 1–7
Heffron RJ, McCauley D (2017) The concept of energy justice across the disciplines. Energy Policy 105 (1):658–667
Hinkley JT, Heenan A, Low A, Watson (2022) Hydrogen as an export commodity—capital expenditure and energy evaluation of hydrogen carriers. Int J Hydrog Energy 47(85):35959–35975
Hjeij D, Bicer Y, Al-Sada MBS, Koç M (2023) Hydrogen export competitiveness index for a sustainable hydrogen economy. Energy Reports 9:5843–5856
Incer-Valverde J, Korayem A, Tsatsaronis G, Morosuk T (2023) "Colors" of hydrogen: definitions and carbon intensity. Energy Convers Manag 291:117294
Institute for Energy Economics and Financial Analysis (IEEFA) (2023) IEEFA submission to the review of the national hydrogen strategy
Jones R (2023) Hydrogen: a "miracle solution", or over-hyped stalking horse for fossil fuels lobby? Michael West Media
Karayel GK, Javani N, Dincer I (2023) Hydropower for green hydrogen production in Turkey. Int J Hydrogen Energy 48(60):22806–22817
Kneebone J (2023) Roundtable: discussion on hydrogen's impact on energy justice. World Energy Justice Congress, Anglett, France
Kolb S, Müller J, Luna-Jaspe N, Karl J (2022) Renewable hydrogen imports for the German energy transition—comparative life cycle assessment. J Clean Prod 373:133289
Kraal D (2017) Review of the petroleum resource rent tax: implications from a case study of the Gorgon gas project. Fed Law Rev 45(2):315–349
Kraal D (2019) Petroleum industry tax incentives and energy policy implications: a comparison betweenAustralia, Malaysia, Indonesia and Papua New Guinea. Energy Policy 126:212–222
Kraal D (2021) Treasury review of the petroleum resource rent tax: gas transfer pricing arrangements. Aust Tax Forum 36(2):161–186
Kraal D (2024) Australian petroleum and coal resources: taxation, emissions and energy justice. In Heffron RJ, de Fontanelle L (eds) The power of energy justice. Palgrave Macmillan, pp 179–185
Kraal D, Heffron RJ (2022) Resource rent tax: its principles, application and need for change in Australia. Aust Tax Forum 37(4):559–599
Kraal D, Odhuno F (2019) Papua New Guinea's extractive resource sector: towards a policy of leaner tax incentives. Port Moresby: National Research Institute of Papua New Guinea. Discussion Paper no. 172
Kraal D, Mulder M, Perey P (2020) Taxation and pricing of natural gas: the Dutch transition to gas market hub pricing and lessons for Australia's integrated gas projects. University of New South Wales Law Journal 43(1):39–81
Langton M (2023) Welcome to country handbook: a guide to indigenous Australia. Hardie Grant
Le TT et al (2023) Fueling the future: a comprehensive review of hydrogen energy systems and their challenges. Int J Hydrog Energy 54:791–816
Lebrouhi BE, Djoupo JJ, Lamrani B, Benabdelaziz K, Kousksou T (2022) Global hydrogen development—a technological and geopolitical overview. Int J Hydrogen Energy 47(11):7016–7048
Merad M (2023) Roundtable: discussion on hydrogen's impact on energy justice. World Energy Justice Congress, Anglett, France
Milne P (2023) Chevron's troubled carbon capture and storage at Gorgon set to worsen in 2023. WAtoday.com.au. https://www.watoday.com.au/national/western-australia/chevron-s-troubled-carbon-capture-and-storage-at-gorgon-set-to-worsen-in-2023-20230711-p5dngj.html
Müller F, Tunn J, Kalt T (2022) Hydrogen justice. Environ Res Lett 17(11):115006

Nuñez-Jimenez A, De Blasio N (2022) Competitive and secure renewable hydrogen markets: three strategic scenarios for the European Union. Int J Hydrogen Energy 47(84):35553–35570

Olabi AG, Abdelkareem M,. Mahmoud S, Elsaid K, Obaideen K, Rezk H, Wilberforce T, Eisa T, Chae K, Sayed E (2023) Green hydrogen: pathways, roadmap, and role in achieving sustainable development goals. Process Saf Environ Prot 177:664–687

Osborne J (2023) Biden commits $1 billion to creating clean hydrogen sales. Spectra. https://spectra.mhi.com/partner-biden-commits-1-billion-to-creating-clean-hydrogen-sales#:~:text=The%20Biden%20administration%20has%20already,least%20one%20of%20the%20hubs

Panchenko VA, Daus YuV, Kovalev AA, Yudaev IV, Litti YuV (2023) Prospects for the production of green hydrogen: review of countries with high potential. Int J Hydrogen Energy 48(12):4551–4571

Pascoe B (2014) Dark Emu: black seeds: agriculture or accident? Magabala Books, Brunswick

Pawar ND, Heinrichs H, Winkler C, Heuser P, Ryberg S, Robinius M, Stolten D (2021) Potential of green ammonia production in India. Int J Hydrog Energy 46 (54):27247–27267

REN21 (2023) Global renewable resources report. https://www.ren21.net/wp-content/uploads/2019/05/GSR-2023_Energy-Supply-Module.pdf

Rigollet C (2023) Roundtable: discussion on hydrogen's impact on energy justice. World Energy Justice Congress, Anglett, France

Rowell J, Bergman N, Al Asedy H, Simpson R (2022) The down low on hydrogen policy and regulation in Australia. Aust Environ Rev 36(6):143–146

Santos Ayllón LM (2022) A justice and responsible research and innovation exploration of marine renewables and green hydrogen in island communities. Science Talks 4:100086

Shyu C-W (2021) A framework for 'right to energy' to meet UN SDG7: policy implications to meet basic human energy needs, eradicate energy poverty, enhance energy justice, and uphold energy democracy. Energy Res Soc Sci 79:102199

South Australia Department for Energy and Mining (2023) South Australia's bill for a world-first hydrogen and renewable energy act. https://www.energymining.sa.gov.au/home/news/latest/south-australia-to-establish-world-first-hydrogen-act

South Australia Parliament (2023) Hydrogen and renewable energy act 2023. https://www.legislation.sa.gov.au/lz?path=/v/a/2023/hydrogen

Sovacool B, Heffron RJ, McCauley D, Goldthau A (2016) Energy decisions reframed as justice and ethical concerns. Nat Energy 1:1–6

Squadrito G, Maggio G, Nicita A (2023) The green hydrogen revolution. Renew Energy 216:119041

Sutton P, Walshe K (2021) Farmers or hunter-gatherers? the Dark Emu debate. Melbourne University Press, Melbourne

United Nations Environmental Program (2019) Towards sustainable desalination. https://www.unep.org/news-and-stories/story/towards-sustainable-desalination

United Nations Framework Convention on Climate Change (2023) The Paris agreement: nationally determined contributions. https://unfccc.int/process-and-meetings/the-paris-agreement/nationally-determined-contributions-ndcs

van der Zwaan B, Lamboo S, Dalla Longa F (2021) Timmermans' dream: an electricity and hydrogen partnership between Europe and North Africa. Energy Policy 159:112613

Walsh SDC, Easton L, Weng Z, Wang C, Moloney J, Feitz A (2021) Evaluating the economic fairways for hydrogen production in Australia. Int J Hydrogen Energy 46(73):35985–35996

Weir J, Sutton S, Catt G (2020) The theory/practice of disaster justice: learning from indigenous peoples' fire management. In Lukasiewicz A, Baldwin C (eds) Natural hazards and disaster justice: challenges for Australia and its neighbours. Palgrave Macmillan, Australia

Diane Kraal is an Adjunct Senior Research Fellow at the Monash Business School, Monash University, Australia. Dr. Kraal's taxation law expertise led to major projects on tax law changes to facilitate greater uptake of electric vehicles in business fleets and regulation for demand management and distributed energy resources. Her latest research book is on decarbonisation. Her publications have applied energy justice in comparing the taxation legislation for gas in Australia and the Netherlands; decision-making on mineral extraction industry tax incentives; and energy policy comparisons for Australia, Indonesia, Malaysia, and Papua New Guinea. She was a member of the Monash delegation to the United Nations COP24, Poland, 2018; Co-Head of the Monash Delegation to COP25, Madrid, 2019; and a delegation member to COP27, Egypt, 2022. She is currently a committee member of the Energy Justice and the Social Contract project, Université de Pau et des Pays de l'Adour, France.

Richard Taylor lectures in sustainability and corporate law at the Monash Business School, Monash University, Australia. Dr. Taylor's research focuses on sustainability, taxation and company behaviour in the resources sector. His articles have explored the effect of tax changes on company taxpayer behaviour in the Lao PDR, Indonesia and Papua New Guinea and the use of grounded theory in qualitative research. He has experience as a mining specialist with the World Bank, Washington DC; and held various industry roles, including Chair of the *Resource Endowment Initiative*, International Council on Mining and Metals. He received an Australian Government Research Training Program (RTP) Scholarship for his Ph.D.

It is Time to Address Nuclear Energy Injustices

7

Ridoan Karim

Abstract

The concepts of "energy justice" and "sustainable development" are pivotal in global policy discussions. Despite the shift towards renewable energy, these sources cannot yet meet global demands affordably. Nuclear technology, with decades of scientific progress, offers a reliable alternative. Innovations like small modular reactors (SMRs) and nuclear fusion have improved safety and environmental impact. However, limited discussions and consensus on nuclear energy as a clean energy source raise questions about its role in sustainable development. Negative public perception, rooted in historical injustices, hinders its acceptance. To address this, deliberate actions are needed to rectify past injustices associated with nuclear power. Sustainable nuclear energy development requires a balanced approach, considering both benefits and drawbacks while adhering to distributional, procedural, and recognition justice principles.

Keywords

Energy justice · Sustainable development · Nuclear energy · Small modular reactors (SMRs) · Nuclear fusion · Historical injustices

7.1 Introduction

In recent policymaking discussions, two buzzwords have gained global attention—"energy justice" and "sustainable development." These terms are becoming increasingly important in ensuring a green future and universal energy security. Although these two concepts are distinct, they intersect significantly in addressing the environmental and social challenges related to energy production and consumption.

Energy justice is all about ensuring that everyone has access to clean, affordable, and reliable energy while considering the social and environmental impacts of energy production and consumption (Jenkins et al. 2016a, b). Sustainable development, on the other hand, aims to meet the needs of the present generation without compromising the ability of future generations to meet their own needs. The United Nations' Sustainable Development Goals (SDGs) recognise the importance of access to clean and affordable energy as a crucial objective that underpins economic growth and social well-being.

The world is moving towards renewable energy sources to ensure universal access to

R. Karim (✉)
Department of Business Law and Taxation,
School of Business, Monash University Malaysia,
Jalan Lagoon Selatan, Bandar Sunway,
47500 Subang Jaya, Selangor, Malaysia
e-mail: ridoan.karim@monash.edu

A. Trakic et al. (eds.), *Energy Justice*, Sustainable Development Goals Series,
https://doi.org/10.1007/978-981-97-6059-6_7

clean and affordable energy. However, despite global efforts, these sources have not been able to meet the worldwide energy demands (Holechek et al. 2022). While renewable energy is environmentally friendly, it is still not affordable for everyone (Balakrishnan et al. 2020). In contrast, nuclear technology, which has been developed over the course of 80 years of scientific progress, offers a reliable and established source of energy (Karim and Lee 2023). Recent technological advancements in nuclear energy, such as the development of small modular reactors (SMRs) and progress in nuclear fusion, have significantly enhanced the safety of nuclear energy production (Karim and Muhammad-Sukki 2023). Surprisingly, discussions regarding nuclear energy as a clean energy system remain relatively limited within the international community. This raises questions about whether the world is overlooking nuclear energy in its pursuit of sustainable development goals.

This chapter attempts to answer the question by exploring the nuclear versus renewable energy debate. It also explores historical injustices related to nuclear energy, examines the intersection of energy justice and nuclear energy from a conceptual standpoint, and finally offers recommendations on how the world can effectively harness nuclear energy to advance sustainable development. When discussing policies, the central focus should be on how to achieve the SDGs in the most equitable way possible. In this regard, the principle of energy justice is crucial (Guruswamy 2010). Additionally, nuclear energy has a significant role to play in achieving the SDGs (Karim and Lee 2023). However, it is essential to use it fairly while upholding the guiding principles of energy justice.

7.2 Nuclear Versus Renewable Energy: The Never-Ending Debate

Renewable energy sources like wind, solar, hydro, and geothermal power have been gaining attention as environmentally friendly alternatives. These sources are sustainable and do not produce greenhouse gas emissions during operation. The advantages of renewables lie in their decentralised nature,[1] potential to create jobs and long-term availability (Karim et al. 2019). Additionally, as the cost of renewable energy technologies decreases, they become more accessible and affordable (He 2020).

Nonetheless, renewable energy sources have their own unique challenges. One of the biggest challenges is that they often rely on weather conditions, making them intermittent. To ensure a continuous power supply, energy storage solutions like the advancement of battery technologies are necessary. Additionally, the development and expansion of renewable infrastructure can be difficult due to land and resource constraints. Each renewable energy technology performs optimally in specific geographic settings. For example, wind turbines require strong, consistent winds and flat terrain to be effective. Solar panels work best in areas with abundant year-round sunlight, while hydroelectric power is most effective in locations with large, fast-flowing rivers. Therefore, the success of renewable energy relies heavily on specific environmental and geographical conditions.

In contrast, finding suitable locations for nuclear energy is generally less challenging than for renewable sources (McCombie and Jefferson 2016). This is why countries like France derive approximately 70% of their energy from nuclear power (Karim et al. 2018). They face difficulties in meeting the specific requirements of renewable energy sources due to their geographical constraints. Moreover, nuclear energy has made significant advancements due to improved fission technology (Karim and Muhammad-Sukki 2023). On the other hand, solar, hydro, and wind power are still in the process of development. Experts in the fields of engineering and

[1] Renewable energy sources are regarded as decentralised sources as they are usually available in many different locations, unlike non-renewable resources such as oil, coal, and gas, which are often found in specific regions. This decentralised nature of renewables allows for a more distributed and flexible energy system, which reduces the dependence on centralised power generation and transmission.

technology view these renewable energy sources as being in their early stages. Conversely, nuclear energy technology has been steadily improving since the 1960s, enabling us to utilise the most advanced and secure atomic technology for generating electricity (Karim and Lee 2023).

While location and resource availability are generally not major concerns for nuclear energy projects, the costs of these projects can vary depending on several factors. Nuclear energy is relatively expensive compared to coal, gas, and oil-based energy production, but it still has a significant economic advantage. A portion of nuclear energy expenses, approximately 31%, is attributed to fuel (World Nuclear Association 2022a, b). This cost is considerably lower than fuel costs associated with natural gas or coal-based energy production, where fuel costs can escalate to as much as 80–90% (Karim and Lee 2023). This characteristic makes nuclear energy less susceptible to price fluctuations of nuclear fuel.

Overall, nuclear energy is a desirable source of clean and environmentally friendly energy, considering its cost-effectiveness in terms of fuel and transportation, high energy density within the fuel, and its ability to provide a consistent source of sustainable energy. Despite these benefits, the global nuclear power industry has faced challenges that have hindered its growth in recent years. These challenges include increasing negative public perception, rising costs, and regulatory hurdles. As a result, many experts recommend limiting the expansion of nuclear energy (Dittmar 2012).

Public concerns regarding the safety of nuclear energy have been increasing over the years due to incidents like Chernobyl, Three Mile Island, and Fukushima. The management of nuclear waste is another factor that has contributed to negative public perceptions of nuclear power over time. Several studies have been conducted to understand the factors that influence public acceptance of nuclear energy, including benefits, risks, trust, and knowledge (Visschers et al. 2011). However, the generalisation of these findings and identification of common factors that affect public perception of nuclear energy remain challenging due to the limitation of many studies that focus on specific regions or countries (Kim et al. 2014). Previous research on public attitudes towards nuclear power has been limited by a binary framework, which only considers acceptance or rejection of nuclear power, failing to account for those who "reluctantly accept" it for various reasons (Karim and Lee 2023).

The acceptance of nuclear energy by the public depends on two fundamental factors: "perception" and "communication." In order to gain public trust in technology, it is crucial to evaluate it objectively and disseminate positive information and knowledge effectively (Visschers et al. 2011). Researchers should consider cultural, social, and historical contexts to fully understand public opinion on nuclear power (Liu et al. 2008). People who live far away from nuclear power plants often view atomic energy in terms of societal benefits rather than individual drawbacks, which is known as "reluctant acceptance" (Corner et al. 2011). Several studies have explored this phenomenon in the context of climate change mitigation, which significantly influences the public's views on nuclear power advancements (Pidgeon et al. 2008). Researchers generally agree that perceived risk and benefit are critical determinants of nuclear power acceptability (Kim et al. 2014).

Despite negative public perceptions, nuclear energy has great potential in fields like medicine, agriculture, energy, and manufacturing. The extensive development of international nuclear law and regulation has resulted in a well-established legal framework that binds states, scientific organisations, and all stakeholders in the nuclear energy industry to ensure the safe and secure use of nuclear materials. Although some aspects of this legal framework may require review and adjustment, it provides a strong foundation for a safe and prosperous future in nuclear energy.

Additionally, the nuclear energy sector can find hope in the advancements in science and technology. Nuclear power plants that rely on artificial intelligence present a potential solution to the threat of meltdowns (Suman 2021). Furthermore, the emergence of next-generation nuclear power plants, such as fusion reactors and molten salt reactors (MSRs), offers the opportunity to achieve 100% safe nuclear energy production (Karim and Lee 2023). These technological innovations could eliminate tragic accidents like Fukushima and Chernobyl, paving the way for unparalleled safety and reliability in the nuclear energy industry.

Hence, the topic of nuclear energy is an important one that deserves informed discussions from the international community. It is essential to consider both the benefits and challenges of nuclear energy while adopting a balanced approach to energy policy. Combining nuclear power with other low-carbon energy sources, such as renewables, can contribute towards a more sustainable and resilient energy future. This approach would enable us to maximise the benefits of nuclear energy while addressing its associated risks and challenges. Furthermore, it is crucial to address historical nuclear injustices in order to clear the path for future advancements. In the upcoming section of this chapter, I will deal with the historical nuclear injustices and explore ways to mitigate them through the conceptual framework of energy justice.

7.3 Nuclear Injustices

Nuclear risks arising from both military weapons development and civilian nuclear power plants have significant justice implications. Nuclear power plants release low levels of ionising radiation and generate substantial amounts of high-level nuclear waste (Iqbal et al. 2021). Justice issues revolve around the siting of these facilities and the possibility of exposing large populations to associated risks (Jenkins et al. 2016a, b). Furthermore, there are also justice concerns associated with the industries that support the mining and processing of nuclear fuels (Jenkins et al. 2016a, b).

Additionally, the nuclear industry is becoming increasingly concerned about ageing and deteriorating reactors (Jenkins and Taebi 2019). Many of these reactors have already exceeded their original 40-year lifespan, which increases the risk of leaks, unplanned discharges, and other malfunctions (Kim et al. 2022). When examining the justice implications of nuclear technologies, it is critical to consider the entire lifecycle, from the inception to the conclusion of nuclear materials and technologies. This highlights the importance of understanding the comprehensive spatial and temporal dimensions of the nuclear hazardscape—an aspect that supporters often overlook when emphasising the absence of carbon emissions from nuclear power plants (Kyne and Bolin 2016).

When discussing energy justice and its implications for nuclear power, I consider three key dimensions. The first of these is distributive justice, which involves the distribution of energy-related issues among different regions and communities, and the criteria used to determine which parties bear the associated risks (Jenkins et al. 2016a, b). The second dimension is procedural justice, which focuses on the fairness or unfairness of the procedures used to impose environmental risks on individuals and locations (Jenkins et al. 2016a, b). Finally, recognition justice is about identifying those acknowledged or ignored when determining the distribution of responsibilities for handling hazards (Jenkins et al. 2016a, b). This aspect is especially important when examining cases of nuclear injustice.

Generally, living in close proximity to nuclear power plants can pose significant risks to individuals due to the complex nature of nuclear technology. These risks can be broadly classified into two categories: those arising from the regular operation of the plants and those resulting from major accidents. In the event of a severe disaster, the surrounding areas may face exposure to highly hazardous radioactive emissions, soil and water contamination, radiation from melted nuclear fuel, and the establishment of exclusion zones where it is unsafe to live, similar to what occurred in the Chernobyl and Fukushima incidents. The Fukushima Daiichi nuclear disaster in 2011 serves as a recent and vivid example of how nuclear power plants can experience catastrophic incidents (Kyne and Bolin 2016). Such events can lead to radiation-related illnesses, the displacement of hundreds of thousands of people, and the long-term contamination of vast areas. Although these catastrophic incidents are infrequent, when they occur, multiple generations must deal with the environmental and health consequences, as the Chernobyl incident has shown (Kyne and Bolin 2016).

During the normal operation of nuclear reactors, there are concerns related to minor radioactive emissions that originate from various sources. These emissions pose environmental and health risks (Kyne and Bolin 2016). As commercial nuclear reactors age, there is a growing apprehension problems such as cooling system leaks, contamination events, plant fires, and what are commonly referred to as "normal accidents"[2] may occur frequently (Kyne and Bolin 2016). This is because the gradual decay of these facilities over time increases the likelihood of frequent malfunctions. People living near nuclear power plants are at risk of exposure to different types of ionising radiation. Radioactive gases are regularly released into the atmosphere from the stacks and steam generators of these plants. The plants discharge around 100 cubic feet of radioactive gases every hour, with a maximum of 22 purges allowed annually for each reactor (Kyne and Bolin 2016). When the materials in the pipes become too dangerous for plant workers to handle, they are released into the surrounding areas. To cool the reactor core, 20,000 gallons of water are required per minute (Kyne and Bolin 2016). This cooling water gets contaminated with radioactive tritium, also known as tritiated water (Kyne and Bolin 2016). Five thousand gallons of tritiated water per minute are discharged into nearby lakes, rivers, or the ocean, while an additional 15,000 gallons are released into the atmosphere as steam (Hamblin 2012).

Potential health implications of exposure to radionuclides encompass: (a) tritium or tritiated water becoming assimilated into body fluids within one to two hours of exposure; (b) plutonium-239, which can lead to blood-related cancers such as lymphoma or leukaemia; (c) iodine-131, which is rapidly absorbed by the thyroid and may potentially cause thyroid cancer; (d) strontium-90, which the body processes in a manner similar to calcium and may accumulate in the breast, increasing the risk of breast cancer; (e) cesium-137, which is absorbed by muscle cells and may contribute to the development of cancer; (f) other radioactive gases that can result in mutations in eggs and sperm (Kyne and Bolin 2016).

The World Nuclear Association has asserted the difficulty of detecting cancer in individuals exposed to less than 100 mSv of radiation (World Nuclear Association 2022b). The US Nuclear Regulatory Commission (NRC) has similarly acknowledged that the effects of low-level radiation exposure are minimal and may not be readily discernible (Kyne and Bolin 2016). However, over the past three decades, numerous studies conducted in both Europe and the USA have consistently established a link between the regular operation of nuclear reactors and cancer, particularly in children (Kyne and Bolin 2016). Moreover, the US National Academy of Sciences claims that there is no safe level of ionising radiation, and even the background radiation from nuclear power plants encountered in day-to-day life can increase the risk of cancer (Kyne and Bolin, 2016). However, the information on the hazards of nuclear power plants in their regular operation is not easily accessible to the public or neighbouring communities (Karim and Lee 2023). Despite the significance of public participation and the right to information in the realm of environmental and energy justice, a significant portion of the nuclear industry has been marked by secrecy and limited public involvement (Karim and Lee 2023). Such secrecy in

[2] In 1984, Charles Perrow, a sociologist from Yale University, authored the book "Normal Accidents: Coping with High-Risk Technologies." Within the pages of this book, Perrow examined complex systems through a sociological lens. His central argument revolved around the notion that our highly interconnected and complex systems inherently carry the potential for numerous unforeseen failures. These failures, he contended, are virtually inevitable and resistant to complete prevention through engineering measures. To describe these occurrences, Perrow introduced the term "normal accidents," which occur when, owing to the system's complexity, multiple failures interact and lead to accidents despite our best attempts to avert them.

policymaking often leads to decisions made without public participation, including the cancellation of nuclear projects.

In 2009, Malaysia's Prime Minister Najib Razak announced a plan to introduce nuclear energy in the country through a small-scale nuclear reactor. Later, the *New Energy Policy 2011–2014* also mentioned nuclear energy development without specifying the projected percentage of its total energy mix. However, in 2018, a new administration took over and the plan was suspended without explicitly mentioning the reasons. It is suggested that public concerns about the potential risks and environmental impact of nuclear energy projects may have played a role. Public opposition and apprehension about nuclear energy projects are common worldwide. Malaysia has the potential to be a model for other countries that are new to nuclear production. When it comes to nuclear projects, it is crucial to learn from the experiences of countries that have attempted to introduce nuclear projects in the past but have failed to do so. Hence, it is important to be transparent with the public and engage them in any future efforts related to such projects. This is where the discussion of procedural nuclear injustices comes into being. Procedural nuclear injustice refers to the unfair or inequitable processes involved in decision-making, governance, and regulation within the nuclear industry (Karim 2023). The process of selecting sites for nuclear power plants has often been criticised for its lack of inclusivity (Wang et al. 2019). In many cases, communities that host these facilities are not adequately consulted or given the opportunity to voice their concerns (Karim and Lee 2023). This lack of engagement can result in the unjust imposition of nuclear facilities on unwilling communities.

Additionally, nuclear power plants are required to have emergency plans in place to deal with accidents or incidents. However, the development and execution of these plans are often criticised for their lack of transparency and inadequate public participation. In the case of the Fukushima Daiichi nuclear disaster in Japan in 2011, procedural injustices were evident in the government's response, including delayed and incomplete information sharing with the affected population (Jobin 2020).

Similarly, the decision-making process for the disposal of nuclear waste has been plagued by procedural injustices. Disputes over the selection of disposal sites have led to prolonged legal battles and conflicts with local communities. The proposed Yucca Mountain Nuclear Waste Repository in the United States is a prime example of procedural issues hindering its development (McCarthy 2019). Yucca Mountain Nuclear Waste Repository, established by the Nuclear Waste Policy Act amendments in 1987, is a proposed storage facility deep within Yucca Mountain. Its purpose is to store spent nuclear fuel and other highly radioactive waste in the United States. This project has sparked intense opposition in the state of Nevada and remains a contentious national issue for Americans. A significant number of Nevadans believe that it is unjust for their state to bear the responsibility of storing nuclear waste, especially when Nevada does not have any operational nuclear power plants (Richter et al. 2022). This sentiment is shared by two-thirds of the state's population, reflecting the ongoing debate over the fairness and equity of this nuclear waste storage project.

Procedural injustices can also occur when regulatory bodies overseeing the nuclear industry are influenced or captured by the interests they are meant to regulate. This can result in a lack of enforcement of safety standards and insufficient oversight. The Fukushima nuclear disaster is an example of such a situation, where regulatory agencies in Japan were criticised for their inadequate supervision of the nuclear plant (Jobin 2020).

Instances of nuclear injustice are not limited to physical or environmental harm caused by nuclear activities. It also involves disregarding and failing to acknowledge the rights, heritage, and contributions of communities impacted by such activities. This form of injustice is more complex, as it encompasses issues of identity, culture, and representation. For instance, Native American communities, such as the Navajo Nation, have experienced recognition nuclear injustice due to the historical extraction of uranium on their lands (Fegadel, 2023). Apart from the health and environmental hazards that accompany uranium mining, these communities have also been deprived of proper recognition and compensation for the devastating effects. The US government failed to address the radioactive contamination and its associated health problems, and for decades, these communities were overlooked, leaving them to suffer alone (Kyne and Bolin 2016).

Recognition nuclear injustice is also evident in the experiences of Pacific Islander communities. The Pacific region has been a testing ground for nuclear weapons, with extensive tests conducted by major nuclear powers (Keown 2018). Affected communities, such as the Marshall Islands, were often relocated and exposed to health risks without their informed consent or acknowledgement of the lasting impacts of nuclear testing (Daborn 2013).

The Chernobyl nuclear disaster is another instance of recognition nuclear injustice. Affected communities in Ukraine, Belarus, and Russia faced both immediate and long-term health consequences. For years, their suffering went unrecognised, with inadequate support, healthcare, and acknowledgement of the extent of their sacrifice. It was only in recent years that efforts have been made to address the lasting impacts and provide recognition to these affected populations.

Communities living in the vicinity of nuclear power plants often experience nuclear injustice. These communities bear the risks associated with power plant operation, including potential accidents and radiation exposure. Yet, their voices may not be adequately heard in decision-making processes related to plant operation, safety measures, or emergency planning.

7.4 Sustainable Development of Nuclear Energy: The Way Forward

The historical injustices associated with nuclear power necessitate affirmative and deliberate actions towards energy justice. Sustainable nuclear energy development involves maximising the potential benefits of nuclear power while minimising its negative environmental, social, and economic impacts. Distributional, procedural, and recognition justice principles are critical when considering sustainable nuclear energy development. These principles ensure fairness, inclusivity, and equity in the nuclear energy sector, making sure that the benefits and burdens of nuclear power are distributed equitably among different communities and stakeholders.

Energy justice pertains to the equitable distribution of benefits and burdens in society (Sovacool and Dworkin 2015). In the context of nuclear energy, it involves ensuring that the advantages and disadvantages associated with nuclear power are fairly distributed among communities and populations. Additionally, energy justice and environmental justice are not distinct but closely related concepts, and they both address issues of fairness, equity, and social responsibility in the context of energy production, distribution, and environmental impacts. Communities living near nuclear power plants or uranium mining facilities should not bear a disproportionate share of environmental risks, such as radiation exposure or contamination.

It is crucial to ensure that the decision-making process for nuclear energy is transparent, inclusive, and participatory to achieve sustainable development. This process requires the involvement of various stakeholders in the site selection, licensing, and emergency planning for nuclear facilities. In addition, regulatory frameworks should be transparent, and information related to safety, environmental assessments, and emergency response plans should be accessible to all (Karim and Lee 2023). Furthermore, the decision-making process should follow the principle of informed consent, especially in cases where indigenous communities or areas with unique cultural significance are affected. Communities impacted by nuclear energy projects should be fairly compensated for their sacrifices, not only economically but also through acknowledgement of their contributions and the hardships they endure.

Sustainable development aims to benefit all members of society, while energy justice focuses on addressing energy-related disparities and ensuring that marginalised communities are not left behind. Hence, the intersection of energy justice and sustainable development lies in their shared goals of ensuring equitable, clean, and sustainable access to energy while considering the social, economic, and environmental well-being of present and future generations. By addressing energy-related issues with a focus on justice and sustainability, societies can work towards a more inclusive and environmentally responsible energy future.

7.5 Conclusion

At the end of the day, it is all about rebuilding the trust of nuclear energy. It is still undeniable that nuclear energy is the most efficient way of producing clean energy. As mentioned earlier, renewables still come with their own set of challenges, whereas nuclear energy has made significant advancements due to improved fission technology. Despite the numerous benefits it offers, many developed countries are hesitant to further expand their nuclear energy capacity. This reluctance is closely tied to concerns about nuclear injustices and the significant public opposition stemming from these injustices. Additionally, historical incidents, such as nuclear accidents and improper disposal of radioactive waste, have eroded public trust in nuclear energy. Addressing nuclear injustices is crucial for rebuilding trust and confidence in the safety and management of nuclear technologies. In order to shift public perceptions of nuclear energy, it is vital to anchor this change in the principles of energy justice. Only then can the world see a sustainable development of nuclear energy that guarantees equitable and environmentally responsible access to energy for all members of society.

References

Balakrishnan P, Shabbir SM, Siddiqi FA, Wang X (2020) Current status and future prospects of renewable energy: a case study. Energy Sour Part A Rec Util Environ Effects 42(21):2698–2703

Corner A, Venables D, Spence A, Poortinga W, Demski C, Pidgeon N (2011) Nuclear power, climate change and energy security: exploring British public attitudes. Energy Policy 39(9):4823–4833

Daborn M (2013) Blown to hell: the health legacies of US nuclear testing in the Marshall islands. Constellations 5(1):26–35

Dittmar M (2012) Nuclear energy: status and future limitations. Energy 37(1):35–40

Fegadel AR (2023) Green victimization of native Americans: uranium mining as a form of toxic colonialism and genocide. Crit Criminol 31(2):1–17

Guruswamy L (2010) Energy justice and sustainable development. Colorado J Int Environ Law Policy 21:231

Hamblin JD (2012) Fukushima and the motifs of nuclear history. Environ Hist 17:285–299

He G, Lin J, Sifuentes F, Liu X, Abhyankar N, Phadke A (2020) Rapid cost decrease of renewables and storage accelerates the decarbonization of China's power system. Nat Commun 11(1):2486

Holechek JL, Geli HM, Sawalhah MN, Valdez R (2022) A global assessment: can renewable energy replace fossil fuels by 2050? Sustainability 14(8):4792

Iqbal J, Howari FM, Mohamed AMO, Paleologos EK (2021) Assessment of radiation pollution from nuclear power plants. In: Pollution assessment for sustainable practices in applied sciences and engineering. Butterworth-Heinemann, pp 1027–1053

Jenkins K, Heffron RJ, McCauley D (2016a) The political economy of energy justice: a nuclear energy perspective. In: Van de Graaf T, Sovacool BK, Ghosh A, Kern F, Klare MT (eds) The palgrave handbook of the international political economy of energy, pp 661–682

Jenkins K, McCauley D, Heffron R, Stephan H, Rehner R (2016b) Energy justice: a conceptual review. Energy Res Soc Sci 11:174–182

Jenkins KE, Taebi B (2019) Multinational energy justice for managing multinational risks: a case study of nuclear waste repositories. Risk Hazards Crisis Publ Policy 10(2):176–196

Jobin P (2020) The Fukushima nuclear disaster and civil actions as a social movement. Asia Pac J 18(9):5392

Karim ME, Karim R, Islam MT, Muhammad-Sukki F, Bani NA, Muhtazaruddin MN (2019) Renewable energy for sustainable growth and development: an evaluation of law and policy of Bangladesh. Sustainability 11(20):5774

Karim R (2023) Prospects and challenges of Islamic finance instruments for low-carbon energy transitions: a legal analysis from an energy justice perspective. J Energy Nat Resour Law 41(2):195–209

Karim R, Lee EYJ (2023) Navigating nuclear energy lawmaking for newcomers: an asian perspective. Springer Nature, Singapore

Karim R, Muhammad-Sukki F (2023) Artificial intelligence (AI) in the nuclear power plants: who is liable when AI fails to perform. In: Taghizadeh-Hesary F, Zhang D (eds) The handbook of energy policy. Springer, Singapore, pp 587–607

Karim R, Karim ME, Muhammad-Sukki F, Abu-Bakar SH, Bani NA, Munir AB, Kabir AI, Ardila-Rey JA, Mas'ud AA (2018) Nuclear energy development in Bangladesh: a study of opportunities and challenges. Energies 11(7):1672

Keown M (2018) Waves of destruction: nuclear imperialism and anti-nuclear protest in the indigenous literatures of the pacific. J Postcolonial Writing 54 (5):585–600

Kim SH, Taiwo TA, Dixon BW (2022) The carbon value of nuclear power plant lifetime extensions in the United States. Nucl Technol 208(5):775–793

Kim Y, Kim W, Kim M (2014) An international comparative analysis of public acceptance of nuclear energy. Energy Policy 66:475–483

Kyne D, Bolin B (2016) Emerging environmental justice issues in nuclear power and radioactive contamination. Int J Environ Res Public Health 13(7):700

Liu C, Zhang Z, Kidd S (2008) Establishing an objective system for the assessment of public acceptance of nuclear power in China. Nucl Eng Des 238(10):2834–2838

McCarthy A (2019) The trail left behind from the nuclear race: radioactive waste. Indon J Int'l Comp l 6:287

McCombie C, Jefferson M (2016) Renewable and nuclear electricity: comparison of environmental impacts. Energy Policy 96:758–769

Pidgeon NF, Lorenzoni I, Poortinga W (2008) Climate change or nuclear power—no thanks! A quantitative study of public perceptions and risk framing in Britain. Glob Environ Chang 18(1):69–85

Richter J, Bernstein MJ, Farooque M (2022) The process to find a process for governance: nuclear waste management and consent-based siting in the United States. Energy Res Soc Sci 87:102473

Sovacool BK, Dworkin MH (2015) Energy justice: conceptual insights and practical applications. Appl Energy 142:435–444

Suman S (2021) Artificial intelligence in nuclear industry: chimera or solution? J Clean Prod 278:124022

Visschers VH, Keller C, Siegrist M (2011) Climate change benefits and energy supply benefits as determinants of acceptance of nuclear power stations: investigating an explanatory model. Energy Policy 39 (6):3621–3629

Wang S, Wang J, Lin S, Li J (2019) Public perceptions and acceptance of nuclear energy in China: the role of public knowledge, perceived benefit, perceived risk and public engagement. Energy Policy 126:352–360

World Nuclear Association (2022a) Economics of nuclear power. https://world-nuclear.org/information-library/economic-aspects/economics-of-nuclear-power.aspx. Accessed 19 Oct 2023

World Nuclear Association (2022b) Radiation and health effects. https://world-nuclear.org/information-library/safety-and-security/radiation-and-health/radiation-and-health-effects.aspx#:~:text=%22From%20an%20epidemiological%20perspective%2C%20individual, presence%20of%20naturally%20occurring%20cancers. Accessed 30 Oct 2023

Ridoan Karim is a Lecturer at the School of Business, Monash University Malaysia, where he teaches and conducts research in the fields of business and cyber law. He is a passionate academic and researcher who has published widely in peer-reviewed journals and presented papers at numerous national and international conferences. His areas of interest include "energy justice", "science, technology and law", "privacy and data protection law", and "Asian and comparative law", among others. Dr. Karim has served as a Consultant and Fellow on projects funded by the Ministry of Education Malaysia, Prime Minister's Office (Law and Institutional Reform), University of Malaya and the Monash Data Futures Institute (MDFI). He earned his Doctor of Philosophy (Ph.D.) from the University of Malaya, Malaysia, as well as a Master of Business Administration (MBA) from the University of Chichester, UK, a Master of Comparative Laws (MCL) from International Islamic University Malaysia (IIUM), and a Bachelor of Laws (LLB) from BRAC University, Bangladesh.

The Role of the Financial Sector in Energy Justice

8

Mirzet Šeho and Sabina Hodžić

Abstract

As the twenty-first century is characterised by a growing awareness of climate change and environmental degradation, the financial sector has started to channel funds into more sustainable energy investments. Therefore, this chapter provides a comprehensive and engaging exploration of how the financial sector, along with financial institutions and financial innovation, can be a powerful ally in the pursuit of energy justice. Furthermore, this chapter presents how financial institutions, instruments, and innovations play a crucial role in shaping sustainable energy systems. Therefore, policymakers, financial institutions, and energy companies need to work together to create regulatory frameworks and investment strategies for the global energy transition. This will not only fulfil economic or environmental but also moral requirements to ensure a just and sustainable future for all generations.

Keywords

Energy justice · Financial sector · Financial institutions · Financial innovation

8.1 Introduction

In the evolving discourse on sustainability and equitable development, energy justice has emerged as a defining challenge of our times and a critical area of focus. Energy justice, encapsulated by the vision of ensuring access to affordable, reliable, sustainable, and modern energy for all (Karim 2023), stands as a beacon of global ambition. Yet, this noble goal remains elusive for millions of people across the globe as they continue to grapple with energy poverty, environmental degradation, and the crippling consequences of an imbalanced energy landscape.

Pursuing energy justice is not just a moral imperative but a practical necessity for sustainable development and a fundamental human right (Huhta 2023). It underscores the conviction that the benefits of energy, from powering our homes to driving our economies, should be extended equitably to every individual, regardless of their socio-economic status or geographic location. Energy justice acknowledges that energy is the lifeblood of modern society (Mulder and Hagens 2008), and those deprived of its benefits find themselves marginalised, economically disempowered, and often relegated to hardship.

M. Šeho (✉)
School of Business, Monash University Malaysia, Bandar Sunway, Malaysia
e-mail: mirzet.seho@monash.edu

S. Hodžić
Faculty of Tourism and Hospitality Management, University of Rijeka, Opatija, Croatia
e-mail: sabinah@fthm.hr

A. Trakic et al. (eds.), *Energy Justice*, Sustainable Development Goals Series,
https://doi.org/10.1007/978-981-97-6059-6_8

At its core, energy justice seeks to address the complex issues of equity and fairness in the distribution, access, and impact of energy production and consumption. Energy justice can be defined as '*the application of rights (both social and environmental) at each component part of the energy system*' (McCauley 2018). It integrates three principal facets: distributional justice, concerning the fair allocation of energy resources and benefits; procedural justice, emphasising the need for inclusive and equitable decision-making in energy governance; and recognition justice, which involves acknowledging and addressing the needs and rights of diverse and often marginalised communities (Jenkins et al. 2016).

At the heart of the pursuit of energy justice lies a critical and often overlooked actor: the financial sector. Financial institutions, including commercial, national, and multilateral development banks, investment firms, and impact investors, to name a few, are far more than intermediaries facilitating monetary transactions. They are influential architects of the global energy landscape. Through resource mobilisation and allocation, investment decisions, lending practices, and risk assessments, the financial sector holds the power to either perpetuate energy injustices or propel us towards a more just, sustainable, and equitable energy future.

The intersection of finance and energy justice is a dynamic and critical area of study, particularly in the context of global efforts to combat climate change and promote sustainable development. This chapter endeavours to provide a comprehensive and engaging exploration of how the financial sector can be a powerful ally in pursuing energy justice, contributing to a future where energy is not only a driver of economic growth but also a pillar of social equity and environmental stewardship.

In the following sections, we delve into the pivotal role of the financial sector in advancing energy justice, highlighting the intricate ways finance has shaped global energy systems throughout history. Through a detailed examination of key financial institutions and instruments, we uncover their essential contributions towards fostering energy justice. A series of compelling case studies, including Germany's Gemeinschaftsbank für Leihen und Schenken (GLS), the Indonesian Tropical Landscape Finance Facility (TLFF), and the European Investment Bank, provide concrete examples of successful financial strategies in action. Further, we investigate the revolutionary potential of financial innovations such as blockchain technology, mobile payment platforms, and crowdfunding in reshaping the energy sector towards greater equity and justice. The chapter concludes by synthesising these insights, underlining the indispensable power of the financial sector in revolutionising energy justice globally.

8.2 Catalysing Transition: Financial Sector's Role in Energy Evolution

8.2.1 How Finance Shaped Energy Systems

To fully appreciate the role of the financial sector in energy justice, it is essential to understand its historical involvement in energy development. The roots of energy financing can be traced back to the Industrial Revolution, a period characterised by rapid industrial growth fuelled mainly by coal. Financial institutions, primarily banks, played a crucial role in providing the capital necessary for developing coal mines and, later, oil and natural gas fields. This era marked the beginning of a symbiotic relationship between the financial sector and energy producers that would significantly shape the future energy landscape.

As the twentieth century progressed, the financial sector continued to invest heavily in large-scale, centralised energy projects. These projects, including massive coal-fired power plants and sprawling oil refineries, were attractive to investors due to their high returns and the growing global demand for energy. However, these investments often overlooked the environmental impact and the social cost, particularly in regions where energy extraction and production took place.

In response to growing concerns over air pollution and the oil crises of the 1970s, there was a notable shift in energy investments. The financial sector began directing funds towards alternative energy sources, such as nuclear and hydroelectric power, as these projects were seen as cleaner and more sustainable alternatives to fossil fuels (Smil 2017). However, they were not without their challenges. For instance, large dams for hydroelectric power often lead to environmental and social issues, including the displacement of local communities and ecological disruptions (Castro-Diaz et al. 2023).

The early twenty-first century witnessed a growing awareness of climate change and environmental degradation. This period marked the beginning of a gradual shift in the financial sector's approach to energy investments, moving towards more sustainable or 'green' energy investments such as wind and solar power. This shift was initially slow, as renewable energy technologies were still maturing and were often more expensive than traditional energy sources (Smil 2017). However, technological advancements, coupled with increasing government incentives and public demand for cleaner energy, began to change the investment landscape. Financial institutions started to recognise the long-term potential and profitability of renewable energy. This shift was further bolstered through international agreements and legislation to emphasise the need for a global green transition to sustainable and affordable energy sources.

8.2.2 The Impact of Finance on the Energy Sector

The financial sector's influence on energy systems is multifaceted, impacting everything from the deployment of resources to the development of infrastructure and technologies. Financial decisions determine which energy projects receive funding, influencing the types of energy sources that are developed and deployed. This has direct implications for accessibility, as the allocation of financial resources can either facilitate or hinder the development of energy infrastructure in underserved regions.

Affordability is another critical aspect of energy justice linked to financial decisions. The cost structures of energy projects, influenced by the types of financing received, ultimately affect the price of energy for consumers. Sustainable financing practices are essential for promoting different energy systems that are crucial for transitioning to a low-carbon future. The integration of environmental, social, and governance (ESG) criteria into investment decisions reflects a growing recognition of the need to align financial practices with sustainable energy goals.

8.2.3 Current Financial Landscape in the Energy Sector

The current financial landscape in the energy sector is a complex and dynamic area where diverse financial instruments and policies converge to shape the development and distribution of energy resources. This landscape is a critical determinant of how energy justice is actualised in different contexts, influencing who gets access to energy, the types of energy produced, and the long-term sustainability of energy systems.

There is a growing emphasis on sustainable and renewable energy financing. This shift is driven by a heightened global awareness of climate change and the urgent need to reduce carbon emissions. Financial institutions, including banks, investment funds, and international finance bodies, have started to integrate environmental criteria into their lending and investment decisions. This trend is evident in the increasing issuance of green bonds by governments and corporations, specifically aimed at funding environmentally friendly projects, including renewable energy ventures.

Another key aspect of the current financial landscape is the role of innovative financial models and mechanisms. Crowdfunding, community-based financing, and impact investing, to name a few, have emerged as alternative funding sources for energy projects, particularly in regions and communities traditionally underserved by conventional financial systems. These models often prioritise social and environmental returns alongside or without financial gains. Furthermore, digital technologies such as mobile payment platforms, blockchain technology, and other fintech solutions revolutionised the ways to transact, mobilise funds, ensure transparency, and reduce transaction costs, making energy systems more efficient, accessible, and affordable. In addition, these technologies have enabled millions of unbanked to be financially included and have the ability to participate in energy projects.

However, challenges in aligning financial flows with energy justice goals persist. Fossil fuel projects continue to attract substantial investment due to their established market presence and proven profitability, raising questions about the financial sector's commitment to a sustainable energy transition and its implications for long-term energy justice. Furthermore, regulatory frameworks and market structures often favour established energy systems, hindering the adoption of innovative financing for sustainable energy solutions. Additionally, the global disparity in financial resource availability means countries with limited financial resources face greater challenges in transitioning to sustainable energy systems. A concerted effort among all stakeholders, including financial institutions, governments, and civil society, is essential to create an enabling environment that reflects the realities on the ground for equitable and sustainable energy financing.

Following the exploration of the current financial landscape in the energy sector, Tables 8.1 and 8.2 present the key financial sector players and instruments and their roles in advancing energy justice.

Table 8.1 Key financial sector players and their roles in energy justice

Player/Institution	Role in energy justice
Commercial Banks	Provide loans and other financial services essential for energy projects. They play a pivotal role in funding both traditional and renewable energy projects, influencing the energy mix and accessibility.
Investment Banks	Specialise in underwriting and managing investments, crucial for structuring large-scale energy projects. They facilitate issuing green bonds and other sustainable investment products, thus promoting environmentally friendly energy solutions.
Insurance Companies	Offer risk assessment and management services for energy projects. Their policies and premiums can significantly influence the economic viability and security of investments, especially in renewable energy sectors prone to high initial risks.
Pension Funds	Control large pools of long-term investment capital. Increasingly, these funds are directed towards sustainable and socially responsible investments, including renewable energy projects, which align with the long-term return expectations and social responsibility goals of these funds.
Sovereign Wealth Funds	Government-owned investment funds with significant financial resources. They can allocate substantial capital towards national energy projects, often prioritising strategic and long-term national energy interests, including sustainability and energy independence.
National Development Banks (NDBs)	NDBs provide critical financing for energy projects, particularly in sustainable energy. They align investments with national policies for energy development. Their role is particularly important in regions where commercial financing is not readily available or is too costly for sustainable energy projects.
Multilateral Development Banks (MDBs)	Provide financial resources and professional advice for energy projects in developing countries. They focus on sustainable development and can influence energy justice by funding projects that improve access to clean and affordable energy in less-developed regions.

(continued)

Table 8.1 (continued)

Player/Institution	Role in energy justice
Venture Capital Firms	Invest in innovative and high-potential start-ups, crucial for new clean energy technologies. Their investments can catalyse advancements in sustainable energy solutions and contribute to the diversification of the energy sector.
Private Equity Firms	Invest capital in private energy companies, including those involved in renewable energy. These firms play a significant role in shaping the energy landscape through strategic investments and can drive the growth of emerging energy technologies.
Crowdfunding Platforms	Enable collective investment from individuals, often in smaller, community-based, or innovative energy projects. This democratises energy financing, allowing broader participation in funding and benefiting local or niche energy initiatives.
Impact Investment Funds	Dedicated to generating a positive social and environmental impact alongside a financial return. These funds are increasingly focusing on renewable energy projects, promoting sustainability, and addressing energy poverty and accessibility issues.
Environmental, Social, and Governance (ESG) Funds	These funds incorporate ESG criteria in their investment decisions, playing a critical role in directing capital towards environmentally sustainable and socially responsible energy projects. This shift is instrumental in promoting long-term sustainability in the energy sector.
International Finance Institutions (IFIs)	Entities like the World Bank and IMF provide financial support and policy guidance for energy projects on a global scale. They play a crucial role in shaping energy policies and financing in developing countries, focusing on sustainable and equitable energy access.
Non-Governmental Organisations (NGOs)	They influence energy policy and provide support at the grassroots level for sustainable energy projects. NGOs may also channel funding, particularly in underdeveloped or rural areas, and advocate for equitable energy policies and practices.
Energy Cooperatives	Community-based organisations that finance and manage local energy projects. They focus on meeting local energy needs, often through renewable sources, promoting energy accessibility and community involvement in energy decisions.
Governmental and Regulatory Bodies	Set policies and regulations that govern and incentivise energy financing, including subsidies, tax incentives, and feed-in tariffs. They can provide direct funding or incentives for sustainable energy projects.

Source Authors' own creation

8.3 Case Studies and Success Stories

Every country's unique natural and economic attributes, shaped by its level of economic development, dictate its energy strategies. A sole focus on energy reinforces these differences between nations. While numerous case studies illustrate this diversity, this chapter will selectively examine a few, aiming to showcase how various nations, irrespective of their location, can leverage their unique opportunities and capabilities to innovate in energy generation. This approach not only facilitates the global energy transition but also broadens the scope for financial investments in sustainable energy initiatives.

8.3.1 Germany's Community Bank for Lending and Giving

With the Climate Action Plan 2050, the German government set ambitious national targets to

Table 8.2 Key financial instruments and their roles in energy justice

Financial instrument	Role in energy justice
Loans	Loans provide funding specifically for projects with a positive environmental impact, such as renewable energy and energy efficiency upgrades, supporting sustainable development, and reducing carbon emissions.
Green bonds	Green bonds raise capital for projects with environmental benefits, including renewable energy, aiding in sustainable energy transitions.
Sustainable and social loans	Loans specifically designed to finance projects with social or environmental impact, promoting energy access, and efficiency.
Equity investments	Equity investments in energy companies, particularly in renewables, drive innovation, and expand sustainable energy solutions.
Venture capital investments	Venture capital funds start-ups and innovative technologies in clean energy, crucial for advancing new sustainable energy solutions.
Crowdfunding	Crowdfunding democratises financing for smaller, community-based renewable energy projects, enhancing local energy access, and independence.
Impact investments	Investments made with the intention to generate positive, measurable social and environmental impacts alongside a financial return.
Carbon credits and trading	Carbon credits and trading mechanisms incentivise reducing emissions, indirectly supporting renewable energy projects and technologies.
Public grants and subsidies	Government-provided financial support for energy projects crucial for kick-starting renewable energy initiatives and research.
Project finance	Long-term, non-recourse or limited recourse financial structures used primarily for large-scale energy infrastructure projects.
Development finance	Finance provided by development banks or international finance institutions for large-scale energy projects, often in developing countries.
Guarantees	Used to mitigate investment risks in energy projects, especially critical for renewable energy initiatives. By assuring investors and lenders against potential losses, they enhance the feasibility and attractiveness of sustainable energy projects.

Source Authors' own creation

reduce greenhouse gas emissions by at least 55% by 2030, 70% by 2040, and 100% by 2050 compared to 1990 levels (BMUV 2016). Additional targets were also set for the energy sector industry, buildings, transport, forestry, and agriculture. To achieve the desired targets, additional financial resources are needed—green funds. These funds depend on the political will, fiscal and monetary policies, and a dedicated national central bank, development, and other banks to stimulate borrowing and lending for green purposes.

For this reason, 'the Gemeinschaftsbank für Leihen und Schenken (GLS)' (Community Bank for Lending and Giving) is a private bank founded in 1974 to support ecological and social investments (Bank 2024). The raison d'être of the bank is to ensure access to needed funds for all in the areas of energy, housing, education, culture, and health. These meet human needs and foster sustainability. The bank is also a member of the Global Alliance for Banking on Values, whose primary focus is on financial flows for activities that deliver economic resiliency, environmental regeneration, and social empowerment for the communities and people they serve (GIZ 2019). At the end of 2017, the bank had over 48,398 members with assets under management of 5 billion euros (Bank 2018). The bank's core activity is lending. There are two types of loans—for savers and investors. For savers, GLS offers savings investments earmarked for specific sectors, e.g. renewable energy, sustainable investments, education, health, and culture. For investors, GLS offers a range of special green funds in which they can invest relatively small amounts starting at EUR 250 (Bank 2018). They relate to environmental and ecological investments. In addition to traditional financial instruments, GLS offers a range of innovative solutions. One example is the crowdfunding platform. All examples have a common feature—long-term financing to achieve ecological sustainability. Moreover, by providing financial support, this example ensures distributional justice concerning the fair allocation of energy resources and benefits.

8.3.2 Indonesian Tropical Landscape Finance Facility

To tackle environmental and social problems, the Indonesian government has set ambitious goals. These goals include reducing the poverty rate from 11% to below 4% by 2025, shifting the current 6% of the total energy mix to 23% by 2025, and increasing carbon emission savings of 29–41% by 2030 (TLFF 2018). Therefore, the TLFF has been established as a multi-stakeholder partnership between the UN Environment Program, the World Agroforestry Centre, BNP Paribas, and ADM Capital. The aim is to finance local projects and companies with a particular focus on green growth, green bonds, and job creation, with a special focus on sustainable agriculture and the renewable energy sector. Furthermore, the objectives are in line with the Paris Agreement, the Sustainable Development Goals, and the priorities of the Indonesian government. In February 2018, TLFF issued the first corporate sustainability bond—PT Royal Lestari Utama (RLU), with a value of USD 95 million (TLFF 2019). It is a joint venture between Michelin (49%) and a subsidiary of the Indonesian Barito Pacific Group (51%) (TLFF 2019). The aim is to finance the production of sustainable natural rubber, the restoration of forested buffer zones, and the implementation of a community partnership programme in Jambi and East Kalimantan provinces. In addition, the TLFF is a platform for granting loans and short grants for capacity building, monitoring, and evaluation activities (GIZ 2019). By ensuring free access to clean water and providing safe waste management at all levels of government, it can be noticed that all three principles of energy justice are addressed.

8.3.3 European Investment Bank

In order to ensure energy and sustainable transitions in Europe and abroad, the European Investment Bank has set up an ambitious goal to become a European Union climate bank (EIB 2024). Nowadays, this is one of the world's leading financiers of climate protection and sustainable development projects. As such, it operates in over 150 countries around the world and has a mandate to work in all developing countries (EIB 2024). In line with its mission and vision and the principles of the Paris Agreement and Sustainable Development Goals, its top priority is to finance climate change mitigation and adaptation projects. Consequently, this will lead to low-carbon and climate-resilient growth around the world and support the transition to a sustainable, low-carbon future that leaves no one behind. Hence, trillions of dollars of investment are needed to achieve the desired goals. For that purpose, the Climate Bank roadmap has been developed. The main objectives of the Climate Bank roadmap are to ensure a just transition for all, support Paris-aligned operations, build strategic coherence and accountability, and accelerate the transition through green finance. To support all stated, in 2022, the bank ensured EUR35 billion, or 56% of the EIB's own resource financing for environmental sustainability projects (EIB 2023). The list of actions that the bank supports is shown in Fig. 8.1.

Based on data from Fig. 8.1, it can be concluded that the European Investment Bank significantly supports all actions in mitigating climate change and adaptation projects. For example, a EUR350 million loan was provided to Danieli, a global leader in steel plant production based in Italy (EIB 2023). This will help promote the circular economy by producing recycled steel at the plants in the Italian province of Udine and in the Sisak province in Croatia. This project also

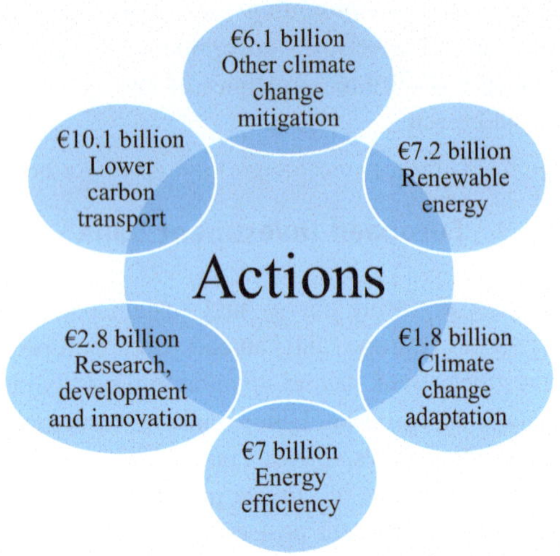

Fig. 8.1 Financial support of the listed actions in 2022. *Source* EIB (2023)

fosters the transition to a circular economy. As a result, it will affect all society actors by ensuring the proper functioning of all three energy justice principles.

8.4 Revolutionising Energy Justice: The Power of Financial Innovation

Financial innovations are crucial in driving economic development (Cheng and Hou 2022) and play a vital role in advancing energy justice by revolutionising the funding, management, and valuation of energy projects. These innovations introduce new financial products, services, and processes that enhance the efficiency and reach of financial markets and institutions, pivotal for facilitating sustainable energy transitions in an inclusive and equitable manner.

The realm of financial innovations is broad, focusing on enhancing accessibility and sustainability within the energy sector. They empower a wide array of stakeholders, from marginalised communities to small investors and enterprises, enabling them to participate in energy markets. Furthermore, these innovations catalyse investments in sustainable and ethical energy solutions. Among the plethora of financial tools—including green loans and bonds, energy efficiency financing, microfinance, carbon credits and trading, and risk guarantee instruments—this section highlights the most impactful: blockchain technology, crowdfunding, and mobile payment platforms, including pay-as-you-go systems. These innovations are at the forefront of creating a fair energy landscape, ensuring clean, affordable, and reliable energy for everyone.

8.4.1 Blockchain Technology

Blockchain technology represents a significant shift in both the financial services industry and the energy sector. It is a distributed ledger protocol that facilitates the secure and transparent recording of transactions and tracking of assets in a peer-to-peer (P2P) network, utilising a series of linked blocks, cryptographically secured and chronologically ordered, to create an immutable and tamper-proof record of data (Gupta 2020). Smart contracts within this framework automate transactions, cutting down administrative and operational costs—key for optimising microgrid and distributed energy resource management, thereby improving energy access.

In the energy sector, blockchain's impact is profound, reshaping financing, energy trading, and asset management. It enables direct electricity trading between consumers through P2P platforms and the tokenisation of energy assets, promoting energy market democratisation, justice, and sustainability.

One of the standout implementations of blockchain in energy is P2P trading platforms. These platforms empower consumers to buy and sell electricity directly amongst themselves, bypassing conventional intermediaries. There are numerous examples that provide these services successfully. One of the leaders in this domain is PowerLedger, an Australian initiative that leverages blockchain to offer a transparent and efficient energy trading platform. By allowing households with solar panels to sell surplus electricity to neighbours, PowerLedger not only maximises renewable energy utilisation but also bolsters local energy economies and curtails transmission losses, boasting over 30 clients across 10 countries and four continents (PowerLedger 2024). Similarly, Brooklyn Microgrid, a project initiated in 2016 in the Park Slope and Gowanus communities of New York, fosters a community-driven model with hundreds of residents trading solar energy within a microgrid setup, showcasing the potential for localised energy autonomy (Microgrid 2024).

Another groundbreaking application of blockchain is the tokenisation of energy assets, exemplified by SunContract, a Slovenian company. This process involves converting energy or energy-producing assets into digital tokens on a blockchain, thus enabling their sale, purchase, or exchange in a transparent and secure manner. SunContract innovates by tokenising solar panels and rooftops, allowing individuals and businesses worldwide to invest in solar farms and earn passive income from the generated electricity (SunContract 2023). This model not only facilitates P2P energy trading but also invites global participation in renewable energy investment. In Slovenia, SunContract has woven a tale of energy self-sufficiency through solar energy. The village of Zavrate, once reliant on the traditional grid, now stands proudly as the nation's first 100% self-powered community (SunContract 2022). Residents now generate their own clean electricity, trade surplus within the village, and enjoy lower energy bills. This is not a one-off marvel; Radeče city centre followed suit, demonstrating the scalability of SunContract's model (SunContract 2022).

The contributions of PowerLedger, Brooklyn Microgrid, and SunContract to energy justice are praiseworthy. By fostering P2P energy trading platforms, these blockchain innovations democratise access to the energy market, allowing households and small producers to sell or exchange surplus renewable energy. This model champions distributional justice by guaranteeing a more equitable distribution of energy resources. Moreover, the use of blockchain technology enhances procedural justice, rendering energy transactions more accessible, equitable, and participatory for a broader range of stakeholders, including those previously marginalised. Furthermore, by enabling investments in renewable energy assets and valuing individual contributions to the energy ecosystem, these platforms advocate for recognition justice, acknowledging the diverse needs and contributions of all participants towards a sustainable energy future. These examples signify a transformative leap towards a more just, inclusive, and sustainable energy landscape, highlighting the pivotal role of financial innovations in achieving energy justice.

8.4.2 Mobile Payment Platforms and Pay-As-You-Go Models

The rise of mobile payment platforms marks a significant evolution in the financial landscape, with profound implications for energy justice. These innovations extend the reach of financial services, making them more accessible, especially in areas traditionally underserved by conventional banking infrastructure. By enabling more straightforward, efficient, and often more affordable financial transactions, mobile payment platforms are reshaping consumer and business engagement with the energy sector.

Mobile payment platforms, such as M-Pesa in Kenya, have revolutionised financial interactions by facilitating direct payments for energy services via mobile devices, overcoming barriers associated with traditional banking. This innovation is particularly impactful in regions with scarce traditional banking services but high mobile phone usage. M-Pesa and similar services have been instrumental in enabling African households to afford solar energy solutions through manageable instalment payments and democratising access to clean energy.

One of the most successful stories in leveraging mobile payments for energy access is M-KOPA Solar, which utilises M-Pesa to provide pay-as-you-go (PAYG) solar systems in Africa. M-KOPA enables customers to make small payments for solar energy, thus enhancing energy access in rural areas and mitigating the energy access gap. As of June 2023, M-KOPA has served over 3 million customers across Kenya, Uganda, Nigeria, and Ghana, impacting over 16 million lives by providing solar home systems to 1 million customers and offsetting 2.1 million tonnes of CO_2 (M-KOPA 2023).

Mobile payment methods enhance financial transaction accessibility and promote transparency and security, advancing procedural justice by ensuring fairness in energy transactions. Additionally, by facilitating investments in renewable energy assets, these financial innovations support the recognition justice, valuing the diverse contributions to a sustainable energy ecosystem. Thus, mobile payment platforms with pay-as-you-go models are crucial in bridging the energy divide, enabling broader and more equitable participation in renewable energy solutions. The integration of fintech into the energy sector represents a significant step towards a more just, inclusive, and sustainable energy future, underscoring the vital role of the financial sector in fostering energy justice.

8.4.3 Crowdfunding

Crowdfunding represents a paradigm shift in how projects secure funding, particularly in the renewable energy sector. This innovative financial model harnesses the collective power of individual investors, both small and large, to pool resources towards a common goal. Unlike traditional financing methods, which rely on a limited number of institutional investors, crowdfunding democratises the investment process, allowing anyone with an internet connection to contribute to projects they believe in. Investments are typically permitted in minimal amounts, thus broadening access to investment opportunities for individuals with limited capital.

In the context of energy projects, crowdfunding can take various forms, including donation-based, rewards-based, debt-based, and equity crowdfunding. Each type offers a different incentive mechanism for contributors, from the satisfaction of supporting a cause (donation-based) to the potential for financial returns (debt and equity). This flexibility makes crowdfunding particularly appealing for renewable energy projects, which often struggle to secure funding through conventional channels due to perceived risks or the innovative nature of the technology.

Some of the most successful examples of crowdfunding in the energy domain include.

- **The Sun Exchange**: the world's first P2P solar cell leasing platform combines solar energy investment with cryptocurrency. The Sun Exchange allows individuals worldwide to buy solar cells and lease them to businesses and communities in sunny emerging markets, earning returns in Bitcoin. This model not only promotes the use of solar energy but also introduces an innovative way to invest in and support renewable energy projects across borders. As of today, The Sun Exchange claims to have over 80 solar projects completed, funded by more than 100 corporate funders and 40,000 individual members from 180 countries; 10 MW+ solar cells sold; and 20 GWh of clean electricity produced (The Sunexchange 2024).
- **Energise Africa**: Targeting the off-grid solar market in sub-Saharan Africa, Energise Africa offers a chance for investors to support solar businesses that install systems in homes lacking access to electricity. The platform provides both social and financial returns, highlighting the impact investment can have on communities' quality of life. They claim to have raised and invested over 32 million pounds, improving the lives of almost one million people and reducing CO_2 by 190,000 tonnes yearly (Africa 2024).
- **Trine**: Focussed on financing solar energy projects in emerging markets, Trine enables individuals to invest in solar projects with a social impact, offering competitive returns. Projects funded through Trine have provided thousands of people with access to clean energy, reducing carbon emissions by more than one million tonnes and fostering local economies (Trine 2024)

Crowdfunding significantly advances energy justice by democratising the financing of renewable energy projects addressing distributional justice through investments that mitigate energy poverty and expand access in underserved regions. This model promotes procedural justice by creating a transparent, inclusive platform that empowers broad participation in energy decisions, fostering ownership and active involvement in the sustainable transition. Furthermore, crowdfunding platforms uphold recognition justice by valuing investments of all sizes and supporting innovative, nontraditional projects, acknowledging the vital role of diverse contributions and community-led efforts in shaping a sustainable energy future.

These highlighted financial innovations are more than mechanisms for funding; they are catalysts for building a just and sustainable energy landscape. The transformative role of finance in realising energy justice demonstrates its capacity to shape a more sustainable and equitable world, emphasising the importance of selecting and focusing on the most impactful innovations in the quest for a fair energy future.

8.5 Conclusion

As global environmental challenges intensify, the intersection of the financial sector and energy justice emerges as a crucial field of study. This chapter underscores the financial sector's critical role in advancing energy justice, illustrating how financial institutions, instruments, and innovations are integral to shaping sustainable energy systems. From historical investments that have influenced the energy landscape to the contemporary shift towards sustainability, finance acts as a pivotal force in the energy sector. The exploration of key financial players and instruments reveals their potential to drive equitable access to energy, while case studies from Germany, Indonesia, and the European Investment Bank highlight practical applications of these principles.

The chapter emphasises the transformative role of financial innovations like blockchain technology, mobile payment solutions, and crowdfunding in democratising energy financing and fostering renewable energy adoption. These technologies ensure accessible, clean, and affordable energy solutions worldwide. Blockchain, for example, revolutionises energy trading and asset management, enabling P2P energy trading and asset tokenisation, which enhance market democratisation and sustainability. Initiatives such as PowerLedger's blockchain-based energy trading, Brooklyn Microgrid's community energy exchange, and SunContract's promotion of energy self-sufficiency illustrate how these technologies underpin equitable energy distribution. They facilitate platforms for households and small producers to trade surplus renewable energy, embodying distributional justice by promoting equitable resource distribution. Furthermore, they advance procedural justice by making energy transactions transparent and inclusive and support recognition justice by valuing diverse contributions to a sustainable energy future. The success of PowerLedger, Brooklyn Microgrid, SunContract, M-KOPA, The Sun Exchange, Energise Africa, and Trine showcases the potential of financial innovations to propel energy justice globally.

As we look ahead, it is clear that the journey towards energy justice is ongoing and requires the sustained commitment of the financial sector. The evolution of financial practices, fuelled by innovation and guided by the principles of sustainability and equity, offers a promising pathway to address the multifaceted challenges of energy access and environmental stewardship. It is imperative that the financial sector continues to adapt and evolve, leveraging new technologies and financial instruments to not only mitigate the impacts of climate change but also to ensure that energy systems are inclusive and equitable.

To amplify the impact of financial innovations on energy justice, collaboration across sectors and disciplines is essential. Policymakers, financial institutions, and energy companies must collaborate to create regulatory frameworks and investment strategies that prioritise long-term sustainability and fairness. By aligning financial flows with the goals of energy justice, we can unlock the potential of green finance, drive the transition towards renewable energy, and ensure that all members of society share the benefits of this transition.

In closing, the financial sector has both the opportunity and the responsibility to be at the forefront of the global shift towards sustainable energy. The examples and strategies discussed in this chapter provide a roadmap for harnessing financial innovation to advance energy justice. As we move forward, the actions of financial institutions, informed by a deep understanding of energy justice, will determine the pace and inclusiveness of the global energy transition. The pursuit of energy justice, underpinned by financial innovation, is not just an economic or environmental imperative but a moral one, ensuring a just and sustainable future for generations to come.

Acknowledgements This work has been supported by the European Union under the BOLSTER project, Grant Agreement, No. 101069586.

References

Africa E (2024) About Trine [Online]. Available: https://trine.com/about/trine. Accessed 23 Feb 2024

Bank G (2018) Die Bank in Zahlen [Online]. Available: https://www.gls.de/privatkunden/gls-bank/zahlen-fakten. Accessed 23 Feb 2024

Bank G (2024) Warum GLS Bank? [Online]. Available: https://www.gls.de/privatkunden/nachhaltig/warum-gls-bank/. Accessed 23 Feb 2024

BMUV (2016) Climate Action Plan 2050 [Online]. Available: https://www.bmuv.de/en/publication/climate-action-plan-2050-en. Accessed 23 Feb 2024

Castro-Diaz L, García MA, Villamayor-Tomas S, Lopez MC (2023) Impacts of hydropower development on locals' livelihoods in the Global South. World Dev 169:106285

Cheng S-Y, Hou H (2022) Innovation, financial development, and growth: evidences from industrial and emerging countries. Econ Chang Restruct 55:1629–1653

EIB (2023) Climate action and environmental sustainability overview

EIB (2024) Climate and environmental sustainability [Online]. Available: https://www.eib.org/en/projects/topics/climate-action/index. Accessed 23 Feb 2024

GIZ (2019) The role of national financial institutions [Online]. Available: https://www.giz.de/en/downloads/giz2019-EN-The-Role-of-National-Financial-Institutions.pdf. Accessed 23 Feb 2024

Gupta M (2020) Blockchain for dummies 3rd ibm limited edition, america. Wiley

Huhta K (2023) Conceptualising energy justice in the context of human rights law. Nord J Hum Rights, 1–15

Jenkins K, Mccauley D, Heffron R, Stephan H, Rehner R (2016) Energy justice: a conceptual review. Energy Res Soc Sci 11:174–182

Karim R (2023) Prospects and challenges of Islamic finance instruments for low-carbon energy transitions: a legal analysis from an energy justice perspective. J Energy & Nat Resour Law 41:195–209

M-KOPA (2023) Impact report

Mccauley D (2018) Global energy justice. In: Mccauley D (ed) Energy justice: re-balancing the trilemma of security, poverty and climate change. Springer, Cham

Microgrid B (2024) Brooklyn microgrid overview [Online]. Available: https://www.brooklyn.energy/about. Accessed 23 Feb 2024

Mulder K, Hagens NJ (2008) Energy return on investment: toward a consistent framework. Ambio 37:74–79

PowerLedger (2024) Clean energy that works for everyone on the planet [Online]. Available: https://www.powerledger.io/. Accessed 23 Feb 2024

Smil V (2017) Energy and civilisation: a history. The MIT Press

SunContract (2022) SunContract supports self-sufficient communities [Online]. Available: https://suncontract.org/suncontract-supports-self-sufficient-communities/. Accessed 23 Feb 2024

SunContract (2023) The perfect harmony: how NFTs seamlessly integrate into SunContract's EcoSystem [Online]. Available: https://suncontract.org/the-perfect-harmony-how-nfts-seamlessly-integrate-into-suncontracts-ecosystem/. Accessed 23 Feb 2024

The SunExchange (2024) SunExchange is powering progress in Africa by delivering world-class solar solutions to high-impact businesses and organisations [Online]. Available: https://sunexchange.com/about-us/. Accessed 23 Feb 2024

TLFF (2018) Tropical landscape finance facility [Online]. Available: http://tlffindonesia.org. Accessed 23 Feb 2024

TLFF (2019) TLFF inaugural transaction: corporate sustainability bond for natural rubber production

Trine (2024) About Trine [Online]. Available: https://trine.com/about/trine. Accessed 23 Feb 2024

Mirzet Šeho is a Senior Lecturer at the School of Business, Monash University Malaysia. He has held various positions with nearly a decade of industry experience before academia. He served as the Head of Financial Controlling at two companies, including a subsidiary of a high-volume, public-listed multinational corporation. In these roles, he led cross-functional teams, fostering advancements in financial management, strategic planning, and internal controls, among other critical areas. In addition to his industry experience, Dr. Šeho is a dedicated and multi-award-winning researcher with a passion for challenging traditional finance practices. He actively publishes in reputable academic outlets and contributes to shaping the future of the field, leveraging his expertise and knowledge to drive innovation and disruption.

Sabina Hodžić is an Associate Professor at the Faculty of Tourism and Hospitality Management, University of Rijeka (Croatia). Her main fields of research are public finance, international taxation issues, local and regional government, digital economy and green development. She lectures on public finance and international taxation and is the author of several papers published in scientific journals. She participated in several academic research projects at home and abroad and was granted a post-doc fellowship from Austria's Agency for Education and Internationalization (OeAD). As a part of professional training, she stayed at Vienna University of Economics and Business (2021), University of New South Wales, Australia (2018), and University of Economics, Slovakia (2016). She is a member of the International Fiscal Association—Croatia. In addition, she is also a member of editorial board of Tourism and Hospitality Management Journal and Croatian Review of Economic, Business and Social Statistics.

9 The Role of Energy Technology Innovation in Solving Energy Injustice

Manjeevan Seera

Abstract

This chapter examines the dynamic interplay of energy technology innovation and energy justice, addressing the global challenge of energy injustice. It begins by defining energy injustice, identifying its key forms as distributive, procedural, and recognition injustices, and demonstrating their global prevalence through examples and statistics. The narrative then shifts to consider how advances in renewable energy technologies, machine learning applications, and smart grid optimizations can address these injustices. Case studies support the role of renewable energy in rural electrification and alleviating energy poverty. The transformative potential of machine learning in improving the efficiency and adaptability of renewable energy systems and smart grids is investigated, acknowledging the challenges and ethical concerns. The chapter concludes with a debate on the intersection of energy technology innovation and energy justice, emphasizing equitable access and distribution of energy resources, supported by policy recommendations, to ensure an inclusive energy transition necessary for sustainable development and equity.

M. Seera (✉)
School of Business, Monash University Malaysia,
Kuala Lumpur, Malaysia
e-mail: manjeevansingh.seera@monash.edu

Keywords

Energy innovation · Energy justice · Renewable energy · Machine learning · Smart grids

9.1 Introduction

The global energy landscape is a complex and intricate web that is essential to everything from simple household functions to large-scale industrial operations. The importance of energy in driving productivity, comfort, and innovation cannot be overstated. However, the benefits of these energy systems are distributed inequitably across the global population, resulting in a phenomenon known as energy injustice. This injustice manifests itself in a variety of ways, including disparities in access to clean energy sources, inequitable energy costs, and disproportionate environmental burdens associated with energy production. Addressing these disparities is not only a matter of fairness; it is also critical for achieving sustainable development, as emphasized by the United Nations in Sustainable Development Goal 7.

Energy injustice is a complicated issue that intersects with larger global issues such as climate change, economic inequality, and social disenfranchisement. It reflects a deeper systemic imbalance in which the most vulnerable populations bear the brunt of environmental degradation

A. Trakic et al. (eds.), *Energy Justice*, Sustainable Development Goals Series,
https://doi.org/10.1007/978-981-97-6059-6_9

while being excluded from energy policy decision-making processes. To navigate the complexities of transitioning to a low-carbon economy, the concept of a 'Just Transition' has been introduced. This transition highlights the importance of policies that are both environmentally sustainable and socially equitable, ensuring that the transition away from fossil fuels does not disproportionately affect disadvantaged communities.

Renewable energy technology innovations, particularly in the realm of Renewable Energy Technology Innovation (RETI), have the potential to address some aspects of energy injustice. In China, for example, RETI has played a significant role in significantly reducing manufacturing carbon intensity, not only locally but also in neighbouring regions, with varying effects depending on distance and industry type. This discovery highlights RETI's potential for mitigating environmental impacts and promoting regional low-carbon development (Xin et al. 2022).

Furthermore, as evidenced by its significant positive impact on China's provinces, RETI's role extends to influencing green development. This effect is inextricably linked with the dynamics of industrial structure upgrading, emphasizing the complex interplay between technological innovation and economic restructuring. Su and Fan's research (2022) sheds light on how RETI, in conjunction with changes in industrial structure, can foster a more sustainable and green economy.

Another critical area where RETI can make a significant difference is in addressing energy poverty. Wang et al. (2022) investigated how RETI can alleviate energy poverty in relation to marketization levels. Their findings show that RETI reduces household energy poverty significantly, with the effectiveness of these interventions varying across regions and marketization contexts. This emphasizes the importance of tailoring RETI strategies to specific regional and economic conditions in order to maximize their impact on energy poverty reduction.

Furthermore, the intersection of green technology innovation and energy consumption structures in high-end equipment manufacturing enterprises provides insights into the long-term performance of the enterprise. Wang (2022) shows that green technology innovation has a positive impact on enterprise performance, with energy consumption structure acting as a moderator. This emphasizes the importance of incorporating green technologies into enterprises' core operational strategies in order to achieve long-term growth and performance.

This chapter will delve deeper into these themes as it progresses, presenting a narrative that intertwines technological innovation with the pursuit of justice in the energy sector. The chapter aims to highlight the transformative potential of technology in creating more equitable energy systems through a series of case studies and scholarly analyses. It will also examine the policy implications of these technologies, advocating for an inclusive approach to energy transition that prioritizes the needs of all segments of society, thus contributing to the global agenda of sustainable development and energy equity.

9.2 Recognizing Energy Inequality

Energy injustice is characterized by disparities in energy access and the negative impacts of energy infrastructure on marginalized communities, and it is a significant impediment to sustainable development. Using contemporary scholarly work, this section delves into the concept, manifestations, and consequences of energy injustice.

9.2.1 Energy Injustice: Definition and Types

Energy injustice manifests itself in three ways: distributive, procedural, and recognition. Inequitable access to energy resources and the unequal burden of environmental harm caused by energy infrastructure are examples of distributive injustice. This type of injustice is exemplified by the fact that certain communities have limited or no access to clean energy technologies and are disproportionately affected by pollution from energy production. When the processes governing energy policy-making exclude certain groups, particularly those most affected by energy decisions, procedural injustice occurs. This exclusion frequently leads to policies that do not reflect the needs or input of marginalized populations. When the specific needs, rights, and cultural identities of communities are not recognized in energy policies, a one-size-fits-all approach that may not suit all groups results.

Lamura's (2022) assessment of injustices in four European Union-mandated National Energy and Climate Plans (NECP) focuses on energy poverty, also known as energy injustice. The study reveals limitations in appreciating energy poverty along the three energy justice tenets, implying a need for broader recognition of structural injustices in order to integrate social and energy goals.

9.2.2 The Effects of Energy Inequality

Energy injustice has a significant social impact, intertwining with issues such as poverty and environmental degradation. Energy poverty, a major component of energy injustice, is associated with insufficient access to essential services, which leads to poor health outcomes and limited economic opportunities. This type of poverty disproportionately affects the most vulnerable people, influencing their daily lives and future prospects. The environmental consequences are equally severe, with marginalized communities frequently residing in areas more vulnerable to pollution and the effects of climate change. These communities bear a double burden: they are less prepared to adapt to environmental changes and are frequently the least accountable for the emissions that cause these changes.

The importance of procedural energy justice is highlighted in Stojilovska's (2023) research on the role of institutions in addressing energy poverty. The study emphasizes the role of the Ombudsman in Europe as an institution capable of uncovering hidden causes of energy poverty and contributing to a socially just energy transition. This approach emphasizes the importance of strong institutional frameworks capable of effectively addressing the multifaceted nature of energy injustice.

9.2.3 Energy Injustice Case Studies

Case studies illuminate the tangible impacts of energy injustice. The research by Barragan-Contreras (2022) delves into the challenges of procedural and socio-environmental justice arising from a large-scale photovoltaic initiative in Yucatan, Mexico. It underscores the critical need for self-determination and the active participation of local communities in energy developments, especially in regions enriched with cultural and ecological diversity. This analysis highlights the necessity for energy policies and projects to be attuned to the nuances of local contexts, ensuring that community engagement is central to the decision-making framework. The case underlines that for energy projects to be just and equitable, they must not only consider the environmental and economic outcomes but also respect and integrate the voices and rights of the communities they impact.

The Kerala case, as analysed by Bedi (2019), encapsulates the multifaceted dilemmas entwined with renewable energy ventures, particularly solar initiatives. Despite solar energy's potential to address India's climate objectives and burgeoning energy demands, the endeavour in Kerala collided with acute land scarcity and entrenched historical injustices. The aspirations linked with the Kasaragod solar project, envisioned as a harbinger of progress, starkly contrasted with the adverse experiences of the local populace, notably the Adivasis. Devoid of formal land entitlements, these indigenous communities bore the brunt of dispossession and livelihood disruption for the sake of development. This scenario accentuates the imperative for an inclusive approach to India's energy transition that marries renewable energy ambitions with the principles of justice and equitable growth. The downscaling of the Kasaragod project, propelled by local resistance and political activism, underscores the potent role of community mobilization in influencing, and occasionally redirecting, the course of national energy strategies.

9.2.4 Resolving Energy Inequity

To address energy injustice, a multifaceted approach that balances technical aspects of energy systems with social equity considerations is required. While investments in renewable energy are critical for reducing emissions and addressing energy poverty, they must be implemented alongside policies that ensure equitable access and address the specific needs of disadvantaged communities. Targeted subsidies and innovative financing models, for example, can help overcome economic barriers to the adoption of clean energy technology. The research highlights community-based energy projects as a promising avenue for empowering local populations and fostering energy autonomy. These projects not only provide access to renewable energy sources but also promote local governance and community participation in energy decisions.

9.3 Energy Technology Innovation's Potential

Energy technology innovations are critical in addressing energy injustice by providing sustainable, equitable, and efficient energy solutions. This section delves into the advancements and how they affect energy equity.

9.3.1 An Overview of Energy Technology Innovation

Renewable energy generation, energy storage, energy efficiency improvements, and the development of smart grid technologies are all examples of energy technology innovation. These innovations are critical to achieving a more sustainable and equitable energy system. According to Ollikka et al. (2017), the integration of ICT with renewable energy solutions addresses the challenges posed by intermittent electricity production from renewable sources. This integration is critical for building a resilient and efficient energy infrastructure capable of adapting to modern societies' dynamic demands and mitigating the effects of energy injustice. Photovoltaic technology, wind turbines, and bioenergy are reshaping the renewable energy landscape, making it more accessible and efficient.

9.3.2 Using Innovation to Address Energy Inequity

Rural electrification is being transformed by renewable energy innovations, particularly in developing countries. Rao and Shrivastava (2016) emphasize the transformative impact of renewable technologies such as solar and wind in India, highlighting how they contribute significantly to socio-economic development and national growth. The proposed nanogrid model in Pakistan uses hybrid renewable energy sources to electrify rural communities, demonstrating the feasibility of such systems in providing affordable and reliable power. This model emphasizes the importance of localized energy solutions in addressing the unique challenges of rural electrification, ensuring equitable and sustainable energy access. The use of these technologies in remote areas not only provides electricity but also opens up new opportunities for economic development and social upliftment.

9.3.3 Energy Technology Innovation Case Studies

Case studies from different regions provide insights into the diverse applications and impacts of energy technology innovations, revealing their potential to address contemporary energy challenges. The work of Nazar (2022) exemplifies this by exploring the role of nanotechnology in enhancing Li-ion batteries, a critical component for advancing energy storage solutions. This research is important for the transition to a low-carbon society, given the increasing reliance on renewable energy sources that necessitate stable and efficient storage systems. Nanotechnology's contribution to improving the electrochemical performance of Li-ion batteries addresses the urgent need for batteries with higher energy densities, longevity, and safety. By focusing on nanomaterials that can endure significant volume changes during redox reactions, nanocoating to stabilize electrode–electrolyte interfaces, and solid-state battery interfaces, this research underscores the industrial and commercial significance of nanotechnology in realizing a sustainable future. The advancements in nanotechnology not only promise to enhance everyday devices but also pave the way for larger-scale applications crucial for a sustainable society, including electric vehicles, drones, and grid-scale energy storage.

Similarly, the study by Kirincic et al. (2018) highlights the unique opportunities and challenges faced by island communities in adopting sustainable energy practices, such as electromobility. Islands, often characterized by their relative isolation and limited resources, present a distinct context for the implementation of sustainable development projects. The research illustrates how modern technologies, particularly electromobility initiatives, can significantly reduce greenhouse gas emissions and foster low-carbon societies within these unique settings. By focusing on integrated actions and community engagement, the case study demonstrates the potential of islands as ideal candidates for sustainable development programmes. These programmes not only aim to achieve long-term environmental goals but also to spur local entrepreneurship and cooperative ownership of infrastructure, thereby creating value for both residents and visitors. Through the lens of island communities, this case study emphasizes the importance of tailored sustainable development strategies that leverage modern technologies to overcome geographical and economic challenges.

9.3.4 Integrating Policy and Innovation for Equitable and Resilient Energy Solutions

Policies play an important role in guiding technological innovations to address energy injustices, ensuring that the advancements in energy technologies translate into equitable benefits, particularly for marginalized and vulnerable groups. Effective policy frameworks are essential in harmonizing the development of domestic renewable technologies and the dynamics of international trade, thereby enhancing domestic innovation and global competitiveness in sectors like solar PV and wind power. It is imperative for policymakers to create an ecosystem that not only nurtures innovation but also guarantees that these technologies are accessible and advantageous across all societal strata.

Looking ahead, the fusion of artificial intelligence and machine learning (ML) with energy technologies is poised to redefine the landscape of energy innovation. These advancements promise to heighten the efficiency of renewable energy systems, refine predictive maintenance, and optimize energy distribution networks. However, the repercussions of climate change on renewable energy sources, such as the potential decline in solar power generation efficiency due to rising temperatures, underscore the urgency for developing climate-resilient energy technologies. The future of energy innovation will be driven by ongoing research into more efficient solar cell materials, improved wind turbine designs, and novel biofuel production techniques, ensuring that technological progress aligns with the imperatives of sustainability and equity.

9.4 ML's Role in Energy Technology Innovation

ML is becoming an increasingly important component of energy technology innovation, particularly in the renewable energy sector. Its applications range from optimizing energy production and distribution to improving smart grid efficiency. This section investigates the transformative impact of ML in renewable energy, such as solar, wind, hydropower, and biomass, as well as its role in decentralizing electricity systems, optimizing smart grids, managing big data in energy, and future challenges and directions.

9.4.1 ML as a Renewable Energy Catalyst

ML is significantly advancing the renewable energy sector, enhancing both the efficiency of energy production and the effectiveness of grid integration. Its applications span various renewable sources, including solar and wind, where it facilitates precise forecasting and optimized energy management. Particularly in wind energy, the application of neural networks for predicting wind patterns has been instrumental. This capability allows for the real-time adjustment of turbine operations, optimizing energy output and reducing maintenance costs. By addressing the inherent variability and intermittency of renewable sources, ML is important in ensuring a more reliable and efficient renewable energy infrastructure.

In wind energy, neural networks have proven effective in predicting wind patterns, enabling the dynamic optimization of turbine operations. This optimization not only maximizes energy output but also significantly reduces maintenance costs, leading to more sustainable and cost-effective wind energy solutions (Abualigah et al. 2022). Similarly, in the solar energy domain, ML algorithms play a crucial role in forecasting solar irradiance. By accurately predicting sunlight availability, these algorithms facilitate the optimal positioning of photovoltaic panels, thus maximizing the energy capture and overall efficiency of solar farms.

Moreover, ML extends its benefits beyond energy production to the intricate domain of energy distribution. Through sophisticated analysis of consumption patterns, ML algorithms can intelligently manage the distribution of energy across the grid. This ensures a harmonious balance between supply and demand, enhancing the reliability and efficiency of the power supply system. In a compelling application within Malaysia, a study utilized genetic algorithms to analyse the impact of local spectral irradiances on the performance of commercial photovoltaic modules (Seera et al. 2021). This research highlights the nuanced relationship between geographic location and solar energy efficiency, demonstrating that even under uniform solar irradiance conditions, the power conversion efficiencies of PV modules can vary significantly based on their location. This insight is critical, as it emphasizes the potential of ML to tailor solar energy systems to specific environmental conditions, thereby optimizing their performance and contributing to the overall efficiency of renewable energy technologies.

9.4.2 ML in Energy Decentralization and Grid Optimization

Advancements in ML is revolutionizing the management of renewable energy sources and smart grid operations by fostering the decentralization of electricity systems and enhancing grid efficiency. In decentralized systems, ML's predictive analytics play a crucial role in forecasting energy generation from local renewable sources, such as solar and wind, which is important for maintaining grid stability in regions with significant renewable integration. These forecasts are facilitated by sophisticated ML models, including neural networks and support vector machines, ensuring accurate predictions of energy outputs.

Simultaneously, ML contributes to smart grid optimization by improving energy efficiency, facilitating effective demand response, and enabling the seamless integration of renewable energies. By analysing consumption patterns, ML allows for dynamic adjustments in energy supply, reducing peak load pressures and optimizing distribution in response to real-time demand. Moreover, ML's capability to anticipate the variable output from renewable sources aids in balancing the grid, thereby ensuring a reliable and sustainable energy supply system. This holistic approach, combining decentralization with smart grid optimization, embodies a more resilient, efficient, and environmentally friendly energy infrastructure, underscored by the insights from Xu et al. (2022) on the importance of ML in enhancing smart grid operations.

9.4.3 Big Data and ML in Energy Management

The integration of big data analytics with ML is revolutionizing energy management systems, especially within renewable energy facilities and smart grid infrastructures. This synergy allows for the extraction of profound insights into various aspects of energy utilization, grid functionality, and opportunities for enhancing efficiency. By harnessing the power of big data analytics, ML algorithms can sift through vast quantities of information collected by smart metres, unveiling patterns in energy consumption. These patterns are important for refining energy distribution strategies and projecting future energy demands, as documented by Tang et al. (2022). The capability to conduct such predictive analyses is indispensable for enlightened decision-making in energy production and allocation, paving the way for more efficient and dependable energy networks.

Moreover, ML extends its utility to the predictive maintenance of energy infrastructure, further bolstering system reliability. Through the continuous monitoring of data from sensors and other diagnostic tools, ML algorithms are adept at foreseeing potential system malfunctions or maintenance requirements. This predictive approach not only minimizes operational disruptions but also enhances the durability and performance of the energy system. By preemptively identifying issues before they escalate, ML contributes significantly to reducing downtime and maintaining a steady, reliable flow of energy, thereby supporting the sustainability and resilience of modern energy landscapes.

9.4.4 Future Prospects and Challenges

The future of energy technology is poised for transformation with the anticipated integration of advanced ML techniques, such as unsupervised and reinforcement learning, as highlighted by Pourpanah et al. (2019). These sophisticated ML methodologies are expected to enhance the precision of energy prediction and management systems, further optimizing the efficiency of renewable energy infrastructures. This evolution promises to leverage the full potential of renewable resources, pushing the boundaries of current energy optimization practices. Yet, the path to harnessing these advancements is not devoid of challenges. The complexities involved in managing vast datasets, safeguarding cybersecurity, and the necessity for specialized expertise are significant hurdles that need to be navigated. Emerging algorithms and cutting-edge technologies, including 5G, are anticipated to play a pivotal role in overcoming these obstacles, facilitating the sustainable integration of ML into energy management systems.

Moreover, the dynamic nature of energy regulations and policies will invariably influence the trajectory of ML applications within the energy sector. The successful adoption and impact of ML technologies hinge on their development and deployment being in harmony with overarching energy and environmental objectives. As the regulatory landscape evolves, it will be imperative to ensure that ML innovations in energy management not only adhere to existing frameworks but also contribute proactively to the global agenda of sustainable and environmentally friendly energy solutions. This alignment is crucial for the long-term viability and acceptance of ML-driven advancements in the energy domain, underscoring the need for a balanced approach that marries technological innovation with regulatory and environmental considerations.

9.5 Crossroads of Energy Technology Innovation and Energy Justice

The relationship between energy technology innovation and energy justice is an important area of investigation in the pursuit of sustainable development. This intersection is about more than just technological advancement; it is also about ensuring that these advancements result in equitable outcomes. As the world grapples with climate change, energy scarcity, and socio-economic disparities, the role of innovative energy technologies grows in importance. However, these technological solutions must be developed and implemented in a way that addresses existing inequities in energy access and distribution. This section investigates how energy technology innovation can serve as a path to justice by bridging the energy divide and ensuring that the benefits of such innovations are distributed equitably across various communities and regions.

9.5.1 Innovation in Energy Technology as a Pathway to Justice

Energy technology innovation is important not only for technological advancement but also for addressing energy injustice. Delina (2021) delves into how renewable energy can revolutionize daily activities such as cooking in rural Thailand, turning them into catalysts for sustainable development. This transcends mere energy access, intertwining social innovation with technological advancements to create environmentally sustainable and socially equitable solutions. The study underscores the need for integrating justice into energy transitions, highlighting that equitable access to energy is fundamental for sustainable development. This aligns with the energy justice discourse advocating for a fair distribution of energy's benefits and burdens.

The investigation reveals the profound impact of renewable energy on local sustainable development through innovations in everyday practices like cooking. The research, using a Grounded Theory approach, emphasizes the importance of equity, accessibility, and affordability in energy justice. It showcases how transitioning to cleaner fuels in rural households contributes to achieving multiple SDGs while addressing energy justice issues. The case study from a rural Thai community illustrates how such transitions can act as sociotechnical innovations, underscoring the role of inclusive social learning and experimentation in fostering just and sustainable development.

9.5.2 ML: Bridging the Energy Gap

ML algorithms are becoming more well-known for their ability to bridge the energy divide. The research by Esquivias et al. (2022) underscores the significant role of ML in optimizing renewable energy systems, which contributes to reducing CO_2 emissions and fostering energy equity, crucial for regions with surging energy demands. The study reveals that economic growth, population density, and technological advancements in renewable energy, alongside the exploitation of natural resources, have historically escalated CO_2 emissions in these economies. However, investments in human capital, adoption of green technologies, and efficient energy utilization can mitigate these emissions. It highlights the opportunity for ML to improve energy demand forecasting, resource allocation, and investment targeting, particularly in underserved areas, thereby promoting a more balanced and sustainable energy landscape. This approach aligns with the broader objectives of environmental sustainability by leveraging technology and human capital development to achieve a reduction in carbon emissions and enhance energy efficiency.

9.5.3 Navigating the Complexities of Technological Innovation and Energy Justice

The intersection of energy technology innovation and energy justice presents a nuanced landscape of challenges and opportunities, as evidenced by recent studies. The exploration of RETI and its impact on industrial green transformation and upgrading (IGTU) in China serves as a compelling case study. Liu et al. (2022) delve into the spatial spillover effects of RETI, revealing its potential to both foster local industrial green advancements and inadvertently hinder progress in adjacent regions. This dichotomy underscores the intricate balance required in pursuing technological advancements while ensuring equitable energy distribution. The study employs an improved entropy weight method to assess the IGTU levels across 30 provinces in China from 2006 to 2020, incorporating kernel density estimation and Moran's I index to elucidate temporal and spatial patterns. The spatial Durbin model elucidates the mechanisms through which RETI influences IGTU, highlighting a "beggar-thy-neighbor" scenario and the significance of regional boundaries in the spatial spillover effects. These findings point to the critical need for tailored policy interventions that consider regional disparities and the multifaceted impacts of technological innovations on energy justice.

Concurrently, the work of Godil et al. (2021) investigates the nexus between economic growth, technological innovation, and renewable energy consumption in mitigating CO_2 emissions within China's transport sector. Utilizing the QARDL approach, the study confirms the dual role of renewable energy consumption and innovation in reducing transport-related CO_2 emissions, juxtaposed with the emission-increasing effect of GDP growth. This paradox highlights the complex interplay between economic development, technological progress, and environmental sustainability. The research emphasizes the pressing need for innovative policies in the transport sector to curtail emissions and foster a sustainable balance between economic growth and environmental health. The backdrop of escalating global environmental concerns and the imperative for sustainable development frames this study, advocating for a transition towards renewable energy as a cornerstone of sustainable economic progress and environmental care.

9.5.4 Ensure Inclusionary Energy Futures

To achieve inclusive energy futures in sub-Saharan Africa, a comprehensive strategy that addresses socio-economic challenges while avoiding exacerbating existing vulnerabilities is required. Akinola et al. (2021) investigate the complexities of transitioning to low-carbon electricity systems in Nigeria, a case study for the region. They emphasize the importance of balancing the deployment of intermittent renewable energy sources with the region's socio-economic realities. The study emphasizes that while Nigeria can meet its NDCs without incurring significant additional costs, reliance on fossil fuel generation for rapid energy scale-up is critical. This approach is critical to ensuring a just and inclusive energy transition in Sub-Saharan Africa, where the socio-economic context differs significantly from that of developed regions. According to the study, while the distributed nature of IRES can democratize energy investment and job creation, it is critical to design clean energy projects that support sustainable development without burdening vulnerable populations.

9.6 Conclusion

The investigation of energy technology innovation as a means of addressing energy injustice, as detailed in this chapter, reveals a complex yet hopeful landscape. Renewable energy technology advancements, ML, and smart grid systems are critical in addressing the multifaceted issue of energy injustice. These innovations are critical in ensuring equitable energy access, increasing efficiency, and promoting sustainable practices. However, the path to energy justice is complicated, necessitating a nuanced approach that balances technological advancement with socio-economic and environmental concerns.

Future efforts must prioritize the integration of social innovation and technological advancements. This integration is critical to ensuring that energy solutions are both environmentally and socially sustainable. Innovative solutions should be tailored to the specific needs of various communities, particularly those in vulnerable and underserved areas. This approach necessitates a thorough understanding of local contexts as well as active community participation in the development and implementation of energy solutions.

The role of policy and regulatory frameworks in directing technological innovation towards addressing energy injustice is critical. Effective policies are required to foster innovation while ensuring equitable access to these technologies.

This includes policies that encourage renewable energy research and development, incentivize the adoption of sustainable practices, and ensure that the benefits of innovation are distributed equitably.

New challenges will emerge as the field advances, particularly in managing large amounts of data, ensuring cybersecurity in smart grids, and the need for specialized expertise in fields such as ML and AI. New algorithms, technologies, and training programmes will be critical in addressing these challenges. Furthermore, the changing landscape of energy regulations and policies will have a significant impact on the future of these applications.

To ensure inclusive energy futures, a comprehensive strategy that addresses socio-economic challenges while avoiding exacerbating existing vulnerabilities is required. Incremental transition strategies focusing on low-carbon development and equitable energy access are critical, particularly in regions such as Africa, where the socio-economic context differs significantly from that of developed regions.

Ongoing research in areas such as advanced solar cell materials, wind turbine blade design, and biofuel production methods will drive innovation in this field. Collaboration between academia, industry, and communities is critical for fostering practical and impactful innovation. This collaboration should also include international collaboration, particularly in the sharing of knowledge and technology to support global efforts to achieve energy justice.

To summarize, energy technology innovation plays a significant and multifaceted role in addressing energy injustice. While technological advancements provide promising solutions, achieving energy justice requires a concerted effort that includes technological, social, and policy dimensions. The way forward entails continuous learning, adaptation, and collaboration to ensure that the benefits of energy innovations are available to all, contributing to a more sustainable and equitable world.

References

Abualigah L, Zitar RA, Almotairi KH, Hussein AM, Elaziz MA, Nikoo MR, Gandomi AH (2022) Wind, solar, and photovoltaic renewable energy systems with and without energy storage optimisation: a survey of advanced machine learning and deep learning techniques. Energies 15(2):578

Akinola O, Chaudry M, Barrett M (2021) Delivering low-carbon electricity systems in sub-Saharan Africa: insights from Nigeria. Energy Environ Sci 14 (8):4406–4422

Barragan-Contreras SJ (2022) Procedural injustices in large-scale solar energy: a case study in the Mayan region of Yucatan, Mexico. J Environ Planning Policy Manage 24(4):375–390

Bedi HP (2019) Lead the district into the light: Solar energy infrastructure injustices in Kerala, India. Glob Trans 1:181–189

Delina LL (2021) Co-producing just energy transition in everyday practices: sociotechnical innovation and sustainable development in the Thailand-Myanmar border. Local Environ 26(11):1325–1340

Esquivias MA, Sugiharti L, Rohmawati H, Rojas O, Sethi N (2022) Nexus between technological innovation, renewable energy, and human capital on the environmental sustainability in emerging Asian economies: a panel quantile regression approach. Energies 15(7):2451

Godil D, Yu Z, Sharif A, Usman R, Khan S (2021) Investigate the role of technology innovation and renewable energy in reducing transport sector CO_2 emission in China: a path toward sustainable development. Sustain Dev 29(4):694–707

Kirincic V, Franković D, Radulovic D, Jakoplić A (2018) Towards a low-carbon island society by strong implementation of electromobility. Presented at the international conference on innovative technologies, IN-TECH 2017, Zagreb

Lamura M (2022) Energy poverty across Europe: problematisation and policy measures through the lens of energy justice. Environ Sci Pollut Res 29:20012–20028

Liu C, Xin L, Li J, Sun H (2022) The impact of renewable energy technology innovation on industrial green transformation and upgrading: beggar thy neighbor or benefiting thy neighbor. Sustainability 14 (18):11198

Nazar L (2022) Nanotechnology for industrial electrochemical energy storage. Electrochem Soc Meeting Abstr 241(7):623

Ollikka K, Kangas H-L, Yukyeong K (2017) Knowledge spillovers from renewable, clean, and smart energy technologies. In: Heading towards sustainable energy systems: evolution or revolution? 15th IAEE

European conference, 3–6 Sept 2017. International Association for Energy Economics

Pourpanah F, Wang R, Lim CP, Wang X, Seera M, Tan CJ (2019) An improved fuzzy ARTMAP and Q-learning agent model for pattern classification. Neurocomputing 359:139–152

Rao GJ, Shrivastava S (2016) Implementation and diffusion modeling of selected renewable energy technologies in India. Int Adv Res J Sci Eng Technol 3(8)

Seera M, Tan CJ, Chong K-K, Lim CP (2021) Performance analyses of various commercial photovoltaic modules based on local spectral irradiances in Malaysia using genetic algorithm. Energy 223:120009

Stojilovska A (2023) Energy poverty and the role of institutions: exploring procedural energy justice–Ombudsman in focus. J Environ Planning Policy Manage 25(2):169–181

Su Y, Fan Q (2022) Renewable energy technology innovation, industrial structure upgrading and green development from the perspective of China's provinces. Technol Forecast Soc Chang 183:121727

Tang W, Wang H, Lee XL, Yang HT (2022) Machine learning approach to uncovering residential energy consumption patterns based on socio-economic and smart meter data. Energy 240:122500

Wang C (2022) Green technology innovation, energy consumption structure and sustainable improvement of enterprise performance. Sustainability 14(16):10168

Wang W, Xiao W, Bai C (2022) Can renewable energy technology innovation alleviate energy poverty? Perspective from the marketization level. Technol Soc 68:101933

Xin L, Sun H, Xia X, Wang H, Xiao H, Yan X (2022) How does renewable energy technology innovation affect manufacturing carbon intensity in China? Environ Sci Pollut Res 29:20012–20028

Xu C, Liao Z, Li C, Zhou X, Xie R (2022) Review on interpretable machine learning in smart grid. Energies 15(12):4427

Manjeevan Seera is an accomplished academic and industry professional, currently serving as an Associate Professor in Business Analytics at the School of Business in Monash University Malaysia. He earned his Ph.D. in Computational Intelligence from Universiti Sains Malaysia and has published over 60 peer-reviewed journal and conference papers in his field. With more than 19 years of experience in academia and industry, Manjeevan has established himself as an expert in his field. Prior to his current role, he served as a Data Analytics Manager at Bank Negara Malaysia and gained extensive experience in the electronic manufacturing industries working with prominent companies like Intel and Flextronics. Manjeevan's research focuses on machine learning principles and applications in finance and engineering, with a particular interest in developing advanced machine learning models for Fintech applications. His current research work centres around the detection and prediction of fraudulent financial transactions.

How to Tax Our Way to Energy Justice

10

Francesco Cannas

Abstract

The purpose of the chapter is to discuss some of the legal issues that currently make many tax systems not entirely suitable to support what is commonly known as the eco- or green transition. The legal analysis focuses on certain concepts, particularly those of horizontal equity and neutrality, which were originally conceived from a predominantly economic perspective and should now be reviewed from a partially different point of view in order to provide a more holistic meaning. The theory underlying the present contribution is that the idea that two persons may be treated differently for tax purposes on the basis of how much they pollute, even if their economic situation is similar, should be elevated to the rank of constitutional value. Its objective is to elaborate conceptual solutions that can subsequently be used as a general model and commencement for reforms either in or outside the EU itself.

Keywords

Environmental taxation · Green taxation · Horizontal equity · EU Green Deal

10.1 Introduction

The word *energy* can have multiple meanings in everyday use and embraces a wide variety of phenomena in the world. Before analysing some of its tax aspects and outlining the scope of research of this chapter, it is appropriate to briefly summarise the meaning of this word.

In very general terms, energy is the ability to perform work (Sellerio 1950) and do something from lifting a weight to heating (in the language of the natural sciences, for example, experts speak of kinetic energy, magnetic energy, etc.). For legal purposes, conversely, it becomes relevant only to the extent that it can be the object of rights, legal relationships, and transactions. Legal scholars have therefore created two categories, specifically that of *natural energy* that can be relevant in both public and private law (Carnelutti 1913) and that of *human energy* that is conceptually difficult—if not impossible—to

The present chapter is the outcome of a period of research at the EW Barker Centre for Law and Business (National University of Singapore—NUS) to which the author offers the deepest gratitude. In particular, thanks are due to Professor Stephen Phua for the guidance and the opportunity to discuss the content.

F. Cannas (✉)
Assistant Professor, Law and Taxation, EDHEC Business School, Promenade des Anglais, France
e-mail: f.cannas@outlook.it

Assistant Professor, Tax Law, University of Turin, Turin, Italy

A. Trakic et al. (eds.), *Energy Justice*, Sustainable Development Goals Series,
https://doi.org/10.1007/978-981-97-6059-6_10

distinguish from human behaviour and is, by itself, irrelevant to law (Messinetti 1965).

Regarding natural energies, which are the subject of this contribution, and particularly electricity and gas, it is often difficult to keep them conceptually separate from their source or means of transport. This is because they are inherently often inseparable for the purpose of human use; thus, the law must ascertain suitable solutions for each case (Biondi 1956). In the field of taxation, for example, EU VAT provides specific rules for identifying the place of taxation of the supply of gas and electricity. This is due to their features that consist of many operators feeding the product into a single network from many places, and it is therefore impossible to identify exactly where the electricity comes from that is used by a specific consumer (De la Feria 2009).

For tax purposes, energy can be relevant in multiple ways, with the primary ones being (i) for energy production and distribution, such as for power generation companies in respect to which tax incentives for research and development of new technologies are particularly important (although it may not always be the case that energy producers and distributors do the research themselves); (ii) for energy use that involves: (ii.a) businesses for which it is a cost that can be deducted in full or in part in some cases; (ii.b) private consumers for whom energy is a basic need and who bear the full cost and taxes levied on it unless exemptions apply; and (iii) for the general public. The latter means those who do not use a certain energy source and for whom the use by others can be a source of pollution of both the air (e.g. when burning fossil fuels to produce electricity) and the landscape (e.g. when installing large wind farms or photovoltaic plants that have a great visual impact). This pollution may even be caused by sectors of the economy that receive subsidies on the basis of political dynamics that have nothing to do with environmental protection (OECD 2018). The above articulation is also reflected in this chapter, focusing mainly on points (i) and (ii).

The taxation of the energy sector is very complex and multifaceted, and the achievement of any politically set goal requires a balance of fiscal policy that considers all the mentioned aspects into account. To remain in the European context, at the end of the last century, the institutional, organisational, and managerial set-up of the electricity (Directive 96/92/EC) and gas (Directive 98/30/EC) industry was profoundly innovated following the implementation of reforms aimed, at least in the declarations of intent, at improving the delivery of services essential for economic development and social welfare. This resulted in a massive liberalisation of the market, privatisation of activities along the entire value chain (production, export, import, purchase, and sale of energy), and an attempt to create a market that is truly European (Travi 1998).

Regardless of the actual achievement of the stated goals and any possible political views on them, whenever an energy policy is established, the tax system must be adjusted accordingly (European Community 1972; Sinisi 2005; Falsitta 2010; Corso 2021).

Currently, the energy sector is experiencing a period of profound change, and there are two major challenges on the horizon. On the one hand, is the United Nations Sustainable Development Goal n. 7 (SDG7) that requires clean and affordable energy to be indiscriminately ensured (Carlsen and Bruggemann 2022). On the other hand, the Ukrainian crisis necessitates a comprehensive review of current energy supplies (European Commission 2022).

The SDG7 is a complex statement with multiple subgoals which concern humanity in general and peremptorily request a generalised effort, especially clean energy production, and others in more limited areas of the world, especially universal access to energy. It should be borne in mind that, among others, as the UN itself states, half of the 10% of the world's population without access to energy is concentrated in sub-Saharan Africa.

The consequences of the war in Ukraine conversely directly affect a smaller number of countries, such as Italy and Germany, which used to purchase 46% and 49%, respectively, of their total gas requirements from Russia. However, the radical change in the supply composition of these large, industrialised countries and OECD members may also have an indirect effect on world prices and supplies, so even more countries could be indirectly affected by these events (IlSole24Ore 2022).

The large production of policy documents on these matters, however, has the consistent feature that fiscal and tax aspects are largely omitted. Thus, it is the scholarly community that must stay abreast of the latest developments and conduct a debate on possible models for implementing the identified collective interests.

The purpose of this chapter is to discuss some of the legal issues that currently make many tax systems not entirely suitable to support what is commonly known as the *eco-* or *green transition*. The legal analysis focuses on certain concepts, particularly those of horizontal equity and neutrality, which were originally conceived from a predominantly economic perspective and should now be reviewed from a partially different point of view in order to provide a more holistic meaning.

The theory underlying the present contribution is that the idea that two persons may be treated differently for tax purposes on the basis of how much they pollute, even if their economic situation is similar, should be elevated to the rank of constitutional value. In this context, the concepts of *environmental equity* and *environmental neutrality* should be introduced in the tax systems.

What is dealt with here are issues that lie at the intersection of politics, major historical changes, and economic interests, with the consequence that any legal analysis, once implemented in the real world, shall take these tangled dynamics into account (Endo 1999).

In addition to this, it must also be considered that the tax system can often play a role that is little more than marginal and certainly cannot be the primary driver of developing a sustainable energy sector because that would be asking too much of taxes. This is why, for example, the revision of certain legal principles may help shape tax incentives for research and development or encourage a more environmentally friendly consumption taxation. However, it is unlikely to foster technology transfer to developing countries lacking infrastructure or ground the removal of subsidies to the most polluting sectors of the economy. This is the ultimate reason why this chapter does not comprehensively analyse, among others, the removal of subsidies to polluting sectors. This is not because it is an irrelevant issue from a tax standpoint but because what is required in that case is the policymaker's decision to remove these subsidies rather than a change in the legal basis on which they are rooted.

From a methodological point of view, this chapter is written from a European and comparative perspective in the sense that it analyses some of the legal issues related to the incentivisation of clean and affordable energy, beginning with those that have emerged from the EU experience. Its objective is to elaborate conceptual solutions that can subsequently be used as a general model and commencement for reforms either in or outside the EU itself.

10.2 The Positive Intervention Through Tax Incentives for the R&D of Newer and Cleaner Technologies: A New Concept of (Horizontal) Equity

To protect the environment, it is necessary to develop new technologies that generate less pollution and are convenient and worthwhile. There has always been a widespread consensus that public intervention in research and development—not only in relation to clean energy but in general—is functional in overcoming certain market failures that may lead to an amount of investment that is less than optimal for the community (OECD 2003). This sector is, in fact, characterised by a high degree of uncertainty and information asymmetry. In general, businesses may have difficulties exploiting the returns of their investments as the results of research often have public good characteristics. It is, therefore, usable also by competitors who did not invest in it. In addition to that, smaller and newly established businesses may also find it more difficult to find sources of funding (Guenther 2017).

With a more specific reference to energy, new and cleaner technologies can be used both directly by consumers and companies in their production processes. Support for research and development falls within what is commonly known as *positive* state intervention, aimed at producing structures that improve environmental quality as opposed to what is referred to as *negative* intervention, which consists of correcting wrong behaviours instead (García and Roch 2016). Stated differently, instead of increasing the price for undesirable behaviour, beneficial tax incentives are tax breaks that reduce the price for conduct that is more desirable from an environmental perspective.

A positive intervention may consist of funding that may or may not be repayable to companies or public institutions such as universities and research institutes to develop new technologies and tax incentives in the strict sense of the word. Although the first of the two categories can also create equality issues and is burdensome for a state budget, the focus here is on the second category as it is a tax instrument in the precise context of the expression.

Tax incentives for research can be of two types: (i) *expenditure incentives*, e.g. through exemptions such as those on excise duties on products for certain development activities or tax credits (discounts) commensurate with the research and development (R&D) expenditures made by businesses; or (ii) *profitability incentives*, e.g. through deductions from the taxable base of a portion of income from the use of certain types of intangible assets and capital gains from their disposal (this is the idea behind certain tax schemes known as *patent boxes* adopted in recent years by many OECD countries).

Compared to other types of subsidies that widely depend on discretionary choices by the public authority, tax incentives are often viewed favourably by some stakeholders because they are considered free-market instruments in the context that the policymaker only sets certain conditions of entitlement and leaves the choice of projects to be invested into private businesses. Their main purpose is, therefore, to increase the amount of overall R&D expenditures and thus the positive externalities that this determines on the production system in terms of innovation and growth (Cabral et al. 2021).

Here, the criticism that has been levelled at some of these instruments is disregarded, such as within the BEPS Project, where some of them were identified as a way to distort competition between states by de facto de-taxing certain intangible assets (OECD 2015) or the fact that, in a global perspective, the issue of justice between states should also be addressed. In fact, these incentives would only burden the budgets of states whose companies are capable of *green* innovation, although it may not always be the case that these are also those that pollute the most. From a strictly legal point of view, the issue arises of effectively restricting the scope of application of these incentives so that they are exploited exclusively for what can be considered *green* research.

The fact is that these incentives have often been applied broadly to all research, but when the policymaker decides to restrict them to technologies that enable the production of *clean and affordable energy*, it becomes necessary to clarify for tax policy purposes what is to be understood by nonlegal expressions such as *green*, *environmentally friendly*, *eco-friendly*, *clean*, etc. (Pitrone 2015). To explain the matter in simple terms, it may be relatively easy to grant a tax incentive for expenditures related to a research laboratory. However, it may be very difficult—first for the tax authorities—to verify that the research taking place there could have a positive impact on the environment in the future (or help to ensure access to energy in those areas of the world where this is not currently the situation).

A technically accurate definition of these terms would also be beyond the research scope of this chapter because it should not be done exclusively by members of the tax community. It would require interdisciplinary work with experts from other branches of knowledge just as a highly technical definition would probably require cooperation between tax authorities and other government agencies for tax audits and litigation before the tax courts after its enactment.

Conversely, adopting a more general definition would again create the problem that is recurrent in most environmental tax incentives, in addition to exposing the tax system to abuses and the risk of tax avoidance. This is precisely because most of them are granted without it being possible to evaluate their true effectiveness either before or after the enactment (European Commission 2020; Easson 2001).

Secondly, limiting a tax incentive to certain subjects, in this case to those favouring the *green energy transition*, makes it potentially selective and, therefore, in some cases, exposes it to the risk of creating market distortions (Miladinovic 2022). This may simply be a political issue in those jurisdictions where the policymaker is entitled to freely decide to favour certain taxpayers, being politically accountable for such a decision in the following elections. In others, such as the EU Member States, this may also create legal issues, particularly with regard to the compatibility of incentives with state aid legislation.

Under the EU legal system, Article 107 of the Treaty on the Functioning of the European Union (TFEU) contains a general prohibition of state aid in the perspective of ensuring that state interventions do not distort competition and trade within the single market (Pérez Bernabeu 2014). It attempts to balance the noninterference of the Member States in free trade with their power to intervene when this is convenient for the community. This is clearly reflected in the structure of the article.

Article 107(1) establishes a general prohibition ('any aid granted by a Member State or through State resources in any form whatsoever which distorts or threatens to distort competition by favouring certain undertakings or the production of certain goods [...]') while the following paragraph (2) makes any state aid compatible with the single market where, among other things, it has a *social character*, helps to recover from the damages caused by a *natural or exceptional event*, or helps the economy of certain specifically identified areas of the Union. In addition to that, Article 107(3) makes it possible to derogate from the prohibition where, upon an evaluation of the measure to be enacted, it emerges that it would be beneficial in promoting the economic development of areas where the standard of living is abnormally low; it would promote the execution of an important project of common European interest or to remedy a serious disturbance in the economy of a Member State; facilitate the development of certain economic activities or areas; or promote culture and heritage conservation to the extent that it is not contrary to the common interest; as well as *any other categories of aid as may be specified by decision of the Council on a proposal from the Commission*.

A very important role in supervising the application of state aid legislation and in balancing the various interests involved belongs to the European Commission, which is a body that acts in the interest of the Union. Additionally, under Article 108(3) TFEU, Member States shall notify any legal frameworks involving state aid prior to entry into force in order for it to decide about its compatibility with the single market. Currently, the categories of aid to be considered as compatible with the common market are listed under the 'General Block Exemption Regulation' (Commission Regulation 2014).

Already in 2010, in planning the exit from the economic crisis effectuated by *subprime mortgages*, the Commission had identified clean energy as one of the pillars of growth. It stated in its Europe 2020 strategy plan that it was necessary '[...] to develop a strategic research agenda focused on challenges such as energy security, transport, climate change and resource efficiency, health and ageing, environmentally friendly production methods and land management, and to enhance joint programming with Member States and regions' (European Commission 2010).

In the following years, the European Commission systematically established some guidelines aimed at evaluating the compatibility with the single market of any proposed aids for energy and the environment. Possible derogations for the energy sector were put under Article 107(3)(c) TFEU, specifically the facilitation of *the development of certain economic activities* (European Commission 2014). As a result, the proper balance was found with any state aid for environmental protection and energy objectives being considered compatible with the single market if it leads to an increased contribution to the Union's environmental or energy objectives without adversely affecting trading conditions to an extent contrary to the common interest. This always considers the other principles that form part of the European legal system, such as the prohibition of restrictive or discriminatory treatment and that of proportionality that obligates an intervention that is not disproportionate to the aims pursued (García and Roch 2016).

By analysing the fundamental principles that underlie the legal framework described, it can be ascertained that the cause of the possible constraints is ultimately to be found in the economic conception of equity. Modern tax systems are, in fact, built on the idea that a *good tax* is a *fair tax*, meaning that taxes shall be based on *horizontal equity* (Elkins 2006).

Even without comprehensively delving into what horizontal equity could mean in theory, which implies, among others, deciding whether to consider the taxpayer's situation before or after the tax levy, how to measure wealth or income, etc. (Simons 1938), there seems to be little or no doubt that the underlying steady idea is that persons should be able to compete on equal terms and, therefore, taxes shall not disturb market dynamics.

From a historical perspective, Elkins (2006) explains the difference between the Libertarian model of equal opportunity and the Liberal conception of equality. The former indicates that individuals should be allowed to compete to the best of their abilities, while the latter results from the feeling that arbitrary social circumstances should not be allowed to influence a person's ability to fully exploit his or her natural talents. Both result in similar tax systems with the difference that, while the first requires that taxes are imposed so as to avoid disturbing the market distribution, the latter requires some adjustments aimed at removing socially created obstacles. The constant that unites them is the strictly economic idea of competition that does not consider other factors such as, as far as it is relevant herein, the environment.

Currently, the economic perspective emerges clearly in the EU legislation on state aid. Indeed, in some cases, there may be a clash between the rules under the TFEU that are aimed at guaranteeing horizontal equity between the economic operators of the single market and the environmental policies of the Union.

In this regard, despite there being complete adherence to the legislation wording, it may also be significant to some extent that the Communication by which the European Commission (2014) establishes the guidelines for deciding which environmental interventions may be compatible with the state aid legislation, no emphasis is placed on the fact that environmental protection is also one of the main objectives established under the founding agreements, such as under Articles 11 and 191 TFEU.

Although this has a purely symbolic value and no practical consequences, and there is no reason to believe that the principles currently in force have not all been taken into due consideration, the Commission's legal reasoning relies on Article 107 TFEU as grounds and thus adopts a strictly economic perspective. It does not expressly mention that environmental protection might instead be considered to be of the same rank as those of economic fairness, thus balancing two principles of the same status.

In the specific case of the EU, it must be borne in mind that its origins are those of a common market, and it is, therefore, reasonable to expect that the perspective adopted is strongly characterised by an economic element. As is elaborated more extensively at the end of the following paragraph, the author advocates for a general increase in the importance of environmental protection with specific reference to the tax system. This should initially be enacted by the lawmaker. As a result, in the context of any legal reasonings, when it comes to applying principles to the facts of real cases and finding a satisfactory compromise, this could gradually provide the interpreters with a more solid legal basis, thereby allowing greater importance to environmental arguments in respect to other common interests.

10.3 The Negative Intervention Aimed at Correcting Polluting Behaviours

10.3.1 A Set of New Principles to Tax Sustainable Energy

In its complexity, the structure of energy taxation often sees the overlapping of various taxes and surcharges on the same item. Under the current European context, to recall one example only, this is the case of excise duties and VAT on electricity and fuel.

It just so happens that these various forms of taxation have little or nothing to do with effective environmental protection and sustainability policies. From a practical point of view, they are levied on producers or distributors and then often shifted onto users, whether they are businesses or private consumers.

If there is a willingness to shape this multiplicity of tax levies to direct them towards greater sustainability, it would be an option to disconnect them at least partially from what is generally intended as the main role of taxes, specifically to generate revenue. Instead, they could be connected to the regulatory activity of the state, i.e. to its power to counteract behaviours that are considered inappropriate by the policymaker (i.e., the one referred to above as *negative intervention*). From a practical perspective, this can be implemented in many ways by either partially modifying existing taxes or creating new *green taxes* with the objective of promoting certain targeted policies as priorities.

In both cases, however, it is essential to rely on a legal basis that allows the tax system to function effectively and does not create areas of uncertainty or potential conflicts with other protected interests.

The next two subsections elaborate and translate the idea into legal terms that the *polluter pays, but two persons who pollute differently shall not be taxed in the same way*. This could be based on the evolution of certain fundamental existing legal principles and, in particular, the promotion of a new concept of horizontal equity and neutrality that are both in the realm of direct and indirect taxes. This could derive and correlate with a full constitutional acknowledgement of the environmental regulatory function of the tax system. There would be a consequent non-application of the ability to pay principle to environmental taxation and the irrelevance of the fact that the impact of taxation on pollution caused by energy consumption might not be precisely measurable.

10.3.2 Constitutional Legitimacy of Regulatory Environmental Taxation and a New Concept of Horizontal Equity

The taxation of energy consumption can be implemented in different ways, ranging from changing some of the rules of income taxation (e.g. denying deductions for the purchase of polluting products) to the introduction of ad hoc taxes aimed at protecting the environment that may be levied on various items and be referred to differently.

The rationale behind all of these forms of taxation is to make the consumer of polluting energy internalise the environmental costs generated by its behaviour. De Sadeleer (2002) explains that *negative externalities* arise when 'the production or consumption of goods or services damages environmental goods without that damage being reflected in their price […]'. From a legal standpoint, this type of taxation is based on what is known as the *polluter-pays principle* that originated as an economic principle—or rather, as the author prefers to say, based on an economic mechanism—and subsequently evolved to become a legal principle (Keating 2006). It was first enshrined in some soft law instruments (CJEU 1973, 1976) and then entered into binding instruments in several legal systems (e.g. Directives 75/439/EEC, 94/62/EC and 75/442/EEC).

When applying this principle, the polluter is responsible for the damage done to the natural environment and obligated to pay for it. According to this legal fiction, if someone decides to pollute, they must pay a sum to restore and compensate for the damage created to the environment and the community. From an economic perspective, they will face higher costs upon engaging in pollution-generating behaviour and will be disincentivised to persist and incentivised to change their sources of energy supply.

Its roots can be found in Principle 13 of the 1972 Declaration of the United Nations Conference on the Human Environment, also commonly referred to as the *Stockholm Declaration*, which reads: 'In order to achieve a more rational management of resources and thus to improve the environment, States should adopt an integrated and coordinated approach to their development planning so as to ensure that development is compatible with the need to protect and improve environment for the benefit of their population'. For that time in history, the acknowledgement of a strong relation between environmental protection and economic development is to be considered a significant achievement.

In the very same period, although a European environmental policy was yet to come, the value of a European environmental policy was declared at a European Council Summit: 'Economic expansion, which is not an end in itself, must as a priority help to attenuate the disparities in living conditions' (1972). Conversely, in the beginning, within the EU law context, specifically under the initial 1957 EEC Treaty (European Economic Community Treaty, also known as one of the Treaties of Rome), the word 'environment' did not appear.

The polluter-pays principle was first included in the founding agreements by virtue of the Single European Act (SEA) 1987, and its wording was subsequently amended several times, reflecting, as can be seen below, its increasing importance and theoretical elaboration.

In 1987, the statement of the principle was contained in Article 130r(2) SEA and reads as follows: 'Action by the Community relating to the environment shall be based on the principles that preventive action should be taken, that environmental damage should as a priority be rectified at source, and that the polluter should pay. Environmental protection requirements shall be a component of the Community's other policies'.

In 1992, the provision on the protection of the environment was contained under Article 130r(2) and then amended by Article G(38) TEU (Treaty on European Union, also known as the Maastricht Treaty) that stated as follows: 'Community policy on the environment shall aim at a high level of protection taking into account the diversity of situations in the various regions of the Community. It shall be based on the precautionary principle and on the principles that preventive action should be taken, that environmental damage should, as a priority be rectified at source and that the polluter should pay. Environmental protection requirements must be integrated into the definition and implementation of other Community policies' (Jans 2011).

In 1997, Article 2 of the Treaty amending the TEU, known as the Treaty of Amsterdam, elevated the concepts of sustainable development and environmental protection to the highest constitutional rank by stating that 'The Community shall have as its task, by establishing a common market.and an economic and monetary union and by implementing common policies or activities [...] to promote throughout the Community a harmonious, balanced and sustainable development of economic activities, a high level of employment [...] a high level of protection and improvement of the quality of the environment [...]'.

Currently, environmental protection and the polluter-pays principle are enshrined in Articles 11 and 191 TFEU. The first states that 'Environmental protection requirements must be integrated into the definition and implementation of the Union's policies and activities, in particular with a view to promoting sustainable development'. While the latter, after having created a connection between the environment and human health in par. 1, states in par. 2 that environmental policies 'shall be based on the precautionary principle and on the principles that preventive action should be taken, that environmental damage should as a priority be rectified at source and that the polluter should pay'.

Building on this legal basis, Directive 2004/35/CE establishes a framework of environmental liability based on the polluter-pays principle with the purpose of preventing and remedying environmental damage. Article 8 of this directive places the responsibility for the costs of the preventive and remedial actions to be taken on the polluting operator. Their justification is that not all forms of environmental damage can be remedied with liability mechanisms, especially those of a widespread and diffuse character where it is impossible to link the negative environmental effects with acts or failure to act of certain individual actors.

Such a fiscal policy and its underlying principles express what is commonly known as the *regulatory function* of the tax system, i.e. the use of tax instruments to pursue purposes other than the *mere* funding of the public budget. As mentioned previously, the consequence of the regulatory function is the application of a tax levy on certain items (which may be related to the production, consumption, or ownership of virtually any items) that are considered negative with a view to the pursuit of objectives that are conversely deemed as positive (Avi-Yonah 2011). This application has a twofold effect; however, it affects items that are already taxed in most cases with the result of creating an increased tax burden that must rely on a solid legal basis so as not to violate the constitutional principle of the ability to pay in the first place (Vanistendael 2014).

Indeed, although ability to pay is commonly intended mainly as a principle founding a progressively higher levy on those taxpayers who have a greater capacity to contribute with the consequence that they contribute a greater share than those with less capacity, virtually all of the theories offering justification for it are based on the concept of equity (Gibson 2012).

Summarising simply and referring to the scholarly literature on this subject for an extensive analysis, one might recall that, according to the Utilitarian Theory, those with a greater ability can contribute more because additional wealth is worth less to them. The consequence is that even a different levy can result in *equal sacrifice* and maximise the welfare of the members of society.

In an Egalitarian view that is more concerned with redistribution and equalisation of the members of the society and is about seeking the best system for those in society who are the most disadvantaged in order to bring their level of welfare consistent with that of everyone else, those with a greater ability can contribute more because they will still be left with *enough* (Murphy and Nagel 2002). Under the Utilitarian conception, the ultimate goal is to increase aggregate welfare.

In both cases, ability to pay can also be stated reversely, specifically that those with the same capacity should contribute the same amount with the consequence that it peremptorily requests the distribution of income in accordance with, among others, *horizontal equity* (Musgrave and Musgrave 1984).

One of the main issues arising from the framework reviewed herein and the polluter-pays principle is that it is widely based on legal fiction, as, in reality, environmental protection has little or nothing to do with the amount of revenue raised through the tax system. Actually, *green taxes* are only effective to the extent that they make the polluter change its behaviour, and the levy must be sufficiently high to achieve this. The higher the levy, the greater the difference is with other taxpayers in a similar situation and thus have a comparable ability to pay and, consequently, the greater the risk of moving away from horizontal equity.

In addition to this, it may also be the case that, regardless of the amount of tax paid, once environmental damage is done, it is no longer repairable or is repairable over a very long timeframe, just as no real restoration is possible for the damaged stakeholders. The result is that the tax has actually failed its main purpose of protecting the environment despite the increased revenue. As a consequence, even the polluter-pays principle becomes difficult to apply in these cases since the additional taxes paid are in fact disconnected or, at most, only minimally connected from any specific environmental policies for the damage caused. When this is the case, it also becomes more difficult to derogate from the ability to pay principle to justify this type of taxation since the latter is a constitutional principle that belongs to the tax realm. It might, therefore, prevail over environmental protection by virtue of the application of the *lex specialis* interpretation criterion.

It is advisable to proceed along two lines to construct a legal framework that allows for the proper promotion of clean energy production and use. The first one is that, according to the rules of each different legal system, environmental protection must be elevated to the highest rank, i.e. constitutional, for tax purposes (in some cases, for example, this may imply including it in specific sections of the written constitution).

This would reflect the fact that there are two types of taxes, i.e. *revenue-generating taxes* and *regulatory taxes*. The former often receive full constitutional recognition and are accompanied by principles such as the ability-to-pay, while the latter are often seen as an exception that receives a special derogation from certain fundamental principles of the legal system. In the light of the importance of the purpose pursued, this would give the second type a full and autonomous legitimacy that is no longer subject to any of the economic criteria governing the generation of revenue.

The second is for the concept of horizontal equity to evolve from a strictly economic conception to a more holistic one that encompasses other aspects of reality with the first and foremost being the environmental dimension. This would make it possible to state that two persons or entities are not in the same situation if they pollute differently, even if their situation is similar economically in terms of income, wealth, assets, etc.

This would also lead to the elevation, expressly or de facto, of the polluter-pays principle to the constitutional level with the consequence that it could be balanced with the ability to pay and thus with (horizontal) equity as the two of them are of equal status. Within this balancing process, the protection of the taxpayer from an excessive tax burden would be guaranteed by other principles, such as proportionality, which would ensure on a case-by-case basis that the measure enacted is not overly excessive in relation to the environmental goal pursued and also that it is not a levy disguised as an environmental tax.

García and Roch (2016) also advocate for the elevation of the value of environmental protection to a constitutional rank. They develop a different legal reasoning, however, and begin with the assumption that environmental taxes are not in contradiction with the ability-to-pay. This is because the latter and the polluter-pays principle are different with distinct functions, and they are compatible for that reason. In essence, there are two different criteria for fairness: (i) a redistribution goal for those taxes depending on the ability to pay and (ii) a restorative nature in the case of eco-taxes. In the author's view, although it is correct to state that the two principles are theoretically compatible, it may be the case that, under the current framework, certain environmental taxes based on the polluter-pays principle are considered contrary to the principle of ability-to-pay. This especially applies when the results are not exactly measurable in terms of environmental impact. This can be exacerbated by the strictly economic approach of the principles underlying the tax system because it can no longer be said that the tax has a restorative nature when the environmental result is not measurable.

Consequently, if a mere indirect and non-measurable aid to environmental protection is not sufficient to make the tax justifiable as restorative, then it should theoretically be justifiable from a redistributive point of view. However, like most environmental taxes, it could not be justified as redistributive and would ultimately succumb vis-à-vis the principle of ability-to-pay.

Regarding the circumstance that environmental protection is, in fact, often disconnected from the revenue generated by environmental taxes, its elevation to a constitutional value, even for tax purposes, could be grounds for the creation of a solid argument in favour of regulatory taxation also when the damage created by the polluter is not repairable by means of the payment of a sum of money. On the contrary, when only revenue collection and ability to pay have constitutional status, environmental taxes are more likely to succumb if not objectively linked to certain difficult-to-measure outcomes.

In the Spanish case of *Transportes Jordi* (2014), the Court of Justice of the EU (CJEU) determined that a tax on the sale of mineral oils was contrary to Directive 92/12/EEC on the movement of products subject to excise duties on the basis that it did not pursue one of the specific purposes among those permitted by the directive. The judges' argument was based precisely on the fact that there was no predetermined allocation of revenue for environmental protection (paras. 32 and 33 of the judgement). Although restricted to a specific directive, this case promotes a standard that can be taken as a model and adapted when designing a tax system to support sustainable energy, although with some modifications. In particular, it specifies that, for a tax to be regarded as having environmental purposes, it must be characterised in one of two ways. First, it must have a predetermined allocation of the generated revenues to environmental purposes. Alternatively, its structure (and particularly the taxable item or the rate) must be designed in such a way that it dissuades taxpayers from acting in a certain way or encourages them to act differently (Wahl 2013).

In the case of this second condition, it would be necessary to make it clear that the levy should be completely disconnected from the possibility of effectively remedying environmental problems. In this way, a tax could be legitimate even if it made only a limited or even merely potential contribution to reducing pollution and would not be exposed to the risk of being considered illegitimate simply because its results are difficult to measure.

10.3.3 The Concept of Environmental Neutrality to Shape Consumption Taxes

In several systems, indirect and consumption taxes are based on the principle of neutrality. This applies particularly to those multi-stage taxes, such as EU VAT, that are intended to be neutral and, because of that, are based on the mechanism of deduction of the input tax paid by the taxable person.

In the specific case of EU VAT, neutrality is one of the declared objectives and main pillars since it aims at allowing the functioning of a single market composed of many states by reducing possible distortions to a minimum (Comelli 2000; Lupi 2000). This is directly reflected in both the preamble and in numerous provisions of the various VAT Directives that have followed one another over time up to the one currently in force as well as in numerous judgments of the CJEU.

Council Directive 2006/112/EEC, currently in force and amended multiple times, establishes the principle of neutrality under recital n.4 where it states that 'the attainment of the objective of establishing an internal market presupposes the application in Member States of legislation on turnover taxes that does not distort conditions for competition or hinder the free movement of goods and services'. This still effectively reflects the nature of what the European (Economic) Community was when, some decades after the end of WWII, indirect taxes were first harmonised. There was a large common market, as defined by the Treaty of Rome (Treaty establishing the European Economic Community), centred on economic cooperation between a number of states adopting a free-market economy (Daniele 2000).

This principle of neutrality, however, while being part of a unitary tax system, cannot be defined in a single way since it keeps several characteristic profiles of the VAT together that are distinct from each other and are, therefore, usually analysed separately.

Terra and Kajus (2011) have comprehensively dealt with these concepts and outlined a distinction between external and internal neutrality. The latter is divided into legal neutrality (equal is treated equally), competition neutrality (the tax burden does not depend on the extent of vertical or horizontal integration), and economic neutrality (absence of interference with the optimal allocation of production).

Currently, the principle of neutrality is built around three main concepts (Herbain 2015). The first is that businesses shall not be charged as the VAT is a tax on consumption. From a practical standpoint, this concept is embodied in a number of mechanisms that make the 'pass through process' along the production chain possible and enable taxing final consumption.

In this context, the CJEU (1996, 2001a, b, 2009) has addressed the neutrality of the VAT system several times with regard to the mechanism of deduction of the input tax paid by the taxable person. The judges have affirmed that: '[…] the deduction system established by the directive is meant to relieve the trader entirely of the burden of the VAT payable or paid in the course of all his economic activities. Thus, the common system of VAT seeks to ensure complete neutrality of taxation of all economic activities […]'.

The second concept is that there shall be no distortion of the single market. This is related to the idea that the facilitation of cross-border trade is one of the most efficient means for increasing global enrichment (Herbain 2015) and that legal and tax systems should affect the decision of businesses as minimally as possible on whether to operate in a certain foreign market. In practical terms, this means that the VAT shall be levied in a manner that does not constitute an obstacle to the flow of goods and services among the Member States and does not distort trade between services and goods supplied domestically and those originating from a different Member State (European Commission 2016).

The last of the three concepts is that of equal treatment that is also referred to as horizontal equity. It was already apparent in the preamble to the first VAT directive for which its eighth recital states that '[…] to result in neutrality in competition, in that within each country similar goods bear the same tax burden, whatever the length of the production and distribution chain […]'. The CJEU affirmed in the case *Investrand* (2007), among others, that 'the common system of VAT […] ensures complete neutrality of taxation of all economic activities' (par. 22) and in the case *Marks & Spencer* (2008) that 'the principle of fiscal neutrality is the reflection, in matters relating to value-added tax, of the principle of equal treatment' (par. 49).

This strong 'economic' view of neutrality has led, among others, to the definition of two goods or services being 'similar' when they are in competition with each other. For instance, there was much debate about whether e-books could benefit from a reduced rate before statutory changes were made (CJEU 2015a, b, c), and one of the main points of discussion was whether they were 'in competition' with paper books (Cannas 2017).

From an environmental perspective, this economic vision has also had rather negative consequences. For example, in *Commission v France* (2001), the court was required to decide whether the reintroduction of a total exclusion from the right to deduct VAT on diesel fuel was contrary to VAT legislation. It explicitly rejected the argument brought by the French Government that the legislative amendment at issue was made for environmental reasons and stated that it cannot justify legislation that breaches the (Sixth) VAT Directive (par. 21).

Similarly, in the 2015 *Commission v United Kingdom* (2015) case, the CJEU stated that the United Kingdom (UK) failed in implementing the VAT Directive by applying a reduced VAT rate to supplies of services of installing energy-saving materials. The court considered that the reduced rate at issue was not adopted by the UK exclusively on the grounds of social interest or, at least, for reasons of principally social interest (i.e. improving the quality of housing and protecting the health of the residents). Instead, it was to serve environmental or energy policy objectives (par. 14). Nevertheless, the court ultimately set aside the piece of UK legislation reviewed on the basis that: (i) the extension of the scope of the reduced rate of VAT at issue to all residential accommodations cannot be described as essentially social (par. 30) and that (ii) goods and services having social or cultural objectives may be subject to a reduced rate of VAT provided they pose little or no risk of distortion to competition (par. 25).

The court's reasoning in deciding these cases has always been formally correct, however, it clearly emphasises that domestic, environmentally friendly divergences regarding the harmonised VAT legislation are not justifiable unless explicitly provided for at the EU level. The consequence is that revenue objectives and the preservation of *perfect competition* within the single market shall always prevail over environmental policies in most cases.

To create a tax system that is conducive to promoting clean energy in addition to elevating environmental protection to the highest constitutional rank, which was already indicated for direct taxes and applied here as well, the concept of neutrality, and thus horizontal equity, should also evolve in indirect and consumption taxes. It is argued here that it is not sufficient for these taxes to be neutral as they shall also be *environmentally neutral.*

From an EU perspective, the recognition of the *environmental neutrality principle* would involve changes to the current VAT Directive commencing with its preamble. In the latter, it should be clearly stated that, in addition to the functioning of the single market, the value-added tax system harmonised at the supranational level also aims to effectively support the environmental objectives of the Union. Of the three concepts of neutrality seen above, it appears to be evident that environmental objectives interfere with the second and the third, i.e. the prohibition of distortion of competition within the single market and that of equal treatment.

Conversely, there appear to be no grounds in this context to advocate for the need to change the concept that 'businesses shall not be charged'. These concepts would remain in the system but changed with the addition of an environmental element and being applied differently. This is because it would be possible to argue that those who produce polluting goods or services are not in a comparable situation as those who produce other polluting ones even though, from the consumers' perspective, they are in competition.

This would thus allow Member States to favour the goods and services with the lowest environmental impact, for example by applying a reduced rate when this is compatible with the other relevant rules. From a legal reasoning perspective, this would be achieved by creating the conditions for 'breaking the currently close link' between the similarity of two services or goods from the consumers' perspective and the *economic neutrality* of the VAT treatment. As a result, two goods or services would not have to be mandatorily considered similar if they pollute differently.

The outlined legal framework would also make it possible to introduce, among others, reduced rates for products for which the specific characteristics should be identified on a case-by-case basis in order to support the production and use of *green energy.*

In conclusion, it can also be briefly stated that such a policy would alter a multiplicity of principles. Remaining within the EU context, an intervention such as that described would, to a certain extent, represent an extension of the *polluter-pays principle* to the area of indirect taxation (Barrett et al. 2017). To use a neologism (Cannas and Fermeglia 2022), it could be declared that the partially new principle of the '*consumer who pollutes pays*' would be established.

10.4 Conclusion

Beyond the more strictly tax aspects, in conclusion, it is relevant to make a brief observation on how everything said in this chapter relates to the concept of '*energy justice*' that is emerging internationally.

This is a principle that is not defined under EU law, although the Union is gradually including it in many of its policy documents. It is therefore what some European scholars have correctly called a '*guiding principle*' (Heffron et al. 2015), which is certainly relevant from a general justice perspective but not binding.

In this context, taxation can only play a relatively marginal role, since at EU level it is only functional to the correct functioning of the Single Market and Member States keep their powers in this area almost intact. The principle of energy justice, therefore, can only *guide* tax policy choices and tax regulation in the EU context, in coordination with Art. 194(1) TFEU, which is the legal basis for the Union's energy policies.

Moreover, Article 194(1) also has rather limited scope due to the legal structure of the Union and has been assessed by the doctrine as unsuitable to effectively guarantee the ambitious goals of (energy) justice set at the international level (Kaschny 2023).

More in general, climate change and the need to rely on clean and affordable energy represent a global challenge for all humanity. There is no doubt that the tax system is relevant in this, and an environmentally friendly tax policy should certainly be prominent in any political agenda. Unlike other challenges of the present time, however, environmental protection is a purely global challenge since pollution knows no borders. Due to this feature, the tax system that is characterised by a domestic and border-constrained dimension can play a relatively limited role and is powerless without large-scale coordinated actions at a global level. For everyone to have access to energy anywhere in the world and for it to not damage the planet, a broad political consensus beyond individual jurisdictions will be needed first. After that, it will be the responsibility of scholars and international organisations with the competence to devise new solutions and promote them in a coordinated manner to create the legal framework and tax structures that will accompany these changes.

References

Avi-Yonah RS (2011) Taxation as regulation: carbon tax, health care tax, bank tax and other regulatory taxes. Account Econ Law Convivium 1(1):1–12

Barrett A, Lawlor J, Scott S (2017) The fiscal system and the polluter pays principle—a case study of ireland. Routledge, London. ebook published on 17 June 2019

Biondi B (1956) I beni. UTET, Torino

Cabral ACG, Appelt S, Hanappi T (2021) Corporate effective tax rates for R&D: the case of expenditure-based R&D tax incentives, OECD taxation working papers n. 54. OECD, Paris. Available at https://www.oecd.org/tax/corporate-effective-tax-rates-for-r-d-ff9a104f-en.htm

Cannas F (2017) Reduced rates and the digital economy: the treatment of (E-)books highlights some possible inconsistencies of the EU VAT system. EC Tax Rev 26(2):96–108

Cannas F, Fermeglia M (2022) Environmental neutrality: redesigning EU VAT neutrality to support the implementation of the European green deal. In: Weishaar SE, Milne JE, Andersen MS, Ashiabor H (eds) Green deals in the making—perspectives from across the globe. Edward Elgar, Cheltenham, pp 62–79

Carlsen L, Bruggemann R (2022) The 17 United Nations' sustainable development goals: a status by 2020. Int J Sust Dev World 29(3):219–229

Carnelutti F (1913) Studi sulle energie come oggetto di rapporti giuridici. Rivista di diritto commerciale 11:354

Comelli A (2000) Iva comunitaria e iva nazionale. CEDAM, Padova, p 302

Commission Regulation (EU) No 651/2014 of 17 June 2014 (the previous GBER, namely Council Regulation (EC) 994/98, was adopted by the European Commission in 2008 and expired on 31 Dec 2013)

Corso G (2021) Il mercato: profili costituzionali. In: Vetrò F (ed) Il mercato dell'energia elettrica a vent'anni dalla liberalizzazione. Marsilio Editori, Venezia, pp 27–48

CJEU (1973) Judgement of 24 October 1973, case C-9/73, *Schülter/Hauptzollamt Lörrach*, ECLI:EU:C:1973:110

CJEU (1976) Judgement of 3 February 1976, case C-59/75, *Manghera e. a.*, ECLI:EU:C:1976:14

CJEU (1996) Judgement of 24 October 1996, case C-317/94, *Elida Gibbs/Commissioners of Customs and Excise*, ECLI:EU:C:1996:400

CJEU (2001) Judgement of 14 June 2001, case C-40/00, *Commission v France*, ECLI:EU:C:2001:338

CJEU (2001) Judgement 22 February 2001, case C-408/98 *Abbey National*, ECLI:EU:C:2001:110, par. 24

CJEU (2007) Judgement of 8 February 2007, case C-435/05 *Investrand*, ECLI:EU:C:2007:87

CJEU (2008) Judgement 10 April 2008, case C-309/06, *Marks & Spencer*, ECLI:EU:C:2008:211

CJEU (2009) Judgement of 29 October 2009, case C-174/08 *NCC Construction Danmark* ECLI:EU:C:2009:669, par. 27

CJEU (2014) Judgement of 27 February 2014, case C-82/12, *Transportes Jordi Besora*, ECLI:EU:C:2014:108

CJEU (2015a) Judgement if 4 June 2015, case C-161/14, *Commission v United Kingdom*, ECLI:EU:C:2015:355

CJEU (2015b) Judgement of 5 March 2015, case C-479/13, *Commission v France* ECLI:EU:C:2015:141

CJEU (2015c) Judgement of 5 March 2015, case C-502/13, *Commission v Luxembourg* ECLI:EU:C:2015:143

Daniele L (2000) Il diritto materiale della comunità europea, 3rd edn. Giuffré, Milano, p 16

De la Feria R (2009) The EU VAT system and the internal market. IBFD, Amsterdam, p 84

De Sadeleer N (2002) Environmental principles: from political slogans to legal rules. Oxford University Press, Oxford, p 21. Reprinted in 2005

Directive (Council Directive) 96/92/EC of the European Parliament and of the Council of 19 December 1996 concerning common rules for the internal market in electricity, OJ L 27, 30 Jan 1997

Directive (Council Directive) 98/30/EC of the European Parliament and of the Council of 22 June 1998 concerning common rules for the internal market in natural gas, OJ L 204, 21 June 1998

Directive (Council Directive) 92/12/EEC of 25 February 1992 on the general arrangements for products subject to excise duty and on the holding, movement and monitoring of such products, OJ L 76, 23 Mar 1992

Directive (Council Directive) 75/439/EEC of 15 June 1975 on the disposal of waste oils, OJ L 194, 25 June 1975

Directive (Council Directive) 94/62/EC of 20 December 1994 on packaging and packaging waste, OJ L 365, 31 Dec 1994

Directive (Council Directive) 75/442/EEC of 15 July 1075 on waste in relation to the cost of the disposal of waste, OJ L 194, 25 June 1975

Directive (Council Directive) 2004/35/CE of the European Parliament and of the Council of 21 April 2004 on environmental liability with regard to the prevention and remedying of environmental damage, OJ L 143, 30 Apr 2004

Directive (Council Directive) 2006/112/EC of 28 November 2006 on the common system of value added tax OJ L 347, 11 Dec 2006

Easson A (2001) Tax incentives for foreign direct investment—part II: design considerations. Bull Int Taxation 55(8):366

Elkins D (2006) Horizontal equity as a principle of tax theory. Yale Law Policy Rev 24(1):43

Endo K (1999) The 1993 white paper on growth, competitiveness and employment. In: Endo K (ed) The presidency of the European Commission under jacques delors. Palgrave Macmillan, London

European Commission (2010) Communication from the Commission 'Europe 2020—a strategy for smart, sustainable and inclusive growth', Brussels, 3.3.2010, COM(2010) 2020 final, p 10

European Commission (2014) Communication from the Commission 'guidelines on state aid for environmental protection and energy 2014–2020', Brussels, 28.6.2014 (2014/C 200/01)

European Commission (2016) Communication 'an action plan on VAT towards a single EU VAT area—time to decide', Brussels', 7.4.2016, COM/2016/148 final, p 5

European Commission (2020) Directorate-general for taxation and customs union, "taxation in support of green transition: an overview and assessment of existing tax practices to reduce greenhouse gas emissions", final report, written by Ecorys and WIFO, Nov 2020, section 4.2 "Assessment of the environmental effectiveness of tax incentives", pp 94–95

European Commission (2022) Communication from the Commission to the European Parliament, the European Council, The Council, the European Economic and Social Committee and the Committee and the Committee of the Regions—REPowerEU: Joint European Action for more affordable, secure and sustainable energy, Strasbourg, 8.3.2022, COM(2022) 108 final

European Community (1972) First summit conference of the enlarged community (Paris), declaration. E.C. Bull 10:15

Falsitta G (2010) Manuale di diritto tributario – parte speciale. CEDAM, Padova, p 744

García EG, Roch MTS (2016) Environment and taxation: state intervention from a theoretical point of view. In: Pistone P, Ezcurra MV (eds) Energy taxation, environmental protection and state aids—tracing the path from divergence to convergence. IBFD, Amsterdam, p 37

Gibson SH (2012) A critical analysis of the ability to pay. Edinburgh Student Law Rev 1(4):1–23

Guenther G (2017) Report 'patent boxes: a primer'. Congressional Research Service, Washington. Available at https://crsreports.congress.gov/product/pdf/R/R44829

Heffron RJ, McCauley D, Sovacool BK (2015) Resolving society's energy trilemma through the energy justice metric. Energy Policy 87:168–176

Herbain CA (2015) VAT neutrality. Promoculture Larcier, Brussels, p 51

Jans JH (2011) Stop the integration principle? Fordham Int Law J 33(5):1533–1547

Kaschny L (2023) Energy justice and the principles of article 194(1) TFEU governing EU energy policy. Transnational Environ Law 12(2):270–294

Keating A (2006) The polluter pays principle in domestic law. Dublin Univ Law J 28:172

Lupi R (2000) Diritto tributario – parte speciale, i sistemi dei singoli tributi. Giuffré, Milano, p 277

Messinetti D (1965) Energia (dir. priv.). In: Calasso F et al (eds) Enciclopedia del diritto, vol. XIV. Giuffré, Milano, pp 867–868

Miladinovic A (2022) Selectivity and the arm's lenght principle in EU state aid law. IBFD, Amsterdam

Murphy L, Nagel T (2002) The myth of ownership: taxes and justice. Oxford University Press, Oxford, p 24

Musgrave RA, Musgrave PB (1984) Public finance in theory and practice, 5th edn. McGraw-Hill International Editions, Singapore, p 218

OECD (2003) Paper 'tax incentives for research and development: trends and issues', Paris. Available at https://www.oecd.org/sti/inno/2498389.pdf

OECD (2015) Report 'BEPS action 5: harmful tax practices, Paris'. Available at https://www.oecd.org/tax/beps/beps-actions/action5/allcontent.htm

OECD (2018) Report 'companion to the inventory of support measures for fossil fuels', Paris. Available at https://www.oecd.org/environment/oecd-companion-to-the-inventory-of-support-measures-for-fossil-fuels-2018-9789264286061-en.htm

Pérez Bernabeu B (2014) R&D&I tax incentives in the European Union and state aid rules. Eur Tax 54(5):182
Pitrone F (2015) Defining "environmental taxes": input from the court of justice of the European Union. Bull Int Taxation 69(1):58–64
Quotidiano ilSole24Ore (2022) Quali Paesi europei dipendono di più dal gas russo? IlSole24Ore, 25 Feb 2022. Available at https://www.infodata.ilsole24ore.com/
Sellerio A (1950) Energia. In: Cosatto M et al (eds) Enciclopedia Italiana, vol XIII. Istituto dell'Enciclopedia Italiana, Roma, pp 967–977
Simons HC (1938) Personal income taxation. The University of Chicago Press, Chicago, pp 30, 16
Sinisi M (2005) La liberalizzazione del settore dell'energia elettrica: logica di impresa e competenza giurisdizionale. Servizi Pubblici e Appalti 2(2):477
Terra B, Kajus J (2011) A guide to the European VAT directives, vol 1. IBFD, Amsterdam, p 283
Travi A (1998) La liberalizzazione. Rivista Trimestrale Di Diritto Pubblico 48(3):645
UN Conference on the Human Environment, Stockholm, Sweden, June 5–16 (1972) Declaration of the United Nations conference on the human environment. In: Report of the United Nations conference on the human environment, princ. 13, at 3, 4, U.N. Doc. A/CONF.48/14Rev.1 (16 June 1972)
Vanistendael F (2014) Ability to pay in European community law. EC Tax Rev 23(3):121–134
Wahl N (2013) (Advocate General), opinion delivered on 24 October 2013, case C-82/12, *Transportes Jordi Besora*, ECLI:EU:C:2013:694, paras. 25 and 26

Francesco Cannas is an Assistant Professor in Tax Law at the University of Turin. He was awarded his Ph.D. by the Vienna University of Economics and Business (WU) and subsequently researched and taught in Belgium, Malaysia, Singapore and Saudi Arabia. His main research interests are indirect taxation, the interplay between taxation and the digital economy, taxpayer's human rights and environmental taxation. Francesco is also a chartered tax advisor with some years of practice in advising international clients.

Part III

Unique Perspectives and Case Studies

Islamic Approach to Energy Justice

11

Jasmin Omercic

Abstract

The Islamic approach to energy justice is rooted in principles like stewardship, justice, and moderation that guide consumer and producer behaviour in reducing and eliminating waste and promoting conservation. Insight into a few principles facilitates understanding how Islam inspires individuals to establish energy justice. The Muslim world has the potential to exemplify the attainment of energy justice, but challenges like frequent conflicts and a sheer lack of cooperation hinder constructive progress. Nevertheless, grassroots efforts in some communities exemplify the Islamic way to energy justice through upholding Islamic notions of consumption and production. Namely, society shall balance consumption habits, optimise resource usage, minimise waste, and uphold environmental stewardship. Likewise, production in Islam demands the lawful only and necessary needs fulfilment, thereby emphasising long-term sustainability. The Islamic approach to energy justice in this chapter exhibits the need for greater awareness among all stakeholders and ignites action through upholding universally appealing Islamic teachings.

Keywords

Sustainable energy · Energy justice · Islamic principles · Islamic production · Islamic consumption · Moderation · Stewardship

11.1 Introduction

Annals of history recorded the relevance of religion in a civilisation's rise and fall as a result of deviating from its teachings. Islam, as one of the Abrahamic religions, known to be a way of life, gave birth to a civilisation connected with nature and dedicated to knowledge creation, preservation, and innovation unlike any civilisation until today (Nasr 2010; Toynbee 1987). It proved to be the major reference point and cause for the awakening of Western civilisation. In the contemporary era of the internet and massive digitalisation, one cannot fail to note the serials of documentaries and other video material featuring the outstanding balance and relationship of Islamic civilisational times with nature. While other civilisations, such as the Persian and Roman, had their distinct pivotal achievements in many fields, the Islamic civilisation mastered them all, especially the arts, which testify to a high touch for aesthetics in congruence with the contextual environment (Nasr 2010; Bakar 2006). Its

J. Omercic (✉)
Assistant Professor of Business Studies,
City University (CU) - Ulster University,
Lusail, Qatar
e-mail: jasmin.omercic@cuc-ulster.edu.qa

A. Trakic et al. (eds.), *Energy Justice*, Sustainable Development Goals Series,
https://doi.org/10.1007/978-981-97-6059-6_11

teachings were all-encompassing and greater than those of any other civilisation until today.

However, Islamic civilisational erudition is incomparable in terms of population and scale to the civilisation in the twenty-first century when humans harness uncounted benefits of scientific discoveries. Quality of life is far better in contemporary times than ever before, but human suffering from the unpredictable consequences backfiring from nature due to mere action following scientific guidelines is ever greater. The Islamic civilisation differed and nurtured a scientific tradition integrated with religious belief and practice. Today, religion has been discarded, nature desacralised, and science greatly appraised to perpetuate Western civilisational development since its re-awakening (Bakar 2006, 2008, 2011). Nevertheless, such was indeed the religious experience at one trajectory of a civilisational reform and transformation amidst confrontation with diverse challenges. It is implicitly not universally appealing, relevant, and the predetermined way forward.

Islam, as a way of life, encompasses religious practice in the form of obligatory and voluntary rituals but also political, economic, legal, environmental, architectural, and other facets of human, flora, and fauna living. Experts of Islam will freely testify how Islamic principles and guidelines have been clearly defined for centuries and open to multiple correct understandings and interpretations due to the dynamism and recognition of every human and other creatures' distinctness (Al-Attas 1978, 2014). Namely, Islam entirely acknowledges all apparent and non-apparent, tangible and intangible, direct and indirect aspects of creation with setup of due balance, perception of sacredness, and integration with nature. It must be reiterated that Islam enables multiple epistemological pathways to the same objective, like energy justice and is a unique religion with textual evidence emphasising the importance of marvelling nature. Islam urges reverence for nature due to reflecting features and meanings of grand names (*asmā' al-ḥusnā wa al-ṣifāt*) Allah SWT (*subḥānahū wa ta'ālā*, Arabic for 'the most glorified, the most high') attributed to himself.

It is evident today that even science confirmed the dynamism of creation and its multifaceted nature after turbulent environmental events. This proves science and Islam compatible and represents the rationale for understanding that there is an Islamic approach to energy justice and that we need to understand it as crucial in the global energy transition agenda. However, the understanding of Islamic teachings is solely and always attributable to Muslims as imperfect representatives of Islam, like adherents of other religions imperfectly represent their religious teachings. Moreover, erudite reading of notable Islamic scholarly contributions like those of Syed Muhammad Naquib Al-Attas allows us to conclude that the core message of Islam resorts to justice (Al-Attas 2014). This makes the discourse on energy justice from the Islamic approach even more notable.

Justice reflects notions of *ādāb* (manners), which in return means the recognition and use of everything vis a vis its ontological nature (Al-Attas 1978). Likewise, energy justice reflects the notions of *ādāb*, which ensures inclusivity in decision-making and access to energy for all. Thus, Islam teaches that everything in this world has a purpose, which, in the ultimate sense, exhibits the ontology of everything. While the true nature of things is immutable, its epistemological application vis a vis the varieties of meanings and usages leads to the misplacement of things. This creates injustice and misalignment of ontology and epistemology. Equally, energy is generated through means that serve another purpose. Through their usage, negative externalities are created while eliminating relatively large segments of society from the benefits of energy usage, distorting the ontology and epistemology of energy production and consumption. It results in energy injustice. This energy injustice reflects the misalignment whereby generated energy over the last few centuries brought development to some and displaced many segments of society from their rightful place, causing an imbalance in nature and the nature of creation. Energy justice, however, refers to fairness in benefits and burdens in processes of energy production and consumption. Implicit is the notion of access to energy.

For a proper understanding of the Islamic approach to energy justice, we review some of the major Islamic principles for energy justice. These principles are greatly relevant for the next section on the Muslim world's potential to attain energy justice. Some challenges, stagnating and preventing the actualisation of that potential, bring us to notions of consumption and production in Islam as a solution to some or all challenges over time. Namely, consumption and production in Islam integrate Islamic principles of energy justice into consumer and producer behaviour, which ultimately positively impacts degrees of waste, conservation, and overall energy justice.

11.2 Islamic Principles for Energy Justice

An exposition of the relationship between Islam and energy justice is viable through an understanding of some fundamental Islamic principles that accumulatively substantiate the genuine Islamic approach to energy justice and its universally appealing nature. These stem from Islam's notions of responsibility as stated in the Holy Qur'ān when Allah SWT said (Yusuf Ali 2001): 'We did indeed offer the Trust to the Heavens and the Earth and the Mountains: but they refused to undertake it being afraid thereof: but man undertook it he was indeed unjust and foolish.' (Al-Aḥzāb, Verse 72)

Hence, humans are caretakers of creation on a very ontological basis and, as such, revelation hints that we can claim a positive relationship between Islam and energy justice (Abdul Jabar et al. 2015). Doing good and preventing bad (*al-amr bi al-m'arūf wa al-nahy 'an al-munkar*) is at the core of Islamic teachings, which enables ultimate justice if the resources available to humankind are adequately distributed.

Starting with the principle of ***stewardship or vicegerency (khīlafah)***, Allah SWT said (Yusuf Ali 2001): 'Behold, thy Lord said to the angels: "I will create a vicegerent on earth. They said: "Wilt Thou place therein one who will make mischief therein and shed blood?- whilst we do celebrate Thy praises and glorify Thy holy (name)?" He said: "I know what ye know not.' (Al-Baqarah, Verse 30)

Indicatively, Islam teaches that humans are stewards of the earth and, as such, shall care for and preserve it. Implicit is the responsibility to utilise natural resources, including energy, in a just and sustainable manner to ensure the welfare of current and future generations. Namely, the Holy Qur'ān emphasises, through the principle of *khalīfah,* the intrinsic nature of man, which shapes his worldview and advocacy for responsible maximisation of energy production and consumption with minimisation of harm to the environment. By Islamic law (*al-sharī'ah al-Islāmiyyah*) and notions of economic efficiency, humankind is principally accountable to God in the hereafter (*ākhirah*) (Khan and Haneef 2022).

Related to the *khalīfah* principle is the principle of ***equitable distribution or simply justice ('adl).*** An earlier reference to Al-Attas identified that justice (*'adl*) is at the core of the Islamic message to humankind. It is fundamental in Islam. Thus, fair and equitable distribution of resources, including energy, is incumbent upon the *khalīfah* to ensure that all people have access to energy, among other necessities (*ḍarūriyyāt*). Therefore, energy justice from the Islamic perspective is intrinsic in the very ontology or purpose of Islam as the religion revealed to Prophet Muhammad, peace be upon him (PBUH) (Salim and Hossain 2016). Equitable distribution of energy would assure everyone's fulfilment of minimum life standards and simultaneously enable greater energy consumption and more privileged access to others if affordable. In other words, competition and freedom to enjoy the bounties (*ni'immāt*) of God are systematic in Islamic practice and thereby clearly denounce the equal distribution of resources, goods, and services, i.e. energy.

Islam emphasises the principle of ***public interest (maṣlaḥah)***, which can be expressed as social responsibility. It ignited the relatively young discipline of Islamic economics (IE) and its practical aspect of Islamic banking and finance (IBF) to develop insurance (*takāful*) and re-insurance (*retakāful*). Related to the principles of stewardship and justice, public interest is given precedence over individual interest. Therefore, say over energy production, access, and consumption are rendered to society rather than the individual. Islam rejects monopoly and any other market structure, such as oligopoly, monopolistic competition or pure competition, if it inflicts harm upon the public (Reda 2013). Hence, safeguarding notions of equity in energy access is the Islamic firm approach in light of the principle of public interest (*maṣlaḥah*). In Islam, the less fortunate are considered to have the right to the resources, goods, and services of the more fortunate, and especially the most fortunate. Likewise, it is obligatory upon every and especially most fortunate individual to allocate a portion for the less fortunate. Alms (*zakāt*) giving is a notable indicator and among the pillars of Islam as a process of cleansing wealth and rendering it long-term *ḥalāl* as well as blocking hoarding (*iḥtikār*) (Younus and Ahmad 2021). It is hereby implicit that the Islamic approach to energy justice incumbently embraces marginalised segments of society and promises public interest or welfare (*maṣlaḥah*).

The principle of ***good character*** (***ḥusn al-khulq***) is notably emphasised in the narrations from the Prophet Muhammad PBUH as follows (Malik ibn Anas 2019): 'Malik reported: The Messenger of Allah, peace and blessings be upon him, said, "I have been sent to perfect good character."' (Al-Muwaṭṭa', Ḥadīth 2588, p. 740)

The purpose of Prophet PBUH in the era of grand ignorance (*jāhiliyyah*) has been to transform the character of the people and distinguish truth (*ḥaqq*) from falsehood (*bāṭil*). Semantically, the *ḥaqq* corresponds to the reality (*ḥaqīqah*) in Islam, like the freedom of choice (*iktiyār*) corresponds to choices leading to goodness (*khayr*) (Al-Attas 2014). Hence, Islam is linguistically a miracle due to the exceptional language of the Holy Qur'ān known to be the word of Allah SWT. Society before the Qur'anic revelation revered the poets due to linguistic mastery, who, in return, marvelled at the unparalleled language of the Qur'ān. Islam apparently leads to the transformation of one's character for the better. Hence, the principle of good character in Islam assures energy justice on notions of equity similar to earlier elaborated principles. It is possible to extend this principle to general environmental ethics, which implies friendly practices and policymaking that safeguard sustainable energy systems and equity (Sayem 2023; Gade 2023).

Another fundamental principle of Islam with strong implications for energy justice is ***moderation*** (***wasaṭiyyah***). In other words, Islam appraises conservation and abhors wastefulness (*isrāf*) and extravagance (*tabthīr*) in consumption and production (Mansour et al. 2017; Kamali 2015). Moderation in the production and consumption of resources promises just outcomes and energy justice. The Holy Qur'ān addresses adherents of Islam as moderates when Allah SWT says (Yusuf Ali 2001): 'Thus, have We made of you an Ummat (community) justly balanced, that ye might be witnesses over the nations, and the Messenger a witness over yourselves;...' (Al-Baqarah, Verse 143)

Moderation (*wasaṭiyyah*) in the verse is translated as 'justly balanced', which relates to earlier principles like justice. The interrelatedness of all the referred principles thus far communicates the integrative Islamic nature and tendency towards balance in all aspects of life. Thus, moderation is a central tenet missing in our contemporary globalised world that propelled a greater quest for energy justice. Simply said, a moderate approach to production and consumption, as taught in Islam, would undoubtedly restore the required balance between humanity and nature as well as enable the fulfilment of the undertaken stewardship covenants with God. Moderation would reduce unnecessary consumption followed by respective production, besides minimising wastefulness in energy production processes. Moderation would enable genuine distribution in the pre-production, during-production, and post-production stages (Mansour et al. 2017; Salim and Hossain 2016; Haneef et al. 2023). Therefore, the Islamic approach to energy justice has the potential, through the moderation principle, to prop up the energy transition agendas in ways unthought conventionally.

Aligned with the notions of energy justice, assuring the inclusivity of marginalised or affected segments of society in decision-making processes is the principle of ***consultation (shūra)*** in Islam. Unlike consensus (*ijmā'*) as a source of knowledge in Islamic epistemology and jurisprudence, *shūra* distinctly refers to non-legal or non-jurisprudential aspects of social living, i.e. energy justice. Justly balanced communities in Islam shall resort to consultation before and during projects of energy provisioning for the short, medium, or long term. While the scholarly consensus exists and endorses consultation as an approach towards greater welfare, the consensus is not a requirement to apply consultation due to matters of non-jurisprudential nature (Al-Raysuni 2011). In addition, we can mention *istiḥsān* (juristic preference) and *istiṣlāḥ* (seeking benefit) as innovative and creative decision-making and operationalised methods aiming for welfare-enriching solutions. The root words of both concepts stem from *iḥsān* (perfection-greatness) or *ḥusn* (goodness), implying the natural inclination of Islamic reasoning towards the good and best practice (*itqān*) (Saleem 2010; Muslimin and Kharis 2020). The Prophet Muhammad PBUH is known to have said: 'Allah loves to see one's job done at the level of *itqān*.' (Saḥiḥ Muslim, Ḥadīth 1976)

Notable is the Islamic approach to working with utmost dedication, consideration, and effort to attain perfection. Extrapolating upon this in terms of energy justice, Islam fosters all efforts to refine production and consumption processes to establish a just balance. As such, involving all concerned stakeholders in energy production and consumption decision-making assures that diverse perspectives are considered, ensuring fairer and genuine energy justice.

Referred principles definitely require innovation, creativity and dynamic collective reasoning (*ijtihād ijtimā'i*). Exploring and implementing sustainable energy solutions through Islamic principles is undeniably a collective task and requires a broad basis to operationalise principles in such a manner to attain the utmost energy justice. Notable to mention is that the essence of all action in Islam towards energy justice also resorts to notions of permissibility (*al-ibāḥah*). Therefore, the application of these principles depends on multiple factors affecting the degree of development of Muslim and even non-Muslim communities. Likewise, it is greatly influenced by cultural, social, political, and other contextual realities requiring filtering as well (Abdul Jabar et al. 2015; Kamali 2015). Non-Muslims have yet to recognise for themselves this unique approach to energy justice through analysis of the implied intentions of the overall Islamic approach and objectives of respective communities. Congruence between the two leads to a logical acceptance of the Islamic approach to energy justice.

Notable contributions to energy justice have already been embraced among an increasing number of non-Muslims through Islamic banking and finance (IBF). *Sharī'ah* compliance is referred to as the benchmark determining the acceptability of certain practices to attain greater welfare, i.e. energy justice. While IBF is subject to criticism, all of it never implies the flawedness of the whole novel industry, indicating the existence of genuine benefits of its distinct Islamic approach (Lone and Ahmad 2017; Hassan 2020). Community-based projects worldwide revealed grassroots initiatives aimed at improving living conditions, including energy access, among other sustainable energy solutions. Thus, advocacy among Muslim organisations is crucial in demonstrating success stories and Islamic harmony with nature via major caretaking of the notions and policies of environmental justice (Kalkavan and Eti 2021). The next section delves more concretely into the state of the Muslim world vis-a-vis energy justice and reveals the potential as well as challenges.

11.3 The Muslim World and Energy Justice: Potential

Resources in the Muslim world are abundant, and energy justice can be attained conveniently under conditions of will and even profitability. It has to be driven by a combination of Islamic principles, as explained earlier. The Muslim world counts dozens of diverse nation-states with varying levels of economic development and governance structures, which strive to be competent and attractive to investments (Abdulai and Siwar 2011; Amhar and Prima 2023). Energy justice represents a great gateway to attain that. For a systematic overview of the potential for energy justice in the Muslim world crossing two billion people, we discuss some indicating factors.

Islamic values stemming from the earlier elaborated principles are a normative basis forming a worldview that unites the Muslim world in diverse ways. These values have a wide scope and multiple right dimensions, which, upon interpretation and application, enable the attainment of objectives aligned with the agenda of energy justice (Haneef et al. 2023; Haneef 1997). For instance, multiple dimensions of the meaning of *khīlafah, 'adl,* or *maṣlaḥah* enable interpreters to agree upon procedures that operationalise steps of project implementation to supply energy generated via hydropower stations to a whole community at an affordable price. As mentioned earlier, the interpretation of these principles depends on the norms, culture, context, etc..

Resource wealth of Muslim nations offers space for innovation and creativity. These resources are oil, natural gas, and renewable energy potential, among others. Energy security and affordability are viable through effective harnessing and managing of these resources for both domestic and global markets. The Muslim world has a great potential to devise energy systems that render them independent and generous to supply affordable and environmentally friendly energy internationally. While this aligns with the global agenda of sustainability and renewable energy transition, it is also profitable and propels nation-states' competitiveness (Kunchu 2016; Abdulai and Siwar 2011).

***Investments in renewable energ*y** projects among Muslim nations can tap into the resource wealth factor and demonstrate how the Islamic approach to energy justice is workable (Abdulai and Siwar 2011). Some Muslim-majority countries can invest in novel renewable energy solar, wind, and hydropower projects to diversify their energy mix and thereby promote cleaner, affordable, and sustainable energy solutions through the Islamic approach. While other renewable energy projects exist already, apparently, they accumulatively prompted discourse on energy justice. Thus, the Islamic approach is not novel in arranging energy transition projects and can benefit from extant appraisals for diverse projects yet to be or already accomplished. However, the novelty of the Islamic approach to investing in renewable energy is its greater consideration and assurance of energy justice in light of earlier elaborated principles (Al-Ketbi and Nobanee 2020). It promises energy access for all as a key aspect of energy justice. The prototype of Muslim-majority countries' collaboration can provide energy access to marginalised and remote communities. For instance, the United Arab Emirates (UAE) and the Kingdom of Saudi Arabia have launched large-scale solar and wind projects contributing to sustainable energy futures and complying with Islamic principles (Xiaohong 2023; Al-Zohbi and Al-Amri 2020).

Muslim nation-states can advocate ***greater technology transfer*** among themselves as a demonstration of capabilities in attaining equity notions of energy justice. The Islamic principles and values enable and advocate such collaboration among Muslims or humanity at large (Kunchu 2016; Amhar and Prima 2023). This has been exemplified in the primetime era of Islamic civilisation. Namely, trade and knowledge-sharing have been key features of the Islamic civilisation and enabled replication and context-based adaptation of workable Islamic approaches to sustainable growth and prosperity (Omercic 2021). In contemporary times, the Islamic approach can be more systematically applied via greater knowledge-sharing, training programs, and joint research initiatives towards understanding notions of energy production and consumption to attain justice and its applications.

Inclusive policy and governance is another major systemic factor that enables Muslim-majority nations to refine governance structures that involve various stakeholders, like local communities in decision-making processes related to energy projects in their area. This participatory approach aligns with Islamic principles like consultation (*shūra*) and is possible through the genuine understanding of the stewardship role of all humanity vis a vis all creation, i.e. environment (Saleem 2010; Kamali 2015; Abdulai and Siwar 2011). This will amplify community energy production and conservation, reduce waste, and promote responsible and sustainable use of resources. It strongly aligns with Islamic principles like *khalīfah*-steward, justice-*'adl*, moderation-*wasaṭiyyah*, etc. These efforts align with Islamic teachings on avoiding extravagance (*isrāf*) and promotion of sustainability.

Islamic economics, banking and finance (IEBF) as an alternative approach to mainstream economics stems from the early decades of Muslim movements for independence in the twentieth century, which incorporates within its operations principles of Islam delineated earlier. IEBF has become a global phenomenon, embraced among non-Muslims too, due to its proven greater resilience in economic and financial shocks stemming from a particular focus on real economic activity, asset-backed rather than asset-based solutions (Omercic et al. 2020). However, this does not imply that the latter is not extant in the IEBF practice. Its relevance and potential to prop up energy justice among Muslim nation-states, especially those with developed IEBF markets like Malaysia, Indonesia, Kingdom of Saudi Arabia, and Qatar, to name a few, is promising based on varying global reports on IBF and Global Islamic Economy (GIE). This brings us to the Organisation of Islamic Cooperation's (OIC's) role in the agenda of energy transition and justice among member nation-states. Again, common Islamic values among OIC member states inspire mutual coordination of efforts towards welfare-enriching objectives for all Muslims and non-Muslims. Unanimous agreement among all states that interest (*riba*) is prohibited and exploitation points to the entrepreneurial spirit of IEBF, which is crucial for the agenda of energy justice (Banna et al. 2020; Khattak and Khan 2023; Hassan and Hippler 2014). It integrates with the principle of stewardship and, thereby, with other principles as well. Proper understanding of these interconnections systematically constitutes the sustainable and socially responsible energy justice approach. The IEBF financing mechanisms can play a fundamental role in renewable energy initiatives. Proactive measures to leverage upon available resources within the Muslim world (natural and human) through IEBF inspire much thought for grassroots projects and address marginalisation issues or exclusion from energy access (Kalkavan and Eti 2021; Amhar and Prima 2023).

Indicatively, the potential for achieving energy justice in the Muslim world is massive yet requires coordinated efforts among high-end and low-end stakeholders, namely the public, private, and third sectors and community on national and international levels. While the coordinated efforts facilitate great cohesion and impact wherever ignited and applied for greater energy justice, contextual circumstances are crucial for consideration to assure relevance and utmost effectiveness without minimising negative externalities.

11.4 Challenges and Examples of Energy Justice in the Muslim World

The Islamic approach to energy justice faces challenges that impact the approach to energy justice, considering earlier Islamic principles and potential for energy justice in the Muslim world. These are important to understand as products of human action rather than the irresolvable status quo, often left unchallenged and preventing the genuine Islamic principles-based approach to energy justice from taking shape.

Therefore, the ***interpretation and implementation*** of Islamic principles present a challenge. While multiple interpretations can lead to energy justice, unawareness of unfavourable interpretations perpetuates the status quo, whose interest is to shape coordinated efforts to their own advantage and thereby buy time before greater energy justice takes place (Kamali 2015; Saleem 2010). The crucial focus in overcoming these challenges should be the notion of justice and the question of whether coordinated parties unconditionally defend or compromise it.

The unfortunate ***economic and political realities*** of the Muslim world stagnate systemic action at government levels, yet general freedoms enable coordinated efforts and action at grassroots levels to improve conditions unvested from status quo interests, i.e. consumer behaviour (Kalkavan and Eti 2021; Sharqieh 2012). Related to this is the unawareness among the population in Muslim-majority nations about their potential impact on energy justice outcomes via consumption primarily and production as well.

For instance, more economical consumption of fossil fuels and community projects can address the challenge of ***convinced economic dependence on fossil fuels***. Gradual transitioning from fossil fuels as major sources of revenue for some countries demonstrates the ability and possibility for more sustainable futures. Qatar and the Kingdom of Saudi Arabia are great examples of nation-states investing in international markets to ensure sustainable revenues for national needs. Many of their investments abroad are directed to profitable businesses aiming for sustainable development goals (SDGs), environment, sustainability, and governance (ESGs), encompassing energy justice (McSparren et al. 2017; Praveen et al. 2020). Investment profits and potential partial, major, or full ownership of major businesses favouring energy justice outcomes enable them to drive their nation-states towards greater energy justice. This ends dependency upon oil per se.

The unfortunate ***geopolitical instabilities*** hinder investments in regions abundant with natural resources while some wars ignited due to natural resources, i.e. Gaza's strip-rich oil reserves. While this deprives Muslim nation-states like OIC of accomplishing major goals and the potential for energy justice, it also defines the rationale for greater coordinated action among most other Muslim nation-states and sets up a pragmatic agenda for energy justice (Sarqieh 2012; Ahsan 2020). The genocide in Gaza stands out clearly and ignited unity among Muslims and even non-Muslims globally.

Lastly, the Muslim world's richness in multiple aspects, including know-how, still reflects inadequate ***technical expertise and infrastructure*** for renewable energy projects, while simultaneous extant know-how and infrastructure lack maintenance (McSparren et al. 2017; Praveen et al. 2020). In other words, greater qualitative aspects of expertise need to be enhanced while maintaining quantity on par.

Having identified a few of the challenges, it is crucial that Islamic principles remain immutable as well as the promising and coherent framework for greater energy justice. We can hereby identify some theoretical and concrete examples to better understand that successful projects or practices demonstrate the unquestionable possibility of Islamic principles and approaches to energy justice. Namely, Islam dedicated distinct attention to the sacredness of nature and all creation as well as the human as the steward of all. Thus, the alignment of Islamic principles with contemporary SDGs, ESGs, and other agendas, of which energy justice is part and parcel, enables the world's communities and institutions to play a unique role in the quicker attainment of energy justice. IEBF activities are already visible globally and joined by non-Muslims as well. The principles of IEBF operations promote value investing without interest (*riba*) and favourable returns (Khattak and Khan 2023). Promoting these kinds of ethical and value investments has become a norm in the twenty-first century and integral to regulatory requirements. Many Islamic bonds (*sukūk*) (estimated to reach $2.4 trillion in 2024) have been issued to fund sustainable, green projects that most foster energy justice (Morea and Poggi 2016a, b; Al-Ketbi and Nomanee 2020; Maulayati and Iswati 2023). However, the landscape of energy justice, sustainability, and IEBF is dynamic and requires constant updates.

Among concrete examples of energy justice projects are the Barefoot College of India training rural women in solar panel installations to assure elementary self-sufficiency (Mininni 2022); SELCO in India providing solar energy microgrids for off-grid villages (Sarker et al. 2022); Green Caravan of Iraq promoting energy efficiency education (Alwash 2023); Morocco's Noor Ouarzazate Solar Complex among the world's largest concentrated solar power plants reducing Morocco's dependence on fossil fuels

(Hamouchene 2023); multiple nation-states energy transition roadmaps for reducing gas emissions (Gielen et al. 2019); commitment of more than 1200 Muslim signatories via the Islamic Declaration on Climate Change (2015) to tackle climate change via earlier discussed Islamic principles (Kaminski 2019); formation of energy transition dedicated ministries and agencies; various energy efficiency and conservation initiatives in mosques, education campaigns, training; establishment of knowledge-sharing platforms like the Islamic Climate Change and the Global Muslim Climate Networks (Ismail et al. 2019; Koehrsen 2021), among others. These examples demonstrate how environmentally friendly and inspiring are the Islamic principles and overall approach to energy justice. It apparently has much to offer if widely embraced among a greater number of Muslims in originating and naturalised nation-states after migration. Many Muslims, native to countries where their parents migrated, have the potential to ignite the Islamic approach to energy justice together with non-Muslims' efforts.

While concise examples of energy production efficiency and consumption are indeed plenty, a distinct brief overview of the consumer and production behaviours in Islam sheds light on why challenges persist and clarifies how initiatives for energy justice could be more effective. Consumption levels are not specifically determined in Islam, but principles like moderation (*wasaṭiyyah*) imply avoidance of wastefulness (*isrāf*), as well as an emphasis on the stewardship role of humans (*khalifah*), which implies strategised consumption behaviour (Islam and Chandrasekaran 2016). Consumption is known to be central to economic growth as part of the gross domestic product (GDP) measure of economic growth, yet the consumerism culture blurred the notions of consumption that is healthy for an economy, its inhabitants, and its environment. Islamic principles like moderation, together with gratitude (*shukr*), lighten up avenues for resource conservation and set a limit to consumerism. Moreover, ethical connotations and social responsibility notions of Islamic principles summon a more calculated consumer behaviour, balancing the interplay of needs versus wants in degrees of higher objectives of Islamic law (*maqāsid al-Sharī'ah*) (Khan 2014; Ilmy and Setiawan 2019). While the importance of these objectives in the Islamic approach to energy justice is beyond the scope, focusing on the objectives, from an analytical point of view, fosters a wiser and more considerate consumer attitude concerning society and the environment. It facilitates the integration of Islamic principles cohesively into consumption and production behaviour to attain greater energy justice. Allah SWT said in the Holy Qur'ān how clearly this balance shall look in practice (Yusuf Ali, 2001): 'And eat and drink but waste not in extravagance. Certainly he (Allah) does not like the prodigals.' (Al-A'rāf, Verse 31) and 'And those who, when they spend, are neither extravagant nor stingy, but are just between that.' (Al-Furqān, Verse 67)

Likewise, the Prophet PBUH is transmitted to have said: 'Eat and drink, but leave off a bit, for that is better for your health. Verily the believers eat in one stomach, while the hypocrites eat in seven.' (Saḥiḥ Bukhāri. Hadīth 5394)

Hence, Islam provides a sound framework for balanced consumption, which has the potential to address earlier raised challenges such as convinced dependence on fossil fuels, interpretations, and implementation, among others. The realisation that humans primarily need to fulfil needs rather than unending wants brings consumption behaviour to a balance (Haneef et al. 2023; Khan 2014).

Islam also encompasses teachings for the production behaviour of firms and individuals. Similar to consumption, production should be guided by ethical and social considerations, responsibility, and notions of equitable distribution. Production in Islam adds another level to distribution, which is personal distribution via *zakāt* (alms) and *sadaqāt* (charities) (Ilmy and Setiawan 2019). In this regard, the Prophet PBUH is recorded to have said: 'The truthful and honest merchant is associated with the prophets, the truthful, and the martyrs.' (Sunan al-Tirmidhī, Ḥadīth 1209)

Likewise, Allah SWT said in the Holy Qur'ān (Yusuf Ali, 2001): 'Give the relative his right, and [also] the needy and the traveller: But squander not (your wealth) in the manner of a spendthrift.' (Al-Isrā, Verse 26)

Among other considerations of production in Islam are productive efficiency of labour and the importance of earning through one's own effort. Production, like any affirmative activity in Islam, is part of worship (*'ibādah*). Again, the Prophet PBUH said: 'Nobody has ever eaten a better meal than that which one has earned by working with one's own hands. The Prophet of Allah, David used to eat from the earnings of his manual labor.' (Saḥiḥ al-Bukhāri, 2072)

Thus, the Islamic teachings that brought about IEBF development and, in light of the energy justice approach, clearly foster innovation and entrepreneurship, which as a result of the prohibition of *riba* (interest), the exploitation and effortless gain is the logical conclusion. In the processes of production in an entrepreneurial venture, there should be great consideration for environmental protection and resource utility maximisation. In other words, quality and excellence in entrepreneurial production processes for greater energy justice should precede profiteering. These production considerations in the Islamic approach to energy justice primarily facilitate community development, grassroots activity or what have conventionally been known as small and medium-sized enterprises (SMSEs)—the backbone of every nation-state economy (Kalkavan and Eti 2021; Hassan and Hippler 2014).

Overall, the Islamic approach to energy justice is unique and faces serious challenges that can be gradually addressed through balanced consumption behaviour and efficient production. Integration of the Islamic principles into these processes of most efficient production and balanced consumption behaviour for energy justice determines the most desirable outcome. Greater awareness about these notions of Islamic consumption and production aspects of energy justice, inclusive of the ethical considerations, sets the speed of energy justice attainment and affects national as well as global markets. The degree to which this will happen depends on the reflective Islamic reasoning (*ijtihād*) integrated with scientific findings because Islam and science are undoubtedly positively related. Namely, science supports the Islamic principles and approach to energy justice.

11.5 Conclusion

The Islamic approach to energy justice is implicitly unique. The exposition of some Islamic principles for energy justice enabled us to grasp how Islam's religious teachings motivate individuals to establish overall justice, including that of energy access for all. It is also implicit that the proper understanding of the principles is a prerequisite for adequate action leading to the attainment of justice. Thus, the multiple correct ways of understanding the Islamic principles for energy justice determine whether Muslim behaviour or even that of non-Muslims will project the genuine Islamic approach to energy justice.

We have seen that the Muslim world has much potential to attain energy justice but also faces challenges that prevent it. Realities in the Muslim world are indeed worrisome due to perpetual conflicts over the decades but also inaction in relation to the potential possible welfare generation unrelated to conflicts that can stem as a result of greater cooperation. While governmental-level cooperation is always more demanding and challenging, grassroots efforts towards energy justice via the application of Islamic principles are still possible. This can be done in Muslim-majority nation-states and that of non-Muslim communities as well. Thus, we resorted to a brief exposition of Islamic views on consumption and production. These impact human consumption habits and identify to what degree current wants are not needs or to what degree thoughtful needs are actually wants only. In other words, balancing consumer behaviour is an objective of the Islamic view on consumption. On a large scale, this leads to greater economising of resources, waste reduction, healthier lifestyle, and genuine environmental stewardship.

Likewise, production views in Islam are aligned with consumption whereby produced is primarily what is lawfully needed and commendable for a better sustainable living. Indicative is a holistic message from the overall Islamic approach to energy justice that a balanced use of resources extracting utmost utility at each stage, inclusive of all stakeholders, is ardently needed in contemporary times to attain energy justice. In other words, the Islamic approach to energy justice is more important today than ever before because of the factual signals that the environmental balance has been seriously shaken and natural laws lead the environment to respond on par. Since human behaviour led to the global environmental crisis and energy injustice in the first place, the Islamic principles, consumption and production views focus on and guide human behaviour towards moderation. This is why the Islamic approach to energy justice, exposed imperfectly in this chapter, becomes ever more significant.

References

Abdul Jabar NH, Ilham Z, Saifuddeen SM, Rahman NNA (2015) Green energy towards sustainability from the Islamic perspective. Int J Sustain Future Human Secur 3(2):31–34

Abdulai AM, Siwar C (2011) Linking sustainable livelihoods to natural resources and governance in OIC countries: prospects and challenges. J Econ Cooper Dev 32(2):97–123

Ahsan M (2020) Desertification in the OIC member countries: factors, challenges and the way forward. Bartın Orman Fakültesi Dergisi 22(2):642–653

Al-Attas SMN (1978) Islam and secularism. Muslim Youth Movement of Malaysia, Kuala Lumpur

Al-Ketbi AY, Nobanee H (2020) Islamic finance and renewable energy: a mini-review. https://ssrn.com/abstract=3538658, https://doi.org/10.2139/ssrn.3538658

Al-Raysuni A (2011) Al-Shūrā: The Qur'anic principle of consultation: a tool for reconstruction and reform. International Institute of Islamic Thought (IIIT), Herndon

Al-Zohbi G, Al-Amri FG (2020) Current situation of renewable energy in Saudi Arabia: opportunities and challenges. J Sustain Dev 13:98–119

Al–Attas SMN (2014) Prolegomena to the metaphysics of Islam. Penerbit UTM Press, Kuala Lumpur

Alwash A (2023) From threat to opportunity: harnessing climate change to build a prosperous future for Iraq and the region. Instituto Affari Internazionali Pap 23:2–17. https://www.jstor.org/stable/resrep51701

Amhar F, Prima EC (2023) Resources of Islamic countries. Islamic Res 6(1):77–82

Bakar O (2006) Environmental health and welfare as an important aspect of civilisational Islam. Al-Shajarah J Int Inst Islamic Thought Civ (ISTAC) 11(1):27–52

Bakar O (2008) Tawhid and science Islamic perspectives on religion and science. Arah Pendidikan Sdn Bhd, Kuala Lumpur

Bakar O (2011) Islamic science, modern science, and post-modernity towards a new synthesis through a tawḥidic epistemology. Revelation Sci 1(03):13–20. https://journals.iium.edu.my/revival/index.php/revival/article/view/36/28

Banna H, Hassan MK, Alam MR (2020) Digital financial inclusion, Islamic banking stability and sustainable economic growth. In: Sarac M, Hassan MK (eds) Islamic perspective for sustainable financial system. Istanbul University Press. https://doi.org/10.26650/B/SS10.2020.017.06

Gade AM (2023) Muslim environmentalisms and environmental ethics: theory and practice for rights and justice. Muslim World 113(3):242–259

Gielen D, Gorini R, Wagner N, Leme R, Gutierrez L, Prakash G, Asmelash E, Janeiro L, Gallina G, Vale G and Sani L (2019) Global energy transformation: a roadmap to 2050. International Renewable Energy Agency (IRENA), United Arab Emirates (UAE)

Hamouchene H (2023) Dismantling green colonialism: energy and climate justice in the Arab region. Pluto Press, London

Haneef MAM, Mohd Nahar A, Mohammed MO (2023) Foundations of Islamic economics. International Institute of Islamic Thought (IIIT), Kuala Lumpur

Haneef MAM (1997) Islam, the Islamic worldview, and Islamic economics. IIUM J Econ Manage 5(1):39–65

Hassan MK, Hippler W (2014) Entrepreneurship and Islam: an overview. Econ J Watch 11(2):170–178. https://ssrn.com/abstract=3263110

Hassan R (2020) Reforming Islamic finance: why and how? J King Abdulaziz Univ Islamic Econ 33(2):67–80

ibn Anas M (2019) Al-Muwaṭṭa' (The Recension of Yaḥyā B. Yaḥyā Al-Laythī (d. 234./848)). Fadel M, Monette C (eds). Harvard University Press, Cambridge, Massachusetts

Ilmy RM, Setiawan I (2019) The concept of production, distribution, and consumption in Islamic economics. Rev Islamic Econ Finance 2(1):41–46

Islam T, Chandrasekaran U (2016) Effect of religiosity on ecologically conscious consumption behaviour. J Islamic Market 7(4):495–507

Ismail R, Saputra J, Aziz AA (2019) Improving environmental performance of the Muslim world: evidence from affluent countries. Int J Energy Econ Policy 9 (3):301–312

Kalkavan H, Eti S (2021) Determining optimal Islamic financing methods for small-scale sustainable energy investments regarding socio-economic welfare. In: Yüksel S, Dinçler H (eds) Strategic approaches to energy management: current trends in energy economics and green investment. Springer. https://doi.org/10.1007/978-3-030-76783-9

Kamali MH (2015) The middle path of moderation in Islam: the Qur'anic principle of wasatiyyah. Oxford University Press

Kaminski JJ (2019) The OIC and the Paris 2015 climate change agreement: Islam and the environment. In: Pal LA, Evren Tok M (eds) Global governance and Muslim organizations. Springer International Publishing. https://doi.org/10.1007/978-3-319-92561-5

Khan F, Haneef MA (2022) Religious responses to sustainable development goals: an Islamic perspective. J Islamic Monetary Econ Fin 8(2):161–180

Khan MF (2014) The framework for Islamic theory of consumer behaviour. J Islamic Bus Manage 4(1). https://platform.almanhal.com/Files/Articles/75756

Khattak MA, Khan NA (2023) Islamic finance, growth, and volatility: a fresh evidence from 82 countries. J Islamic Monetary Econ Fin 9(1):39–56

Koehrsen J (2021) Muslims and climate change: how Islam, Muslim organisations, and religious leaders influence climate change perceptions and mitigation activities. Wiley Interdisc Rev Clim Change 12(3):1–19. https://doi.org/10.1002/wcc.702

Kunchu JAB, Sarmidi T, Md Nor AHS, Yussof I (2016) New evidence on natural resource curse for OIC and non OIC countries. Pertanika J Soc Sci Human 24 (4):1767–1792

Lone FA, Ahmad S (2017) Islamic finance: more expectations and less disappointment. Invest Manage Fin Innov 14(1):134–141

Mansour MS, Hassan KH, Bagheri P (2017) *Sharī'ah* perspective on green jobs and environmental ethics.' Ethics Policy Environ 20(1):59–77

Maulayati RR, Iswati S (2023) Sukuk trend analysis: bibliometric and systematic literature review from 2011–2021. Jurnal Ekonomi Syariah Teori dan Terapan 10(4):377–390. https://doi.org/10.20473/vol10iss20234 pp377-390

McSparren J, Besada H, Saravade V (2017) Qatar's global investment strategy for diversification and security in the post-financial crisis era. Centre on Governance, University of Ottawa, Ottawa

Mininni GM (2022) The Barefoot college 'eco-village approach to women's entrepreneurship in energy. Environ Innov Soc Trans 42:112–123

Morea D, Poggi LA (2016) Islamic finance and renewable energy: an innovative model for the sustainability of investments. In: International annual conference (AEIT), 1–7

Morea D, Poggi LA (2016) Islamic finance and renewable energy: an innovative model for the sustainability of investments. In: International annual conference: sustainable development in the Mediterranean area, energy and ICT networks of the future. Institute of Electrical and Electronics Engineers

Muslimin JM, Kharis MA (2020) *Istiḥsān* and *istisḥāb* in Islamic legal reasoning: towards the extension of legal finding in the context of Indonesia. Al-Risalah 20 (2):163–179

Nasr SH (2010) Islam in the modern world: challenged by the west, threatened by fundamentalism, keeping faith with tradition. Harper One, New York

Omercic J (2021) Integration of knowledge (IoK) methodological reasoning of Islamic economics (IE) as the wisdom of humanity: a heterodox economic approach. Int J Islamic Econ Fin (IJIEF) 4(2):233–262

Omercic J, Haneef MAM, Mohammed MO (2020) Economic thought, foundational problems of mainstream economics and the alternative of Islamic economics. Int J Islamic Econ Fin (IJIEF) 3(2):171–198

Praveen RP, Keloth V, Abo-Khalil AG, Alghamdi AS, Eltamaly AM, Tlili I (2020) An insight to the energy policy of GCC countries to meet renewable energy targets of 2030. Energy Policy 147:111864. https://doi.org/10.1016/j.enpol.2020.111864

Reda A (2013) Islam and markets. Rev Soc Econ 71 (1):20–43

Saḥiḥ Bukhāri, Hadīth 2072. Accessed 14 Feb 2024. https://sunnah.com/bukhari:2072

Saḥiḥ Bukhāri, Hadīth 5394. Accessed 14 Feb 2024. https://sunnah.com/bukhari:5394

Saḥih Muslim, Hadīth 1976. Accessed 14 Feb 2024. https://muslimmatters.org/2016/09/07/the-forgotten-sunnahs-ihsan-itqaan-and-self-reliance/

Saleem MY (2010) Methods and methodologies in fiqh and Islamic economics. Rev Islamic Econ 14(1):103–123

Salim R, Hossain MZ (2016) Equitable distribution of wealth and resources in Islam. In: Choudhury MA (ed) God-conscious organisation and the Islamic social economy. Routledge, New York

Sarker T, Dey S, Yousaf A, Mishra A (2022) Social enterprises and SDGs: a case study of SELCO Solar Light Pvt. Ltd., India. In: Ward A, Steckler E (eds) World scientific encyclopaedia of business sustainability, ethics and entrepreneurship. World Scientific, Singapore

Sayem MA (2023) Islam and environmental ethics: a Qur'anic approach. In: Kwan SS-M, Chow W-Y (eds) Asian spiritualities and social transformation. Springer Nature, Singapore

Sharqieh I (2012) Can the organization of Islamic cooperation (OIC) resolve conflicts? Peace Conflict Stud 19(2):219–236

Sunan Al-Tirmidhī, Hadith 1209. Accessed 14 Feb 2024. https://www.abuaminaelias.com/dailyhadithonline/2019/06/02/honest-traders-prophets-truthful/

Toynbee AJ (1987) A study of history: abridgement of volumes I–VI. Oxford Paperbacks, Oxford

Xiaohong D (2023) Islamic finance on investment in green energy projects: a comprehensive study of UAE. J Xi'an Shiyou Univ Nat Sci Edn. 19(10):675–692

Younus TS, Ahmad R (2021) Zakat management and economic sustainability: United Arab Emirates model. In: Sarea A (ed) Impact of zakat on sustainable economic development. IGI Global, Pennsylvania

Yusuf Ali M (2001) The Holy Qur'ān: translation and commentary. Islamic Vision Ltd., London

Jasmin Omercic is an Assistant Professor of Business Studies at City University (CU) in partnership with Ulster University of the United Kingdom in Lusail, Qatar. Previously, he was an Assistant Professor at the Department of Economics of the Kulliyyah of Economics and Management Sciences (KENMS) of the International Islamic University Malaysia (IIUM). At IIUM, he also served as a Research Fellow at the Centre of Islamic Economics (CIE). Jasmin's research interests include alternative economics, banking, and finance approaches leveraged on technology and thought with a future and sustainability-oriented mindset. He has published multiple papers and presented at various conferences and seminars on related and evolving topics. Jasmin was a Country Correspondent for the prestigious Islamic Finance News (IFN) and a Malaysian Institute of Management (MIM) Certified Trainer at the Student Affairs and Development Division (STADD) of IIUM, designing and delivering leadership and management trainings. He aims to advance practice with distinct quality control and empower the next generation of leaders and thinkers.

12 Challenges in Addressing Energy Injustice in ASEAN

Mohamed Hanipa Maidin

Abstract

The phrase energy justice has become the buzzword in international human rights law. The central theme of energy justice is essentially attaining and securing equality between social, economic and political participation in the energy system. Ergo, this chapter seeks to critically discuss the ongoing challenges faced by ASEAN in confronting energy injustices in the region. It also ventures into the discussion of the theory or concept of energy justice. In discussing the electrification process in ASEAN, this chapter also examines the development and the challenges of such a process in almost each and every state in ASEAN. Given the status of Singapore as a developed nation, it is deliberately excluded from the discussion. The demand for ASEAN to swiftly shift into renewable energy has been growing day by day. At the end of the day, ASEAN's enormous challenge is its huge dilemma in picking and choosing between two major priorities: either to succumb to international pressures by completely moving to renewable energy thus disregarding the people's plight or to stick and maintain the status quo by absolutely ignoring such international pressures.

Keywords

Energy justice · ASEAN · Electrification · Renewable energy

12.1 Introduction

Whenever we talk about energy, our mind will invariably be directed to a global resource scarcity *vis-à-vis* population growth in an increasingly unpredictable social and environmental climate. Nowadays, the phrase "energy justice" has become the buzzword in international human rights law. The central theme of energy justice is attaining and securing equality between social, economic and political participation in the energy system. Its ultimate aim is to make the whole system more ethical (University of Sussex 2023).

In accomplishing energy justice, especially in the Association of Southeast Asian Nations (ASEAN), all the key players in energy sectors have to take notice and, in turn, recognise the presence of all the stakeholders who are supposed to have a say when energy systems are being designed and deployed and what the implications of such systems are and to whom (University of Sussex 2023). Energy injustice will invariably creep in when energy poverty is pervasive. Hence, to attain a full realisation of energy justice, ASEAN has to address the scourge of energy poverty as well. It is not

M. H. Maidin (✉)
International Islamic University Malaysia (IIUM), Kuala Lumpur, Malaysia
e-mail: hanipamaidin12@gmail.com

A. Trakic et al. (eds.), *Energy Justice*, Sustainable Development Goals Series,
https://doi.org/10.1007/978-981-97-6059-6_12

disputed there is a sharp dichotomy between energy poverty in developed countries and developing countries. The inherent nature of energy poverty in developing countries is, nonetheless, far more chronic.

The energy poverty in developing countries is not only confined to a lack of access to modern energy sources needed to provide modern energy services but it is also exacerbated by the want of infrastructure, well-functioning energy services markets and sufficient income to afford modern energy sources. Except for Singapore, it is safe to say that other countries in ASEAN are still categorised as developing nations. Be that as it may, the issues of energy injustice are still prevailing in ASEAN.

In this note, I seek to critically discuss the ongoing challenges faced by ASEAN in confronting energy injustices in the region. In penning this note, I intend to structure the chapter in the following order. Firstly, it is important to get to know ASEAN at great length especially its historical development, its mission and vision. Hence, in the first part of the chapter this note shall focus its discussion on ASEAN in general.

In the second part of the chapter, I shall venture into the discussion of the theory or concept of energy justice and the significance of such a concept. Thirdly, I shall discuss electrification in ASEAN in which I will touch upon the development as well the challenges of electrification in each state in ASEAN. I purposely exclude Singapore in my discussion on electrification in ASEAN as Singapore, in my view, does not really encounter any major problem as far as electrification in ASEAN is concerned.

Thereafter, in the fourth part of the chapter, I attempt to embark on the discussion of the issue of energy transition in ASEAN and its impacts on countries in ASEAN, especially on the problem of inequality. In this part, I shall also touch on the renewable energy (RE) in ASEAN including a discussion on solar energy. Finally, I will highlight several key challenges faced by ASEAN in effectively translating the theory of energy justice into practice.

12.2 ASEAN—Its Historical Development, Mission and Vision

Initially, especially at its infancy stage, i.e. prior to the eruption of World War II, countries in Southeast Asia (SEA) were not officially recognised as a region. *Au contraire,* they were merely identifiable by the country's own name. Hence, they were not seen as a country from a particular region (Keling et al. 2011).

Therefore, Burma, Malaya or Sumatra, for instance, which had been colonised by superpowers (Allied Forces) were only called Southeast Asia. (Keling et al. 2011). In fact at the Quebec Conference which was held in August 1943, the word Southeast Asia was employed in which the allied parties had placed Malaya, Sumatra, Thailand and Burma under the Southeast Asia Command (SEAC I). The idea was to protect the colony in the Southeast Asian region (Keling et al. 2011).

In July 1945 during the Potsdam Conference, SEAC 1 was duly expanded by the Netherlands East Indies (Indonesia) comprising several other countries of Indochina such as Southern Vietnam and Cambodia. The new identity was then identified as SEAC II. During World War II, these Southeast Asian countries were categorised as the Allied Forces Zone under the command of Lord Mountbatten (Turnbull 1999).

According to Nathan, the idea of establishing ASEAN came into picture with a few bilateral and multilateral agreements. These ideas were first mooted in the Southeast Asia Friendship and Economic Treaty (SEAFET) after Tunku Abdul Rahman's official visit to the Philippines in January 1959 (Nathan 1988). Tunku Abdul Rahman was the first Prime Minister of Malaysia.

SEAFET was very limited in its scope whereby it was confined to economy, trade and education relationships only. Nonetheless, the idea had duly inspired the establishment of ASEAN. SEAFET, however, was a failure due to persistent disagreements among the Southeast Asian countries (Saravanamuttu 1983).

On a positive note, however, the failure of SEAFET had led to the establishment of another regional organisation, i.e. Association of Southeast Asia (ASA) which was formed on 31 July 1961. And three countries namely Malaya, Thailand and Philippines agreed to form this regional pact. It would be fair to say that the primary objective of ASA is to create peace and regional stability. Nonetheless, ASA was also designed to cultivate cooperation in the field of economic, social science and culture as well as to provide training facilities and research (Pollard 1970).

Unfortunately, ASA was doomed to failure as well. Such a failure was often attributable to the emergence of conflict especially between Malaya and the Philippines. The Philippines decided to pull out from ASA as a result of Malaya's proposal to include and claim Sabah into Malaysia. Such a proposal was vehemently objected to by the Philippines (Pollard 1970).

And with the failure of ASA, another regional organisation came into existence, namely an association known as MAPHILINDO comprising Malaysia, Philippines and Indonesia. It was essentially a non-political confederation of the three Southeast Asian countries in the Malay Archipelago. MAPHILINDO was slated to create cooperation in the field of economy, culture and social science (Arnfinn Jorgenson Dahl 1982). This association also carried another mission, i.e. to end the dispute between Indonesia, Philippines and Malaysia entailing territorial issues. Once again MAPHILINDO also descended into the same fate like its predecessors when each country was preoccupied in championing their respective national interests (Patmanathan 1980).

Despite hiccups here and there but to its credit, SEA, however, never easily gave up in trying very hard to revive the relationship between the regional states. A conference attended by SEA leaders was, therefore, held in Bangkok in 1967. The conference successfully produced the Declaration of Bangkok on the 8th of August 1967 leading to the establishment of ASEAN.

As far as ASEAN is concerned, its founding fathers were Indonesia, Malaysia, Philippines, Singapore and Thailand. On 7 January 1984, Brunei Darussalam decided to join ASEAN, and on 28 July 1995, it was followed by Vietnam. Lao PDR and Myanmar officially joined ASEAN on 23 July 1997. Two years thereafter, Cambodia finally joined it on 30 April 1999, making up what is today the ten Member States of ASEAN.

ASEAN's visions and missions are duly embodied in the ASEAN Declaration which, inter alia, to accelerate the economic growth, social progress and cultural development in the region through joint endeavours in the spirit of equality and partnership in order to strengthen the foundation for a prosperous and peaceful community of Southeast Asian Nations and to promote regional peace and stability through abiding respect for justice and the rule of law in the relationship among countries of the region and adherence to the principles of the United Nations Charter.

Since its inception ASEAN has been faithfully adhering to the principle of sovereignty, non-interference, consensus and unity in diversity. And all those principles are expressly embodied in the ASEAN Charter (The ASEAN Charter). And those principles, to some scholars, seem to define the ASEAN Way (Mariam 2018).

12.3 Energy Justice and Its Significance

Unlike climate justice which is focussing on addressing root causes of climate crisis and distributional inequality of climate impacts, energy justice, conversely, places its emphasis on the need for an equitable supply of affordable and sustainable energy for all (Iges 2023).

Relatively speaking, energy justice is a nascent phenomenon, and, in essence, it primarily seeks to apply justice principles to a plethora of energy-related issues such as energy policy, energy production and systems, energy consumption, energy activism, energy security, the energy trilemma political economy of energy and climate change (Jenkins et al. 2016a, b; Prehoda 2020).

In ventilating the issue of energy justice, scholars have added new adjectives to the word justice. The three most common adjectives which have been prefixed to the word justice are distributive or distributional justice, recognition-based justice and procedural justice. Hence, it is safe to say that the most frequently used energy justice framework is the tenet-based one that includes distributive, procedural and recognition-based justice (Van Uffelen 2022).

Distributional or distributive justice essentially entails unequal allocation of environmental benefits and ills as well as uneven distribution of their associated responsibilities (Kirsten Jenkins et al. 2016a, b). The need to incorporate distributional considerations into energy has been deeply ingrained in a plethora of international climate agreements through concepts such as common but differentiated responsibilities and intergenerational equity.

The procedural justice is often linked to the right to be duly included in the decision-making process. In other words, all the relevant stakeholders ought to be involved in the decision-making process. This form of justice is rooted in the need to act in a fair and equitable manner before any policy is to be carried out (Bhullar 2023a, b).

The case of Charanka Park in India is often cited as *locus classicus* in highlighting the potential for marked inequity in the distribution of benefits and costs of such "environmentally good" projects. Some also argue the procedural justice was not duly observed in that case.

A well-researched article by Yenneti and Day (2016) merits profound study and contemplation as the article was written with the primary objective of identifying and understanding the distributional justice concerns in the implementation of the Charanka solar park from the perspective of the Charanka Village community.

As a result of a long-fought activist campaign, the East Kalimantan government of Indonesia adopted provincial regulations strengthening oversight over coal mining and clean-up operations in 2013. Unfortunately, the environmental and social impacts of coal production on Javanese fisherfolk residing close to the coal power stations remain unaddressed. The procedural justice was clearly violated when local residents were not fully informed about the construction of the power plant and its potential side effects from the get-go. Consequently, no legitimate channels were provided for them through which to appeal and make their voices heard. As a consequence, a number of protests took place, with demonstrators demanding the right to access relevant information about planned construction. Indonesian media portrayed this as a challenge to the economic status quo (Iges 2023).

Recognition-based justice is often described as a theory of social justice that emphasises the recognition of human dignity and of difference between subaltern groups and the dominant society and it has a distinct status compared to distributive and procedural justice as its meaning seems the least tangible to grasp (Honneth 2004).

Generally speaking, scholars tend to refer to theories developed by Nancy Fraser and Axel Honneth whenever they seek to canvass the discussion on recognition-based justice. Nancy Fraser places recognition justice in the cultural realm. In essence, Fraser invariably characterises her stance as the "status model of recognition". Hence, she views recognition as concerned with the cultural status order, or patterns of cultural values and cultural values, in her estimation, are a hierarchy in nature. To illustrate her point, she often cites this example, namely executives are often valued more than homeless people (Blunden no date).

To sum up, recognition-based justice in the eyes of Fraser refers to the cultural status order of society. And misrecognition would rear its ugly head when such patterns are institutionalised in a way that prohibits the participatory parity of certain groups. Fraser argues that misrecognition is inherently erroneous in that it prevents people from participating equally "as peers in social life". In her theory, she calls such a phenomenon as "participatory parity" (Bozalek et al. 2020).

Honneth's recognition theory, on the other hand, is based on the normative values through which identity is established in social relations. Through social interactions, not only the individuals who are duly shaped, the normative mechanisms upon which society is founded are equally shaped too. Be that as it may, Honneth articulates a theory of recognition in which social justice can be achieved through the relationships we have with each other, subject to subject. He also argues that our unique human interdependence is intersubjective recognition, and has been institutionalised in society, in the sphere of family, civil society and the state (O'Brien 2013).

12.4 Electrification in ASEAN

Generally, electrification is defined as the process of increasing access to affordable electricity. And in assessing these experiments, the equity of that process ought to be taken into account. For that reason, equitable electrification is always known as precondition of socially and geographically inclusive development (De Langre no date).

Electrification in ASEAN has to do with the issue of energy vulnerability. Energy justice will, in turn, be linked to energy vulnerability and energy/fuel poverty, such as the problem of accessibility and affordability of energy services (Bhullar 2023a, b). As far as the ASEAN countries are concerned, reaching full electrification has been more challenging since those who remain without access to the national grid live in the most difficult places to reach—often in remote communities or on smaller islands (Al-Saffar 2017). It is said that "still around 65 million people across the ten Association of Southeast Asian Nations (ASEAN) countries are without access to electricity" (Al-Saffar 2017). Even if people are connected to electricity in these remote locations, such a connection is only made possible via hefty costs, not to mention polluting diesel generators (Al-Saffar 2017).

Despite an expansion of electricity access in recent years, it is estimated that around 2500 villages are still without access to electricity in Indonesia (Setyowati 2021). In 2022, the electrification rate in Indonesia was at 99.63%. The Indonesian government envisioned a 100% rate by 2023. Nevertheless increasing the electrification ratio invariably becomes an enormous task due to the fact that Indonesia has more than 17,000 islands and many communities are isolated from one another (Indonesia: electrification rate 2022|Statista 2023). Even the Indonesian government duly acknowledged that more than 500,000 Indonesian households had no access to any form of electricity by May 2021. And the scourge of electricity shortages are relatively common occurrences in remote areas of Indonesia, particularly in eastern Indonesia, where there are many small islands and relatively few residents (Yuxuan and Renaldi 2022).

Taking into account the Indonesian geography, building and expanding the power supply has always been a major challenge. In order to attain access to a stable and adequate electricity supply in Indonesia, the networks of countless islands need to be interconnected for nationwide coverage. However, logistical hurdles remain to be the biggest challenge in providing electricity in rural areas especially in eastern Indonesia. Most of the electricity networks across the archipelago are owned and operated by the state-owned electricity provider PLN (Perusahaan Listrik Negara or State Electricity Company). It is a corporation which is owned by the Indonesian government and has exercised a monopolistic character trait as far as electric power distribution in Indonesia is concerned. It generates the majority of the country's electrical power, and it is responsible for supplying electricity to the grid throughout Indonesia (Indonesia Investments 2024). Unfortunately, PLN was reported to have incurred gigantic debts due to expensive coal power contracts (Yuxuan and Renaldi 2022).

In his recent parliamentary reply, the Malaysian minister of rural and regional development, namely Datuk Seri Dr Ahmad ZahidHamidi, revealed that at least 137 Orang Asli (indigenous people in Malaysia) villages across the country still do not have any electricity supply (The Star 2023). Therefore, his ministry is currently studying ways on how to get electricity to these Orang Asli villages across peninsular Malaysia. In Malaysia, the Orang Asli Development Department (JAKOA) is developing a plan to provide electricity through the national grid to all Orang Asli villages in Peninsular Malaysia (The Star 2023). The minister's aforementioned reply, unfortunately, merely reaffirms a whole raft of studies depicting that poor and marginalised social groups are often disproportionately and severely implicated by the sting of energy production facilities.

To its credit, access to electricity in Thailand was reported at 100% in 2021 Such a data was revealed via the World Bank collection of development indicators, compiled from officially recognised sources in February of 2024 (TRADING ECONOMICS no date). Hence, it would be fair to say that Thailand has been very successful in developing its electrical power generation, transmission grid and distribution network. In 1976, when Thailand launched its Accelerated Rural Electrification Program, only 21% of Thai homes had access to power. However, in 1996, Thailand managed to reach 98% and by 2016 it achieved essentially full access at 99.99% (Admin 2023).

The Thailand transmission grid is operated by the state-owned Electricity Generating Authority of Thailand (EGAT). Despite being able to provide a full access of electricity to its entire population, Thailand is facing another obstacle. In Thailand, power generation remains the largest source of CO_2 emissions in Thailand, followed closely by the transportation and industrial sectors. Under such circumstances, the Thai government has duly set its targets to attain carbon neutrality by 2050 and net zero emissions by 2065. And to accomplish these goals, the Thai government has developed its "Thailand Energy Infrastructure Policy 4D1E" with the pillars of Digitalization, De-centralization, De-regulation, De-carbonization, and Electrification (Admin 2023).

According to Mariam Sherman, World Bank Country Director for Myanmar, Cambodia and Lao PDR, Myanmar has the lowest electrification rate in Southeast Asia with only 50% of households connected to the public grid. (World Bank Group 2020b) and 80% of the rural people having no access to grid electricity (Burma-Energy 2023). Hence among the ASEAN countries, Burma has the lowest electrification rate. The World Bank estimated Myanmar needs to double its current installed power generation capacity over the next five to seven years to achieve universal electricity access by 2030 (World Bank Group 2020b).

In 2014, the Government of Myanmar adopted the National Electrification Plan (NEP) to achieve universal access to sustainable electricity services by 2030. To its credit, the World Bank has shown its analytical support to NEP. Hence, the NEP has delivered electricity access to 2 million people and to schools, rural health clinics and community centres by extending the public grid in over 5000 rural villages and delivering Solar Home Systems and renewable energy mini-grids in 7200 villages throughout the country (World Bank Group 2020b).

In May 2020, the World Bank's Board of Executive Directors approved a $350 million credit from the International Development Association (IDA) to increase the output and efficiency of power generation and improve the resilience of Myanmar's electricity system to climate change and disasters. In addition to that, the Board equally approved $110 million in additional financing for the Essential Health Services Access Project, implemented nationwide since 2015 (World Bank Group 2020b).

Though the NEP has envisioned 100% nationwide electricity access by the year 2030, the political turmoil in Myanmar in which military coup defines its political landscape has been the main source of obstacles hindering the mission of the NEP to successfully meet its 2030 target (World Bank Group 2020b).

Be that as it may, as far as the electrification in Myanmar is concerned, the guaranteed access to a stable and adequate electricity supply has always been extremely demanding. Among the formidable challenges faced by the country are the suspension of CNTIC VPower's LNG plants in Yangon, the impairment of transmission lines connected to Baluchaung hydropower plants, the seasonal depletion of water reserves and crippled human resources due to the civil disobedience movement across government ministries all contributed to the growing power outages across the country. To rub salt to the wound, the rise of global oil and fuel prices, devalued local currency (MMK), the shortage of U.S. dollars in the local market and the slowdown of foreign investment have negatively impacted the energy sector. Energy projects approved before the military takeover have been suspended due to the political and economic turmoil in the country (Burma-Energy 2023).

To sum up, we may conclude that Myanmar is a good template illustrating the devastating impacts of political turmoil and uncertainty to the economic development and sustainability of even good and sound policies to its population.

According to data shared by Statista Research Department in November 9, 2023, in 2022, almost all households in urban areas in Vietnam had electricity, while the share of households living in rural areas with access to electricity was slightly lower, albeit still relatively high, at 99.5%. With this impressive data, apparently Vietnam may be said to have reached almost 100% electrification. Anyway the percentage of Vietnamese households with electricity access had been gradually increasing year on year in the past decade (Statista 2023). Hence, the real challenge faced by Vietnam is not really in providing access to electricity but in ensuring an affordable access to electricity would be highly and practically sustainable.

In Vietnam, state-owned Vietnam Electricity (EVN) had owned all power plants in the country since 2006, thus monopolising electricity transmission. It also owns and operates all transmission lines and substations. Despite the government's initiative in making policy changes for over the last 11 years to create a competitive energy market, it has yet to complete the task. Hence, EVN remains the only purchaser of electricity in the state. This means whenever suppliers increase prices due to higher costs, EVN will have to absorb them since it does not have the authority to increase retail tariffs. Ergo, in 2022, for instance, EVN reported a substantial loss amounting to VND 26 trillion (VnExpress 2023).

Herein the dilemma, though EVN's monopoly has been substantially thinning out and it now only accounts directly or indirectly for around 37.6% of the country's installed capacity of 80,000 megawatts, the government still needs to retain a monopoly over transmission to ensure national energy security and prevent private companies from taking over and raising prices. Consequently, EVN has to pay for the remaining 62% of power supplied, mainly to gas and oil giant Petrovietnam, coal mining company Vinacomin and a number of private companies (VnExpress 2023).

Another stumbling block is that despite some changes in the laws which allow private companies to invest in transmission, so far only one company has done so, connecting its solar plant with the national grid. A slew of major encumbrances such as land acquisition problems, low profit margins and red tape tend to deter private companies from entering the transmission business.

In the meantime, EVN has been consistently making big losses. When the Ministry of Industry and Trade conducted a review in 2022 on EVN's operations, it found out that the cost of electricity production had actually by increased more than 9% since 2021, but the government eventually approved a retail price hike of only 3%, leaving EVN selling power at a loss of VND112 per kilowatt-hour.

When EVN is not financially healthy, it goes without saying that it will not be able to invest in new power plants or ensure transmission stability. This is yet another predicament usually faced by many ASEAN countries like Vietnam. On the one hand, a state-owned electricity provider like EVN has to bear enormous losses in transmission business but on the other hand the government has to exercise restraint in allowing EVN to increase electricity tariffs in ensuring that people do not have to pay hefty electricity bills. In a way, by resorting to this populist approach the government reluctantly needs to ensure the affordability of energy services which is one of the key elements of energy justice. The question is how long could EVN and the Vietnam government sustain such persistent losses?

As of December 2022, the statistics depicted the only regions with a 100 level of electrification in the Philippines were the National Capital

Region (NCR) and CARAGA. Meanwhile, the electrification rate of the Autonomous Region in Muslim Mindanao (ARMM) was at 44.9%. About 880 thousand households in the country remain unserved (Statista 2024).

Before the Philippine government launched the Accelerated Barangay Electrification Program (ABEP) in 1999, barangay (i.e. village, suburb or other demarcated neighbourhood; a small territorial and administrative district forming the most local level of government) electrification level only stood at 76.9%, having energised only 32,281 out of 41,945 total barangay coverage. The ABEP was subsequently renamed as the "O'Ilaw Program", and under this programme, the Independent Power Producers (IPPs) were strongly recommended to participate as part of their corporate social responsibility. The programme finally led to the creation of the Foundation for Rural Electrification and Economic Development (FREED) as a mechanism to mobilise private funds for rural electrification (Expanded rural electrification | Department of Energy Philippines, no date).

As un-electrified communities are mostly populated in rural areas, rural electrification has been given a top priority by both the Philippine government and the private sectors. Be that as it may, the Expanded Rural Electrification Program (ER Program) was duly established and an ER Program Team was also created in April 2003, thanks to the enactment of the Electric Power Industry Reform Act of 2001 or "EPIRA". The ER Program entails the electrification of sitios (territorial enclave that forms part of a barangay) and households.

The ER Programme was slated to achieve 100% barangays electrification by 2008 and 90% household electrification by 2017. And under ER Programme, the participation of private sectors has been highly significant and crucial in ensuring the implementation of rural/missionary electrification would be successful. We may say that via this ambitious programme, the Philippine government, to its credit, envisioned greater access to electricity services.

Nonetheless, despite such an ambitious forecast, the number of households without electricity in the country is still haunting. In 2023, the Department of Energy (DOE) revealed in the Senate that the number of un-electrified households has increased to 1.8 million from 800,000 within five years. And population growth has been the fall guy (Panti, GMA Integrated News 2023).

Both Cambodia and Laos do not only share common historical experiences (French colonialism, civil war, mass destruction, communism and reconstruction) but they also share the same starting point in electrification, i.e. only 15% of households had access to electricity in 1995. Despite sharing the said common denominator, Laos, nonetheless, managed to reach 87% of households electrified in 2015, while Cambodia lagged at 35%. And the average price of electricity is 3–4 times higher in Cambodia than in Laos. Laos has a single tariff structure for the whole country, while Cambodia has hundreds. Electricity is therefore affordable and accessible throughout Laos whereas in Cambodia, power is relatively affordable and accessible in the main cities, while rural areas are either not served or can scarcely afford it. And both are also two small low-income countries in Southeast Asia and over 85% of their populations lived in rural areas (De Langre no date).

Cambodia and Laos have chosen two different institutional models for their electrification. Though Cambodia is often said to have opted for market liberalisation, weak state utility and decentralisation, its system is, however, an amalgamation of the private sector-dominated with an element of the public sector. In Cambodia, all electricity providers must obtain licences, including Electricité du Cambodge (EdC). The Lao system, on the contrary, is overwhelmingly state-dominated, though elements of the private sector are still visible and present. Nevertheless the interests of private firms are subordinated to the needs of Electricité du Laos (EdL). Thus, we may say that Laos has a more monolithic system, with a strong monopolistic state-owned utility that uses and directs the foreign private sector in specific capital-intensive subsectors and projects (De Langre no date).

De Langre argues that the success story of Laos in electrification has to do with its policy in favouring the validity of the developmental state model. He also argues that Cambodia's failure is due to its policy of early deregulation and privatisation (no date).

There has been a debate whether electrification is best done with large or little state intervention. If electrification is best to be determined by the market via deregulation, this question would arise: Can the market per se ensure a socially optimal allocation of resources on its own? The proponents of state intervention often argue that strong state intervention, with a commitment to social objectives, is more likely to achieve accessibility and affordability targets; a view shared by both the West and the USSR in the post-war period. On the other hand, the advocates of market- oriented electrification argue the underlying assumption behind the Electricity Law, which remains the legal foundation of the current system, is that competition is the best way to increase accessibility and affordability (De Langre no date).

From the foregoing discussion, we may fairly deduce that electrification has two dimensions that are seemingly incompatible: affordability and accessibility. On that score, increasing one should decrease the other. Being relatively poor countries like many ASEAN countries, it seems that investing in infrastructure is a costly venture that might lead to price hikes causing burdensome to the ordinary people. On the other hand, keeping prices would make large capital investments less likely. Under such circumstances affordability and accessibility seem to interact in a zero-sum game, and this relationship could be expected to hold even stronger in capital-poor countries like Cambodia, Laos, Burma, Indonesia and the Philippines.

Despite such a grim outlook, there is still a silver lining when a recent study has projected countries across Southeast Asia would, in all likelihood, achieve universal electrification by the early 2030s (IEA 2017).

12.5 Energy Transition in ASEAN *Vis-à-Vis* the Issue of Inequality

Tony Mazzocchi—a *Vis-senior official in the US Oil, Chemical and Atomic Workers (OCAW) union—was said to have coined the phrase "a just transition" for the first time in the late 1970s (Gambhir et al. 2018). Faced with the prospect of the closure of a number of industries due to environmental sustainability concerns, Mazzocchi mooted the idea of a just transition strategy. In essence, this strategy accepted the closures, yet it proposed a government to fund a scheme by providing wages, retraining and relocation support for workers as well as financial support and technical assistance for the development of alternative industries and jobs for displaced workers (Gambhir et al. 2018).*

In recent times, the term "just transition" has often been associated with the phenomenon of low-carbon transition, which entails a gradual shift from fossil fuels towards low-carbon energy sources. But, scholars in this field argue that the transition to low-carbon energy could possibly regenerate the problem of inequality as this transition does not operate in a vacuum (Bhullar 2023a, b). Though the energy transition phenomenon seems to be inevitable, Southeast Asian countries, however, badly need to effectively address the potential scourge coming from such an energy transition, as it tends to reproduce patterns of inequality that have long plagued the fossil fuel-based energy system.

ASEAN's substantial reliance on fossil fuels for economic and industrial development has been invariably presented as a *fait accompli*. Hence, it would be highly logical for ASEAN to face enormous challenges regarding the future energy lands due to energy transition. The tremendous impact coming from this transition should never be taken lightly, let alone underestimated (Rufaidah 2023).

One thing is for sure a slew of ASEAN countries would face big challenges to adopt energy transition. After all, according to the International Energy Agency's (IEA) report "Coal in Net Zero Transitions", the energy transition has led to a significant decline in employment in the coal industry. (Dong et al. 2023) Moreover, the new employment opportunities provided by clean energy may not be enough to absorb all those who have become unemployed due to the energy transition Moreover, the new employment opportunities provided by clean energy may not be enough to absorb all those who have become unemployed due to the energy transition.

To cite one example of a possible threat of energy transition to the Indonesian population if the country is fully determined to pursue the strategy of energy transition without a proper plan of action in place. The imminent challenge is, therefore, how could ASEAN design and execute policies which promote a more equitable and inclusive energy transition.

12.6 The Development of Solar Energy in ASEAN

It is an undeniable fact that the ASEAN region is richly endowed with diverse renewable energy sources and one of them is solar energy. Solar energy is believed to be the most reliable, abundant and free source of energy for a carbon-free economy. And it is also considered as the most reliable and promising form of renewable energy as it is inexhaustible, sustainable and practically unlimited. For that reason, solar energy has emerged as a highly promising candidate due to its remarkable growth rate (Summary of Inflation Reduction Act provisions related to renewable energy|US EPA 2023).

It is interesting to note here that the amount of energy the earth receives from the sun in a single day surpasses our yearly energy requirements. This fact implies that our planet does not face any energy crisis. Due to their geographical location, most ASEAN countries exist near the equator and they receive significant amounts of irradiation from the sun throughout the year. Despite such advantages, the real challenge for ASEAN is, therefore, how to convert this abundance of solar energy into usable electrical power (Summary of Inflation Reduction Act provisions related to renewable energy|US EPA 2023).

By virtue of a surge in demand for clean or renewable energy particularly in facing the scourge of emissions, solar energy has started to enter mainstream policy in most ASEAN member states. And Vietnam has become the front-runner in embarking on solar energy and followed by Thailand. According to McKinsey & Co, Thailand, Vietnam and Malaysia commutatively account for about 10% of global solar cell and module production. Hitherto Southeast Asia has annual capacity to produce 69 GW worth of solar power modules and the components to make them (Citraningrum 2023).

As ASEAN is still wanting in its capability to produce polysilicon or wafers, the region—at least for the time being—still needs to rely on China—which is known to be the hub of polysilicon or wafers—in importing those necessary items. China's dominance in many clean technologies has not, however, escaped the attention of the United States. Hence, the emergence of geopolitics of energy and the race has now been shifting towards countries and regions with a competitive edge in clean technologies (Citraningrum 2023).

In America, the U.S. government has adopted climate legislation to incentivise domestic manufacturing, most importantly the Inflation Reduction Act. In essence, the Act is offering funding, programmes and incentives to accelerate the transition to a clean energy economy and will likely drive significant deployment of new clean electricity resources (Summary of Inflation Reduction Act provisions related to renewable energy|US EPA, 2023).

Despite the fact solar power generation is beginning to take off in ASEAN and global investors are willing and ready to pump in the money by investing into the solar projects in the region, ASEAN is unfortunately still facing the problem of effective and workable coordination. Solid grid interconnectivity between member countries, increase in storage capacity and power system flexibility to accommodate more solar energy are still lacking (Citraningrum 2023).

12.7 Key Challenges in Translating the Theory into Practice

Whereas in OECD countries, the transition from fossil fuel dependence to a system based on cleaner energy has been rapidly carried out, ASEAN seems to continue to rely on fossil fuels for the daily survival of its population. In the process, ASEAN also needs to find the best ways to use fossil fuels so as not to harm the environment. This fact clearly cements this painful reality: coal use will continue to be the second largest energy source in ASEAN, especially in power generation after gas, in the foreseeable future to meet fast-growing electricity demand (Phoumin et al. 2021).

However, the increase in coal use for power generation in ASEAN countries will only lead to the widespread construction of coal-fired power plants, hence resulting in increased greenhouse gas (GHG) and CO2 emissions if the best available clean coal technology (CCT) is not employed (Phoumin et al. 2021).

Hence, we can clearly see how ASEAN countries are being trapped in a catch-22 situation. On the one hand, they are being pressured to move to renewable energy—hence abandoning fossil fuel—to decrease greenhouse gas (GHG) and CO_2 emissions, but on the other hand, they may still need to resort to unpopular decisions by abandoning fossil fuel, closing the coal mines and halting the construction of new coal-fired power plants. In such circumstances, there will be an inevitable collision between two priorities, namely the protection of the environment and the protection of job security for many of their citizens whose earnings depend on working in such coal-fired power plants and coal mines.

It goes without saying that such a transition may have unintended and immediate economic impacts on local workers, and many of whose livelihoods depend upon polluting industries such as fossil fuels. These pressures can compound other vulnerabilities facing workers and communities, including environmental degradation and social problems arising from the pursuit of unsustainable economic growth.

In Indonesia alone, it is reported that coal miners will be at risk of losing their jobs as the nation shifts towards clean energy, which involves the early closure of coal mines. A recent Global Energy Monitor report estimates that around 30,000 workers in the coal sector could face potential layoffs from 2020 to 2040 (Baskoro 2023). A key player in the coal mine business is reported to have said: "Coal mine closures are inevitable, but economic hardship and social strife for workers is not" (Baskoro 2023).

In fact, the impacts of the closure of coal mines or coal plants do not only hit the mine workers. According to the World Bank, such a closure may trigger domino effects as well, whereby it may cause tremendous effects to workers along the coal value chain. It may equally hurt local economies reliant on mine workers' earnings and fragment community well-being and social capital. In addition to that, it may also squeeze public finances (World Bank Group 2022a).

And worse still, in some communities, closing mines can create a persistent, destabilising demand shock as displaced workers struggle to transition to new jobs in which only a few alternatives would be available. And workers may be unwilling to accept lower-paying options or move to regions with greater labour demand (World Bank Group 2022a).

In a comprehensive and detailed report prepared by the International Labour Organization (ILO) entitled "A Just Transition in Southeast Asia: The Impact of Coal Phase-out on Job", the report, inter alia, reveals that "in Indonesia, the Philippines, and Viet Nam, job losses due to mine closures in coal-dependent regions and the loss of indirect jobs in their industrial ecologies will have a profound negative effect on the labour markets, economies, and livelihoods of local communities" (ILO 2022). The same report, therefore, recommends that concerted efforts ought to be made to duly mitigate or reduce the devastating repercussions of coal phase-out on affected populations via just and equitable policies from those three affected countries (ILO 2022).

12.8 Conclusion

Though ASEAN seems to have unanimously shared a regional understanding that the transition to renewable energy has to be translated into reality within this decade, any attempt to translate such a shared vision into reality still remains unclear and blurred. Yes, many ASEAN states may have, on paper, openly proclaimed that they had duly devised a whole raft of strategies in their endeavour of shifting from heavy reliance on fossil fuel to reliance on renewable energy, yet how they are going to implement such strategies still remains in the dark. Worse, some of such strategies, with due respect, have even become a laughing stock.

To be fair, in reality, ASEAN is in fact facing a herculean task in embarking on a low-carbon transition, i.e. by shifting—in a gradual and piecemeal approach—from fossil fuel to low-carbon energy. One of such enormous challenges is the ongoing dilemma of picking and choosing between two major priorities: either to succumb to international pressures by completely moving to renewable energy thus disregarding the people's plight or to maintain the *status quo* by completely ignoring such international pressures. For whatever prejudices one may harbour against ASEAN, this regional association, in my view, deserves a modicum of respect for still believing that, at the end of the day, pursuing energy justice still boils down to safeguarding and protecting the people's welfare and interest.

References

Admin (2023) Thailand electrical infrastructure 2023. Tractus Asia, 6 June. https://tractus-asia.com/blog/thailands-electrical-infrastructure-status-2023-tractus/

Al-Saffar A (2017) Bringing electricity to all corners of Southeast Asia—analysis, IEA. IEA. Available at https://www.iea.org/commentaries/bringing-electricity-to-all-corners-of-southeast-asia

Baskoro FM (2023) 30.000 coal workers could lose jobs as Indonesia shifts to clean energy. Jakarta Globe, 17 Oct 2023. https://jakartaglobe.id/business/30000-coal-workers-could-lose-jobs-as-indonesia-shifts-to-clean-energy

Bhullar L (2023a) Meanings of energy justice in the low-carbon transition. In: De Gruyter eBooks, 65–80. https://doi.org/10.1515/9783110752403-011

Bhullar L (2023b) Meanings of energy justice in the low-carbon transition. In: Bellantuono G, Godden L, Mostert H, Wiseman H, Zhang H (eds) Handbook of energy law in the low-carbon transition, 65–80. De Gruyter, Berlin, Boston. https://doi.org/10.1515/9783110752403-011

Blunden A (no date) Nancy Fraser's status model of recognition by Andy Blunden. https://ethicalpolitics.org/ablunden/works/recognition-status.htm#: ~ :text=To%20redress%20this%20injustice%20still,on%20a%20par%20with%20the

Bozalek V, Hölscher D, Zembylas M (2020) Nancy Fraser and participatory parity. https://doi.org/10.4324/9780429055355

Burma-Energy (2023). https://www.trade.gov/country-commercial-guides/burma-energy

Citraningrum M, Tumiwa F (2023) ASEAN can be a power in solar energy with teamwork. https://asia.nikkei.com/Opinion/ASEAN-can-be-a-power-in-solar-energy-with-teamwork

Dahl AJ (1982) Regional organization and order in South East Asia. Macmillan Press Ltd., London

De Langre G (no date) Electrification models and inclusive development. https://www.cairn.info/revue-l-europe-en-formation-2015-4-page-183.htm

Dong K et al (2023) Revisiting energy justice: Is renewable energy technology innovation a tool for realizing a just energy system? Energy Policy 183:113820. https://doi.org/10.1016/j.enpol.2023.113820

Expanded Rural Electrification|Department of Energy Philippines (no date). https://www.doe.gov.ph/expanded-rural-electrification

Gambhir A, Green F, Pearson PJG (2018) Towards a just and equitable low-carbon energy transition. Grantham Institute briefing paper no 26 (Aug 2018). Imperial College London. https://www.imperial.ac.uk/grantham/publications/briefing-papers/towards-a-just-and-equitable-low-carbon-energy-transition.php

Honneth A (2004) Recognition and justice. Acta Sociologica 47(4):351–364. https://doi.org/10.1177/0001699304048668

IEA (2017) WEO-2017 special report: energy access outlook—analysis—IEA. IEA (n.d.). https://www.iea.org/reports/energy-access-outlook-2017

IGES (2023) Just and Sustainable transitions for a net-zero Asia: emerging issues and solutions. IGES [Preprint]. https://doi.org/10.57405/iges-12716

Indonesia: electrification rate 2022|Statista (2023). https://www.statista.com/statistics/865133/indonesia-electrification-rate/

International Labour Organisation (2022) A just energy transition in Southeast Asia: the impact of coal phase-out on jobs, 1 May 2022. https://www.ilo.org/asia/publications/WCMS_845700/lang–en/index.htm

Indonesia Investments (2024) Perusahaan Listrik Negara—PLN|Indonesia investments. https://www.indonesia-investments.com/business/indonesian-companies/perusahaan-listrik-negara-pln-soe/item409.

Jenkins K et al (2016a) Energy justice: a conceptual review. Energy Res Soc Sci 11:174–182. https://doi.org/10.1016/j.erss.2015.10.004

Jenkins K et al (2016b) Energy justice: a conceptual review. Energy Res Soc Sci 11:174–182. https://doi.org/10.1016/j.erss.2015.10.004

Keling MF et al (2011) The development of ASEAN from strategic approach. Asian Soc Sci 7(7). https://doi.org/10.5539/ass.v7n7p169

Mariam S (2018) Human rights the "ASEAN way": exploring the possibilities for a regional ADR and adjudicative body in Southeast Asia. https://digitalcommons.lmu.edu/cgi/viewcontent.cgi?article=3048&context=llr

Nathan KS (1988) Malaysia: reinventing the nation. In: Alagappa M (ed) Asian security practice: material and ideational influences. Stanford University Press, New York

O'Brien T (2013) Awakening to recovery and Honneth's theory of recognition. ResearchGate [Preprint]. https://www.researchgate.net/publication/320211657_Awakening_to_Recovery_and_Honneth's_Theory_of_Recognition

Onecms (2023) What is energy justice? University of Sussex. University of Sussex. Study online, 18 July 2023. https://study-online.sussex.ac.uk/news-and-events/what-is-energy-justice/

Panti LT, GMA Integrated News (2023) DOE: Households with no electricity increased from 800K to 1.8M in five years. GMA News Online, 29 Sept. https://www.gmanetwork.com/news/topstories/nation/883667/doe-households-with-no-electricity-increased-from-800-000-to-1-8-million-in-five-years/story/

Patmanathan M (1980) Reading in Malaysia foreign policy. University Malaya Press, Kuala Lumpur

Phoumin H, Kimura F, Arima J (2021) ASEAN's energy transition towards cleaner energy system: energy modelling scenarios and policy implications. Sustainability 13(5):2819. https://doi.org/10.3390/su13052819

Pollard VK (1970) ASA and ASEAN, 1961–1967: Southeast Asian regionalism. Asian Surv 10(3):244–255. https://doi.org/10.2307/2642577

Prehoda E (2020) Energy justice and U.S. energy policy: case study applications exploring U.S. energy policy through an energy justice framework. https://doi.org/10.37099/mtu.dc.etdr/243

Rufaidah R (2023) Investment in clean energy in ASEAN countries: prospects and challenges—ASEAN Centre for Energy. ASEAN Centre for Energy, 30 Nov 2023

Saravanamuttu J (1983) The dilemma of independence two decades of Malaysia's foreign policy 1957–1977. Penerbitan Universiti Sains Malaysia, Penang

Star Online (2023) More than 130 Orang Asli villages still have no electricity. The Star, 2 Nov 2023. https://www.thestar.com.my/news/nation/2023/11/02/more-than-130-orang-asli-villages-still-have-no-electricity

Setyowati AB (2021) Mitigating inequality with emissions? Exploring energy justice and financing transitions to low carbon energy in Indonesia. Energy Res Soc Sci 71:101817. https://doi.org/10.1016/j.erss.2020.101817

Statista (2023) Households with electricity access in urban and rural Vietnam 2010–2022. https://www.statista.com/statistics/1326662/vietnam-households-with-electricity-access-by-urbanization-level/

Statista (2024) Household electrification rate Philippines 2022, by region. https://www.statista.com/statistics/1446585/philippines-household-electrification-rate-by-region/

Summary of Inflation Reduction Act provisions related to renewable energy|US EPA (2023). https://www.epa.gov/green-power-markets/summary-inflation-reduction-act-provisions-related-renewable-energy

Trading Economics (no date) Thailand—access to electricity (% of population)—2024 data 2025 forecast 1990–2021 historical. https://tradingeconomics.com/thailand/access-to-electricity-percent-of-population-wb-data.html

Turnbull CM (1999) Regionalism and nationalism. In: Tarling N (ed) The Cambridge history of Southeast Asia. Cambridge University Press, Cambridge

Van Uffelen N (2022) Revisiting recognition in energy justice. Energy Res Soc Sci 92:102764. https://doi.org/10.1016/j.erss.2022.102764

VnExpress (2023) Vietnam electricity and the conundrum of establishing a competitive market. VnExpress International—Latest News, Business, Travel and Analysis From Vietnam, 19 Sept. https://e.vnexpress.net/news/business/economy/vietnam-electricity-and-the-conundrum-of-establishing-a-competitive-market-4652366.html

World Bank Group (2022a) Global perspective on coal jobs and managing labor transition out of coal. World Bank, 5 Apr 2022. https://www.worldbank.org/en/topic/extractiveindustries/publication/global-perspective-on-coal-jobs-and-managing-labor-transition-out-of-coal

World Bank Group (2020b) Myanmar: power system efficiency project brings country closer to universal electricity access. World Bank, 30 May. https://www.worldbank.org/en/news/press-release/2020/05/29/myanmar-power-system-efficiency-project-brings-country-closer-to-universal-electricity-access

Yenneti K, Day R (2016) Distributional justice in solar energy implementation in India: the case of CharankaSolar park. J Rural Stud 46:35–46. https://doi.org/10.1016/j.jrurstud.2016.05.009

Yuxuan, Renaldi (2022) Coal fever in Indonesia. https://earthjournalism.net/stories/coal-fever-in-indonesia

Mohamed Hanipa Maidin is a former Deputy Minister of Law and ex-member of Parliament for two terms. In 1994, he was called to the Malaysian Bar to be duly licensed as an advocate and solicitor. In 1997, he set up his own legal firm under the name of Mohamed Hanipa and Associates. He was an active litigation lawyer for almost 25 years in practice and being a lawyer, he had appeared in various public interest litigations in Malaysian courts including the Federal Court. Several of his cases were reported in media as well as in various legal journals in Malaysia. He obtained his Bachelor of Laws [LLB (Hons)] and Bachelor of Laws [LLB (Shariah)] from the International Islamic University Malaysia (IIUM). He is also a passionate writer and many of his articles have been published in media in Malaysia and abroad.

Efficacy of Citizen Suits Against Energy Polluters in the USA

13

Julio C. Colón

Abstract

Energy production is all but a necessity in the modern world. The energy needs of the USA are largely met by coal-fired plants. These operations generate significant amounts of waste. In attempting to resolve environmental harms from these enterprises, citizens in the USA are able to bring the matter before the courts, but how effective are such lawsuits? Dozens of cases have come before the courts, and commentators publish on the state of the law regarding environmental statutes and their interpretation. This paper specifically describes three major statutes—the Clean Air Act, the Clean Water Act, and the Resource Conservation and Recovery Act—and studies how citizen suits fared in the courts. Ultimately, it concludes that the statutes' provisions, which preserve the right to common law remedies available to citizens against environmental polluters, may be the most effective part of each statute in protecting the environment.

Keywords

Coal · Coal ash · Pollution · Water pollution · Air pollution · Ground pollution · Environmental litigation · EPA

J. C. Colón (✉)
Killeen, TX, USA
e-mail: juliocolon@utexas.edu

13.1 Introduction

Most of the energy in the USA is produced from fossil fuels, and many power plants are fired by coal (US Energy Information Administration 2023). In 2022, 19.5% of all power plants in the USA were fueled with coal, second only to natural gas fuel plants (US Energy Information Administration 2023). However, coal is not the cleanest fuel, and its use produces chemicals that harm human health, such as sulphur dioxide gas, which causes respiratory illnesses, and solid wastes, namely fly ash, which can be suspended in the air as a particulate, or find its way into groundwater used for drinking (US Energy Information Administration 2023).

Given the potential health and environmental harms of pollution from coal-fired power plants, the considerable amount of environmental litigation which challenges the operation of coal-fired power plants is unsurprising. Gerrard (2009) indicated that in 2009, half of all judicial or administrative cases seeking to reduce greenhouse gasses concerned such enterprises. There is no indication that such litigation will slow down. Millions of tonnes of coal ash are generated every year from electricity production in the USA (Evans 2011:3). This massive accumulation of waste has inflicted major catastrophes on the US population. For example, in 2008, a dam retaining a large coal ash impoundment in the state of Tennessee broke and released over 1 billion

A. Trakic et al. (eds.), *Energy Justice*, Sustainable Development Goals Series,
https://doi.org/10.1007/978-981-97-6059-6_13

gallons of coal ash sludge into nearby rivers, destroying homes, vehicles, trees, and wildlife in the process (Korb 2011:1183–1184). The residents affected by that disaster sued the Tennessee Valley Authority, the company responsible for the operation, alleging negligence, trespass, and other claims [Korb 2011:1184 (citing '*Crichton v. Tenn. Valley Auth.*', 2010)].

Even in less tragic cases, private citizens are pursuing environmental justice in the courts. As individuals, communities, and non-profit organisations, environmental advocates are suing energy polluters under common law theories and various federal statutes, by which courts have jurisdiction to hear certain claims regarding the operations of energy producers under governmental license. The question explored in this paper is the efficacy of those citizen suits against energy producers regarding environmental and health-related claims. Along these lines, the ability of citizens to directly affect the operations of energy producers using the courts goes directly towards the matter of whether the procedures in place for the redress of public concerns are truly just.

13.2 Air, Land, and Sea: Litigating Against Energy Polluters

After the electrification of the USA, concerns about environmental consequences quickly followed. While the benefits were shared by most, the consequences were visited on communities closest to the pollution. So community members allied, forming coalitions of their own as well as allying with non-profit organisations formed for the purpose of preserving the environment.

A framework for legal battles already existed. This framework is derived principally from three federal acts which, among other purposes, serve as a basic guideline for enterprises whose businesses discharge or emit pollutants into the environment; they are the Clean Air Act (42 USC § 7401 et seq.), the Clean Water Act (33 USC § 1251 et seq.), and the Resource Conservation and Recovery Act (42 USC § 6901 et seq.). Enforcement of the acts is typically the responsibility of the US Environmental Protection Agency (EPA) and, at times, other federal agencies (42 USC § 7603 Emergency powers; 33 USC §1319 Enforcement; and 42 USC § 6928 Federal enforcement). However, each of these statutes provides citizens with the legal means to file suit in federal court against alleged polluters (42 USC § 7604 Citizen suits; 33 USC § 1365 Citizen suits; and 42 USC § 6972 Citizen suits). These three statutes and the case law stemming from citizen suits employing those laws are at the heart of this chapter. The environmental standards, jurisdictional provisions, and the case law produced during the last few decades will be described in order to make determinations regarding the potency of legal advocacy for deterring environmental polluters in a modern world where electrification is all but an absolute necessity.

But beyond federal law, there still exists a nonstatutory option for environmental advocates making claims against energy polluters, and that is the common law. The USA is a common law jurisdiction, and hence, several common law torts are effective in this class of lawsuits. This section will also consider how torts such as nuisance, trespass, and other causes of action fare in the courts.

Each of the following subsections describes the applicable law, then proceeds to look at the relevant provisions, and analyses relevant case law with respect to citizen suits. Because the majority of electricity in the USA is generated from coal-fired plants, the analysis will focus on those cases which deal with coal fly ash as a pollutant, in air, land, and water.

13.3 The Clean Air Act

The Clean Air Act (CAA) is a federal statute that seeks to protect public health and regulate emissions by means of regulated National Ambient Air Quality Standards (NAAQS) (42 USC § 7409). These NAAQS are set by the EPA and are published in the Code of Federal Regulations (40 CFR Part 50 for current National Primary and Secondary Ambient Air Quality Standards). Under these standards, a certain amount of pollution is permitted. For example, for particulate matter, according to 40 CFR § 50.7, the NAAQS for 'particles with an aerodynamic diameter less than or equal to a nominal 2.5 μm' are within acceptable standards when 'the annual arithmetic mean concentration, as determined in accordance with appendix N of this part, is less than or equal to 15.0 μg per cubic metre'. Keeping in mind relevant NAAQS, the EPA and authorised state agencies can issue permits to steam-generating electric plants and other stationary sources of air pollutants (42 USC § 7651 g; 42 USC § 7661a).

Of course, the above standards come with powers that permit the EPA to enforce them, such as 42 USC § 7603 permitting the EPA to bring suit in federal court in order to enjoin a person from generating pollution which presents 'an imminent and substantial endangerment to public health or welfare, or the environment [...]'. Similarly, the CAA empowers citizens to enforce the law in two ways.

Firstly, if the EPA does not object to a permit issued by a state permitting agency, then any person may raise concerns during the public comment phase of the permitting process, and such person may request that the EPA object to the issuance of the permit if after its own review, it declines to object to the issuance of the permit (42 USC § 7661d(b)(2)). Secondly, 42 USC § 7604(a) provides that any person can file a civil action in federal district court against (1) a person or governmental body (including the US government) if they are alleged to have violated an emission standard under the statute or promulgated by the EPA or state agency; (2) or against the Administrator [of the EPA] for failure to perform nondiscretionary duties; or (3) or against those who begin certain major constructions without the necessary permits or who are in violation of said permits. Citizens acting under this provision are required to give notice at least 60 days before commencing suit (42 USC § 7604(b)). When the court determines it is appropriate, it may award litigation costs, including attorney and expert fees, upon issuing a final order (42 USC § 7604(d)). A similar provision, 42 USC § 7607, provides, 'A petition for review of action of the [EPA] Administrator in promulgating [NAAQS ...] may be filed only in the United States Court of Appeals for the appropriate circuit [...]'.

With regard to the citizen suit provisions, it is necessary that the party seeking fees prevail in some respect. For example, in *Ruckelshaus v. Sierra Club* (1983), respondents, the Sierra Club and Environmental Defense Fund, sued the EPA, petitioner, under 42 USC § 7607(b), alleging that the EPA's private industry contacts tainted its rulemaking ability—a claim that the appellate court ultimately rejected. Notwithstanding its failure on the merits of the case, respondents filed for and were awarded approximately $90,000 between them (*Ruckelshaus v. Sierra Club* 1983:682). The US Supreme Court overturned this decision, citing, among other principles, the 'American Rule', which provides that 'the prevailing litigant is ordinarily not entitled to collect a reasonable attorneys' fee from the loser' (*Ruckelshaus v. Sierra Club* 1983: 684).

Case law regarding 42 USC § 7604 for violating emissions standards, permits, or derogating mandatory tasks is sparse but growing. Importantly, it is already the rule in some jurisdictions that a 42 USC § 7604 suit attacking the validity of a permit under the CAA is akin to a collateral attack on a validity-issued permit, thus effectively ruling that construction projects with permits must be allowed to become operational and only sued under § 7604 if it can be alleged that it is then violating the terms of the permit or an emission standard. For example, in *CleanCOALition v. TXU POWER* (2008:479), 'Appellants interpret the phrase "without a permit" to mean "without a permit that complies with the CAA." However, we decline to rewrite the plain language of the statute'.

More interesting case law exists, perhaps with regard to the CAA on how courts interpret 42 USC § 7604(e), which provides: 'Nothing in this section shall restrict any right which any person (or class of persons) may have under any statute or common law to seek enforcement of any emission standard or limitation or to seek any other relief (including relief against the Administrator or a State agency'.

There are three important principles that can be derived from the many appellate cases on the matter of common law actions which may compete with provisions of the CAA. The first is that § 7604(e) exists as a savings clause and therefore expresses an intent to preserve common law rights in state courts in the face of plausible preemption by the federally imposed CAA, preemption meaning that federal laws typically abrogate state laws where the two would be in conflict. For example, in *Bell v. Cheswick Generating Station* (2013), a group of plaintiffs filed a class action suit against Cheswick Generating Station, a coal-burning power plant, alleging common law trespass, nuisance, and negligence due to the buildup of fly ash and coal dust. The problem was so serious that plaintiffs alleged that the particulates made them 'prisoners in their [own] homes' (*Bell v. Cheswick Generating Station* 2013:192). The trial court dismissed the plaintiffs' case based on the reasoning that the CAA preempted state common law claims (*Bell v. Cheswick Generating Station* 2013:193). The appellate court reversed, reasoning that there was little distinction between the CAA savings clause and that of the Clean Water Act, which the US Supreme Court in *International Paper Co. v. Ouellette* (1987) had already stated does not preempt state law claims in the source state (*Bell v. Cheswick Generating Station* 2013:196–197).

The second important principle is that between interstate parties, the common law applied must be derived from the state, which is the source of the pollution. The *Bell v. Cheswick Generating Station* court acknowledged this principle as it rejected the argument that allowing common law state claims could result in an inconsistent standard across states, noting that the *International Paper Co. v. Ouellette* court already determined this very issue in its decision, writing, 'First, application of the source State's law does not disturb the balance among federal, source-state, and affected-state interests' (*Bell v. Cheswick Generating Station* 2013:197). It also cited *North Carolina v. Tennessee Valley Authority* (2010), which reversed an injunction of the trial court in favour of North Carolina against several power plants in neighbouring states (*Bell v. Cheswick Generating Station* 2013:196). The trial court, in that case, had applied the law of North Carolina rather than those of the source states, and this was one of the most convincing arguments in favour of the appellant power companies, as the appellate court stated, 'the injunction improperly applied home state law extraterritorially, in direct contradiction to the Supreme Court's decision in *International Paper Co. v. Ouellette,* 479 US 481, 107 S.Ct. 805, 93 L.Ed.2d 883 (1987) (*North Carolina v. Tennessee Valley Authority* 2010:296)'.

Lastly, the point of contention that an activity given license by the state, such as the issuance of a permit under a CAA-sanctioned state programme, does not necessarily mean that state common law is preempted by the state regulation which enabled permitting the activity, and the question itself turns on the law of the state wherein the pollution occurs. This point was certainly at issue in *North Carolina v. Tennessee Valley Authority* (2010:296), as it stated itself in its holding, 'it is difficult to understand how an activity expressly permitted and extensively regulated by both federal and state government could somehow constitute a public nuisance'. However, in making this observation, it is important to note the law of the source states in question at the time, which provided in Alabama that 'there can be no abatable nuisance for doing in a proper manner what is authorised by law', and in Tennessee that '[a]n activity that is explicitly licensed and allowed by Tennessee law cannot be a public nuisance' (*North Carolina v. Tennessee Valley Authority* 2010:309–310). Although, this is not always the case.

Regarding this principle, the court in *Freeman v. Grain Processing Corp.* explained that an activity explicitly sanctioned by state law does not necessarily preempt common law claims. In that case, residents living near a corn milling operation made claims of nuisance, trespass, and negligence against the mill due to the odours and fumes emitted from the facility (2014:63). The trial court dismissed their claims based on the argument that the common law claims were preempted by the CAA and implementing Iowa state statutes (*Freeman v. Grain Processing*

Corp. 2014:63). The Iowa Supreme Court reversed the decision with respect to common law being preempted by both the CAA and Iowa statutes, and sufficing to mention the court's reasoning as to the latter, it was noted that while no cases were on point as to whether the Iowa environmental statute in question preempted common law claims, the court had already 'made clear that a lawful business, properly conducted, may still be a nuisance [...]' (*Freeman v. Grain Processing Corp.* 2014:84, 88).

In close, it appears as though the permitting requirements of the CAA are protective of energy polluters in a sense because, although the process of obtaining permits is certainly resource and time-intensive, it is just part of the cost of the business of producing and selling energy. Also, the provision for civil suits in 42 USC § 7604 may be too scientifically based to be beneficial for the average citizen. Consider, for example, the information required to properly allege that the chemical composition of the emissions from a facility is violating NAAQS. By necessity, such an allegation also requires sophisticated measurement and testing. Hence, what can be assumed about the resources required to prove the case? On the other hand, a common law claim which only alleges the deprivation of use and enjoyment of private property is more straightforward to allege and to prove. Therefore, it is accurate to describe the pursuit of common law claims against polluters as a 'more immediately enforceable deterrent to uncontained emission of airborne particulates' (DeBellis 2014:616).

13.4 The Resource Conservation and Recovery Act

The Resource Conservation and Recovery Act (RCRA) empowers the EPA to oversee the cycle of solid waste from its creation to disposal for the purpose of protecting the health of both citizens and the environment (See 42 USC § 6901). Among other practices, Congress passed the act in order to prevent the dumping of solid waste, whether hazardous or nonhazardous (See 42 USC § 6907(a)(3). In order to do so, RCRA gives the EPA authority to promulgate guidelines for various solid waste management activities, including state solid waste management plans for coal combustion residuals (42 USC §§ 6945(d), 6946).

These guidelines make state participation in these programmes enforceable by making certain federal aid contingent on participation (42 USC § 6947). RCRA distinguishes between solid waste, in general, and 'hazardous waste' (42 USC § 6903). RCRA defines 'hazardous waste' as solid waste which can significantly contribute to mortality and illness, or pose a substantial hazard to human health or the environment (42 USC § 6903(5)).

RCRA enforcement mechanisms for violations of its provisions regarding hazardous waste include allowing the EPA to issue civil penalties and suspend permits, and as well as possible criminal penalties for more egregious violations (42 USC § 6928). For solid waste in general, the EPA may commence enforcement actions under § 6928 against coal combustion residuals units with permits to operate if the permit issuing state requests or the EPA finds that federal enforcement is necessary in order to bring the coal combustion residuals unit into compliance with the terms of the permit (42 USC § 6945(d)). Such permits are issued by states with the approval of the EPA, which ensures that they are in line with the guidelines published in 40 CFR § 257 (42 USC § 6945(d)).

Further enforcement measures include 42 USC § 6972, RCRA's provision for citizen suits, which allows individuals to bring civil actions in US district court when (1) a person (including the US government or other governmental agency) 'is alleged to be in violation of any permit, standard, regulation, condition, requirement, prohibition, or order [under RCRA]'; (2) a person (including the US government or other governmental agency) past or presently involved in transportation, generation, storage, treatment, or disposal of solid or hazardous waste 'which may present an imminent and substantial endangerment to health or the environment'; or (3) the Administrator (of the EPA) for failure to perform a nondiscretionary duty. When appropriate, the court may order litigation costs, including attorney and expert fees, to the 'prevailing or substantially prevailing party' (42 U.S. C. § 6972(e)).

Early on, citizen groups have utilised strategies alleging RCRA violations in order to advocate in favour of cleaner energy. Coal combustion residuals, namely fly ash, are not hazardous waste under the RCRA (see 42 USC § 6945(d) and 40 CFR § 257.50); thus, the more stringent testing and recordkeeping relating to hazardous waste as described in 42 USC § 6922 and 40 CFR § 262.11 were less frequent issues in cases brought by environmental advocates against energy polluters. However, an energy-producing enterprise similar to coal-fired plants was not immune to RCRA's hazardous waste requirements. In *Chicago v. Environmental Defense Fund* (1994), the US Supreme Court had occasion to decide whether the residuals from a municipal garbage incinerator which produced energy were excluded as hazardous waste under RCRA, namely 42 USC § 6921(i), which provides: '[…] the mass burning of municipal solid waste shall not be deemed to be treating, storing, disposing of, or otherwise managing hazardous wastes for the purposes of regulation under this subchapter, if […] such facility—(A) receives and burns only— (i) household waste (from single and multiple dwellings, hotels, motels, and other residential sources) [..]'.

After initially failing to prove its case at the district court, the Environmental Defense Fund prevailed on appeal, and the appellate court overturned the decision, stating, 'ash generated from the incinerators of municipal resource recovery facilities is subject to regulation as a hazardous waste under Subtitle C of RCRA' (*Chicago v. Environmental Defense Fund* 1994:331). At the time of appeal to the US Supreme Court, even the EPA itself had aligned with the position of the City of Chicago that municipal waste combustion was excluded as hazardous waste (*Chicago v. Environmental Defense Fund* 1994:331). Ultimately, the US Supreme Court upheld the decision of the appellate court because 42 USC § 6921(i) did not contain an exception for the end product – the ash itself (*Chicago v. Environmental Defense Fund* 1994:334–335). The Court makes the crucial observation that '[…] the waste the facility produces (as opposed to that which it receives) is not even mentioned' (*Chicago v. Environmental Defense Fund* 1994:334), and surprisingly it did not make the § 6921(i) phrase 'receives and burns' equal to 'receives and produces'. Somewhat influential in its reasoning was the omission of the term 'generating' from the § 6921(i) statement 'treating, storing, disposing of, or otherwise managing' even though 'hazardous waste generation' is separately defined under the statute (42 USC § 6903(6)).

One crucial distinction between the citizen suits provision of RCRA in comparison with the CAA and Clean Water Act is that it permits civil actions against polluters whenever the alleged actions 'may present an imminent and substantial endangerment to health or the environment' (42 USC § 6972(a)(1)(B)). For this reason, commentators consider RCRA a crucial statute in environmental advocacy because it permits suits whenever the harm continues, even if the conduct has long since ended (Wohlers 2022:337). And while it is a potent tool for environmental advocates, proper allegation is essential.

As noted in *Chicago v. Environmental Defense Fund*, courts will be unwilling to read even slight assumptions into the law. For example, in *Little v. Louisville Gas and Elec. Co.* (2014), plaintiffs complained of coal fly ash from electricity producer Louisville Gas and Electric Company. The pollution was such that property owners began to notice a thin film of dust covering their homes, and they filed suit in US district court alleging violations of the CAA, RCRA, and common law negligence, nuisance, and trespass (*Little v. Louisville Gas and Elec. Co.* 2014:796–797). Among its various claims, one of the plaintiffs' counts under RCRA alleged that 'Defendants' handling of coal combustion residuals 'may present an imminent and substantial endangerment to health or the environment', in violation of 42 USC § 6972(a)(1)(B) (*Little v. Louisville Gas and Elec. Co.* 2014:807). Defendants moved to dismiss this claim because it reflected a collateral attack on the facility's permits as it sought an injunction against fly ash from escaping the site, and even imminent and substantial endangerment to health or environment is not sufficient to enjoin activity authorised under its permit. Defendants also argued that it was better redressed under the CAA (*Little v. Louisville Gas and Elec. Co.*2014:813).

Regarding its first argument, indeed, RCRA plain language does prohibit citizen suits that seek to 'enjoin the issuance of a permit' (42 USC § 6972(b)(2)(D). However, plaintiffs argued that their suit was not seeking to attack the issuance of the permit but to bring the defendants' conduct in line with the terms of their permits (*Little v. Louisville Gas and Elec. Co.* 2014:813–814). Ultimately, the court was swayed by the evidence that the EPA-sanctioned state programme which issued the permits contained within its statute a provision for those objecting to the terms of a permit to file 'with the district a petition for a hearing', and because plaintiffs circumvented that process it was indeed essentially a collateral attack on permit decisions (*Little v. Louisville Gas and Elec. Co.* 2014:814).

Two points are important to note. Firstly, the court made the point that the 'imminent and substantial endangerment' case needed to be filed after the pursuit of the remedies within the state statute, which enables the permitting process. Secondly, while plaintiffs defended their claim by explaining that it was actually attempting to bring defendants in line with their permits, the proper provision for that was not 42 USC § 6972 (a)(1)(B) 'imminent and substantial endangerment', but rather § 6972(a)(1)(A), which specifically allows citizen suits for 'violation of any permit'. Lastly, it is important to note that the court drew an analogy to CAA case law and determined that RCRA's savings clause also does not preempt state common law claims against polluters (*Little v. Louisville Gas and Elec. Co.* 2014:817).

In summary, three ideas stand out from analysing the case law involving RCRA. The first is that the standards promulgated under the statute envision enforcement by environmental advocates as evidenced by the ability of citizens to not only bring suit due to violations of permits and standards but also due to pollution that presents an 'imminent and substantial endangerment' to human health and the environment, even when the actions leading to the pollution have ceased (42 USC § 6972(a). As stated in *Utility Solid Waste Activities Group v. EPA* 2018:424, 'Enforcement is left to states' own policy decisions and to the initiative of people bringing citizen suits to enforce the federal standards'. The second important point is that state solid waste plan statutes promulgated in correspondence with RCRA (42 USC §§ 6941-6949a) will be considered by trial courts when deciding if a claim under § 6972(a)(1)(B) truly presents an imminent and substantial threat of harm. Perhaps the reasoning is that the more expedient means of resolving a matter derived under the state solid waste plan would be used if the matter was truly imminent, but in any case, it is best to exhaust state-enacted measures before pursuing an RCRA claim in the US district court. Lastly, it is important to understand the RCRA savings clause in 42 USC § 6972(f), which reads: 'Nothing in this section shall restrict any right which any person (or class of persons) may have under any statute or common law to seek enforcement of any standard or requirement relating to the management of solid waste or hazardous waste, or to seek any other relief (including relief against the Administrator or a State agency)'.

It is likely that courts will read this provision in the same way as the CAA savings clause, and thus, plaintiffs will have common law claims available to them, which ought to be alleged along with any RCRA claim. However, the applicable law would be that of the source state, and hence, local jurisprudence regarding the preemption of common law claims by state statutes would apply.

13.5 The Clean Water Act

The Federal Water Pollution Control Act, better known as the Clean Water ACT (CWA), was created in order to regulate the discharge of pollution into US waters and to regulate their quality (see 33 USC § 1251 et seq.). Officially, Congress's goal was to eliminate polluting discharges into the nation's waters by the year 1985 (33 USC § 1251(a)(1)). Although the CWA contains a myriad of research and construction-related points, the heart of the statute for environmental advocates are the standards, enforcement, and permitting provisions (33 USC §§ 1281-1377a).

Among the important standard-making provisions are § 1311(f), prohibiting 'discharge any radiological, chemical, or biological warfare agent, any high-level radioactive waste, or any medical waste, into the navigable waters'; the § 1312(a) empowerment of the EPA to promulgate effluent limitations for the purpose of preserving water quality standards for the 'protection of public health, public water supplies, agricultural and industrial uses, and the protection and propagation of a balanced population of shellfish, fish and wildlife, and allow recreational activities in and on the water' (also 33 USC § 1312 Information and Guidelines); §1319 enforcement powers, which permits the EPA to pursue civil and criminal penalties for violations of certain provisions of the statute or permits issued under it; and finally, § 1341 and § 1342 which establish a certification and permitting system for the discharge of pollutants into navigable waters. Also relating to enforcement in extraordinary circumstances, the EPA may seek an injunction against polluters in the US district court when evidence shows that pollution presents an 'imminent and substantial endangerment to the health [or livelihood] of persons' (33 USC § 1364(a)).

In achieving its purpose, the CWA also empowers states to create their own permit plan to ensure compliance with CWA effluent discharge and water quality standards (33 USC § 1342). The statute provides that the federal permitting programme is suspended for states with EPA-approved programmes, and it ensures continued compliance from states by the EPA's ability to issue corrective measures to the state in order to bring its programme back in line with CWA requirements (33 USC § 1342(d)).

As for citizen advocacy, the CWA contains a similar provision as the statutes previously discussed. 33 USC § 1365(a) states that a citizen may file suit in US district court whenever a person or agency is alleged to be in violation of a standard under the statute or a state or federal order issued under the statute. The court may order attorney and expert fees related to the litigation whenever appropriate (33 USC § 1365 (d)). Also similar to the CAA and RCRA, other statutory and common law remedies are not preempted by the CWA (33 USC § 1365(e).

Three concepts or terms are recurring matters regarding CWA claims by ordinary citizens under § 1365. The first is the meaning of the term 'navigable waters', which is defined as 'the waters of the United States, including the territorial seas' (33 USC § 1362(7)). Because the power of the federal government to legislate in this area is derived from its authority over US waters, the rules regarding the definition of navigable waters are a point of contention in the case law. Second is the definition of 'point source', which 33 USC § 1362(14) defines as 'any discernible, confined and discrete conveyance, including but not limited to any pipe, ditch, channel, tunnel, conduit, well, discrete fissure, container, rolling stock, concentrated animal feeding operation, or vessel or other floating craft, from which pollutants are or may be discharged'. Because the CWA controls the discharge of pollutants into navigable waters from a point source, the precise definition of this term has also been litigated. And the third recurring issue was that of standing. The CWA provision for citizen suits is not completely unbridled, and § 1365(b) contains a notice requirement and a restriction on lawsuits when the EPA has already commenced action against the polluter. And, of course, there has been litigation as to the extent of the § 1365(e) reference to rights under 'common law to seek enforcement of any effluent standard or limitation or to seek any other relief'.

In enforcing the CWA, federal government agencies have taken an expansive definition of the phrase 'waters of the United States' (33 USC § 1362(7)). Some examples are the (1) Army

Corps of Engineers asserting jurisdiction over property whose vegetation required saturated soil and was near a lakeshore (*United States v. Riverside Bayview Homes* 1985); (2) requiring a permit to discharge materials over isolated ponds that formed over trenches because migratory birds were present (*Solid Waste Agency of Northern Cook County (SWANCC) v. US Army Corps of Engineers* 2001); and (3) irrigation canals that divert and drain water to various creeks (*Headwaters, Inc. v. Talent Irrigation Dist.* 2001).

In 2006, the US Supreme Court attempted to reconcile the various interpretations of what was to be considered 'waters of the United States'. In *Rapanos v. United States*, the Court was called upon to determine whether 54 acres of water-saturated land owned by John A. Rapanos were 'waters of the United States' as asserted by US regulators (2006). In that case, Rapanos wanted to develop the land but was held up by 12 years of litigation and permit decisions threatening civil and criminal penalties under the CWA, all for a parcel which was 11 to 20 miles away from the nearest body of navigable water (*Rapanos v. United States* 2006:720–721). Ultimately, the court decided in favour of Rapanos, but in a plurality decision which asserted two different rules. In the first opinion written by Justice Scalia and joined by three other justices, the reasoning was based on the proposed rule that 'the phrase "the waters of the United States" includes only those relatively permanent, standing or continuously flowing bodies of water "forming geographic features" that are described in ordinary parlance as "streams[,]... oceans, rivers, [and] lakes."' (*Rapanos v. United States* 2006:739). However, in his opinion concurring in judgement, Justice Kennedy reasoned that wetlands not adjacent to waters that are navigable can then only be considered navigable waters under the CWA if they 'possess a "significant nexus" to waters that are or were navigable in fact or that could reasonably be so made' (*Rapanos v. United States* 2006:759). Hence, because the Court could not decide by a majority as to why Rapanos won, the decision failed to clarify the matter for lower courts.

Following *Rapanos v. United States*, the EPA made various rules under its CWA administrative authority. The 2015 rule adopted by the EPA essentially adopted Justice Kennedy's opinion: 'waters of the United States means: [...] (7) All waters in paragraphs (a)(7)(i) through (v) of this section where they are determined, on a case-specific basis, to have a significant nexus to a water identified in paragraphs (a)(1) through (3) of this section' (80 Fed. Reg. 37,054, 2015:37,105).

Later, during the administration of President Trump, a new rule was adopted which broke waters down into two groups, jurisdictional waters and non-jurisdictional waters, with the latter group importantly excluding groundwater and ephemeral streams, among other forms of surface water (85 Fed. Reg. 22,250, 2020:22,338). During this period, it was suggested by some commenters that states take the initiative to protect waters within their borders (Wilkerson 2019:462).

After various legal challenges to the Trump era rule, in 2023, the Supreme Court would eventually reach a consensus on a definition of 'navigable waters' under the CWA in *Sackett v. EPA*. In that case, Sackett was prohibited from building on his land by the EPA (*Sackett v. EPA* 2023:1325). According to the EPA, the property contained wetlands covered by the CWA 'because they were near a ditch that fed into a creek, which fed into Priest Lake, a navigable, intrastate lake' (*Sackett v. EPA* 2023:1325). On appeal from trial court, the Ninth Circuit appellate court noted that it believed that Scalia's decision did not completely reject the 'significant nexus' theory of Kennedy's opinion, and in any case, Kennedy's understanding of significant nexus was already the standard of the Ninth Circuit in determining waters covered by the CWA (*Sackett v. U.S. Envtl. Prot. Agency* 2021:1088–1089). In doing so, it then reasoned that the EPA was reasonable in its decision because the wetlands were adjacent to a covered tributary which 'provide important ecological and water quality benefits [...] "especially important in maintaining the high quality of Priest Lake's water, fish, and wildlife."' ('*Sackett v. U.S. Envtl. Prot. Agency*', 2021, p. 1093). However, on reviewing the case the Supreme Court decided that the 'significant nexus' theory was no longer the proper standard, based on its interpretation of the correct influence of the EPA in comparison with the states' role, and the plain meaning of the phrase 'wetlands adjacent thereto' as inclusive of waters covered by the CWA under 33 USC § 1344(g)(1) ('*Sackett v. U.S. Envtl. Prot. Agency*', 2023, pp. 1338–1340). In doing so, it ultimately adopted the plurality opinion from *Rapanos v. United States* that the CWA covers only waters which 'as a practical matter indistinguishable from waters of the United States' (*Sackett v. U.S. Envtl. Prot. Agency* 2023:1327). This requires firstly that 'adjacent' waters such as wetlands border a 'relatively permanent body of water connected to traditional interstate navigable waters' and secondly that there be a 'continuous surface connection' making it difficult to determine where one water ends and the other begins (*Sackett v. U.S. Envtl. Prot. Agency* 2023:1341).

As noted from the previous sections, the major source of energy industry pollution, coal ash, is a threat to the water beneath us as the surface impoundments used to dispose of the ash leach chemicals beneath the surface and into groundwater used for, posing serious health risks to citizens. The *Sackett v. U.S. Envtl. Prot. Agency* decision nearly makes the CWA completely ineffective for cases which address this issue because groundwater cannot meet the definition of 'waters of the United States' (33 USC § 1362(7). Other litigation strategies against energy polluters employing the CWA require understanding the term 'point source'. According to the CWA, any 'discharges of pollutants from a point source [... that] would interfere with the attainment or maintenance of that water quality [...]' are unlawful except with a permit from the EPA (33 USC §§ 1311(a), 1312(a)). According to 33 USC § 1362(14), a 'point source' is defined as: '[A]ny discernible, confined and discrete conveyance, including but not limited to any pipe, ditch, channel, tunnel, conduit, well, discrete fissure, container, rolling stock, concentrated animal feeding operation, or vessel or other floating craft, from which pollutants are or may be discharged'.

Under § 1362(12), 'discharge of pollutants' is also clearly defined, meaning 'any addition of any pollutant to navigable waters from any point source'. This is important because if pollutants were not discharged from a 'point source', then the CWA citizen suits provision would not apply as there would not be an alleged 'effluent standard or limitation' violation under the CWA (33 USC § 1365(a)).

Courts wrestled with the proper definition of 'point source', and the law was in contradiction. For example, in *Yadkin Riverkeeper v. Duke Energy Carolinas*, citizens attempted to enforce provisions of the CWA against an energy producer, alleging that the coal ash residuals deposited in nearby lagoons violated its permits and polluted navigable waters via groundwater seepage (2015:437). Among other arguments, the energy company claimed that the coal ash lagoons and groundwater were not point sources (*Yadkin Riverkeeper v. Duke Energy Carolinas* 2015:443). The trial court disagreed, reasoning that (1) because the coal ash lagoons were designed in such a way as to impound the coal ash, they were confined and discrete and (2) because the lagoons were leaking pollutants, which were allegedly arriving at navigable waters, that was a form of conveyance, hence making them a 'point source' (*Yadkin Riverkeeper v. Duke Energy Carolinas* 2015:444). In *Upstate Forever v. Kinder Morgan Energy Partners*, the appellate court made a similar conclusion regarding pollutants seeping through groundwater into navigable waters, noting that there does not have to be a direct conveyance between the point source and the navigable waters, and thus diffusion through groundwater did not rule out an ongoing violation (2018:648).

However, other cases were not as amenable to CWA citizen suits, refusing to budge in order to read an understanding into the statute that was not explicitly written. In *Kentucky Waterways Alliance v. Kentucky Utilities Company*, the plaintiff alleged that coal ash produced by energy producer Kentucky Utilities Company seeped through the bottom of man-made ponds through the groundwater and contaminated a lake (2018:928). Plaintiff alleged that both the groundwater was a point source, and the pond was a point source, with groundwater being hydrologically connected to the navigable waters (*Kentucky Waterways Alliance v. Kentucky Utilities Company* 2018: 932–933). The appellate court rejected this reasoning on the basis that 'point source' is itself defined as a conveyance, and the facts did not show that the means by which the pollutants travelled to the lake were in form 'discernible', 'confined', or 'discrete' (*Kentucky Waterways Alliance v. Kentucky Utilities Company* 2018: 933–934). The Fourth Circuit appellate court made similar conclusions in *Sierra Club v. Virginia Elec. & Power Co.*, ruling that the power company's coal ash landfills and settling ponds were not 'point sources' because even though 'arsenic from the coal ash stored on Dominion's site was found to have reached navigable waters', the landfills and ponds were not created to convey anything, and what did reach navigable waters was generalised and diffuse, rather than discrete or channelled (2018:410–412).

Finally, in 2020, the Supreme Court addressed this issue in *County of Maui, Hawaii v. Hawaii Wildlife Fund*. In that case, wastewater treated by county facilities was treated and discharged into wells, where it eventually made its way into the Pacific Ocean through groundwater (2020:1466). Hawaii Wildlife Fund brought a citizen suit under the CWA, claiming that this was a discharge of pollutants requiring a permit (*County of Maui, Hawaii v. Hawaii Wildlife Fund* 2020:1466). Maui argued that the statute created a rule leaving no room for interpretation, that '[a] point source or series of point sources must be "the *means of delivering* pollutants to navigable waters."' (*County of Maui, Hawaii v. Hawaii Wildlife Fund* 2020:1470). However, the Court disagreed and ruled in favour of Hawaii Wildlife Fund, noting that while the statute does not presume to give the EPA general authority over groundwater, the notion that a pollutant's travel through any amount of groundwater nullifies its emancipation from a 'point source' was too restrictive an interpretation (*County of Maui, Hawaii v. Hawaii Wildlife Fund* 2020:1472–1474). Rather, it determined that discharges of pollutants reaching navigable waters would require permits under the CWA if it amounted to the 'functional equivalent' of a direct discharge from the point source into navigable waters, based on factors such as: '(1) transit time, (2) distance travelled, (3) the nature of the material through which the pollutant travels, (4) the extent to which the pollutant is diluted or chemically changed as it travels, (5) the amount of pollutant entering the navigable waters relative to the amount of the pollutant that leaves the point source, (6) the manner by or area in which the pollutant enters the navigable waters, (7) the degree to which the pollution (at that point) has maintained its specific identity. Time and distance will be the most important factors in most cases, but not necessarily every case' (*County of Maui, Hawaii v. Hawaii Wildlife Fund* 2020:1476–1477).

Accordingly, the discrete conveyance requirement of a 'point source' is contextualised in time and distance, and consequently, pollution that reaches navigable waters within a relatively modest time and distance argues in favour of the pollutants being discharged from a point source. Finally, there is a limitation related to the CWA and common law actions. In *International Paper Co. v. Ouellette*, the Supreme Court held that the CWA's saving clause preserves common law claims such as nuisance inasmuch as the law applied is that of the state from where the discharge of pollutants originates (1987:497–499).

In summary, recent cases have clarified important aspects of the law regarding environmental advocacy by means of CWA citizen suits. As it stands, wetlands are neither completely covered nor excluded by the law, and groundwater is completely excluded. However, any time that pollutants find their way into navigable waters such as interstate rivers, lakes, and oceans, the specific facts regarding the journey of the pollutants must be analysed to determine whether advocates can argue that there is a functional equivalent of a direct discharge into navigable waters from a point source. If so, then the area between the point source and navigable waters, be it land or another medium such as groundwater, may receive protection indirectly as a product of any injunction over the discharge into navigable waters. And finally, it is important to allege common law claims that may apply to the pollution, such as nuisance and trespass, as a strategy to seek maximum relief in courts.

13.6 Conclusion

Energy production is a massive endeavour requiring high amounts of capital for infrastructure, fuel, and human resources, as well as large investments in acquiring permits and keeping the operation legal. There will always be pollution, and because of our need for electricity, there must be a compromise regarding the consequences. However, justice requires that such consequences be distributed fairly, rather than cast entirely on a future generation or another neighbourhood. Fortunately, citizens and environmental advocates do have some recourse against energy polluters in courts.

At the federal level, the CAA, RCRA, and CWA include provisions which permit citizen suits against energy polluters for violations of their permits, or for violations of the standards within the statutes themselves. Each statute also includes a provision that allows citizen suits against the regulator for not enforcing standards.

Environmental advocates are still limited by the statutes in what can actually be challenged. Importantly, there can be no collateral attacks against the issuance of a permit. Therefore, advocacy must begin at the permitting stage. In these instances, state law is used because it may contain procedures for challenging the issuance of permits, which must be exhausted before seeking recourse under the citizen suit provisions of the federal statutes. After the issuance of a permit and the exhaustion of challenges during the public hearing process and potential state remedies, possible challenges remain. In particular, advocates may challenge the energy producer's compliance with the terms of the permit in regard to monitoring, discharge amounts or processes, or other specifics.

Of the three statutes discussed in this chapter, the citizen suit provision of RCRA arguably presents the most advantage for plaintiffs because it permits citizens to make claims under the act for violations which present imminent and substantial endangerment to health or the environment, while the CAA and CWA reserve that kind of action to the EPA itself. Of course, the success of such a claim relies on the proof the plaintiff is able to make available, as is the case with other statutes. This is one of the drawbacks of the citizen suit provision—which is that a level of sophistication outside of the reach of most citizens is required to make allegations under the federal statutes. After all, compliance with the standards promulgated by the EPA and contained within government-issued permits requires specialised knowledge to properly analyse. Hence, there is reason to question whether the law can perform according to its potential.

Notwithstanding these drawbacks, the savings clause in each of the three statutes permits the plaintiffs to make common law claims against energy polluters even when their activity is under a permit issued by federal law. Courts have interpreted this clause in each case to mean that the federal statute will not preempt state common law claims, and therein lies the best hopes for procedural justice with regard to the need of those affected by pollution resulting from energy production. Because the manner of energy production may change in ways that could make the CAA, RCRA, and CWA less effective for environmental advocates, the ability to make common law claims permits us to hope that there will continue to be access to justice in the courts as the future arrives.

References

DeBellis EA (2014) 'Bell v. Cheswick' and what the clean air act's savings clause means for fenceline communities. Ecol Law Q 41(2):613–620. http://www.jstor.org/stable/44321217

Evans L et al (2011) State of failure: how states fail to protect ourhealth and drinking water from toxic coal ash, 3. Available at https://earthjustice.org/wp-content/

uploads/stateoffailure_2013-04-05.pdf. Accessed 18 Dec 2023
Gerrard MB (2009) Coal-fired powerplants dominate climate change litigation. New York Law J 242(61)
Korb B (2011) Holding our breath: waiting for the federal government to recognize coal ash as a hazardous waste. J Marshall l Rev 45:1177
US Energy Information Administration (2023) Electricity explained: electricity in the United States. Available at https://www.eia.gov/energyexplained/electricity/electricity-in-the-us.php. Accessed 18 Dec 2023
Wilkerson T (2019) Sierra Club v. Virginia Electric & Power Co.: how a clean water act misinterpretation may open the floodgates to future groundwater polluters. Vt J Envtl l 21(3):442–466
Wohlers K (2022) The particle problem: using RCRA citizen suits to fill gaps in the clean air act. Mich Law Rev 121(2):325–352

Julio C. Colón began his legal career at the University of Texas School of Law where he was conferred a Juris Doctor. During law school, he was active in the international law society and published an article in the Texas International Law Journal, which was well-received by scholars internationally. Mr. Colón's first work in environmental law was as a legal intern for the Texas Commission on Environmental Quality. Since then, he has worked as a public servant and professor. He is currently a solo practitioner attorney and continues to work in the area of environmental law, assisting non-profit organisations working for the protection of wildlife and wildlife habitat.

14 Examining the Paradoxical Nexus: Energy Justice Amidst Australia's Growing Fossil Fuel Exports and Its Implications for Climate Justice

Evgeny Guglyuvatyy

Abstract

Australia's vast renewable energy potential positions it well for a clean energy revolution. Historically, however, the government has prioritised fossil fuel industries, thereby impeding the development of effective climate change policies and emission reduction efforts. Recent policy initiatives to increase renewable energy production, promote the transition to clean energy, and advance energy justice. Nevertheless, Australia's continued export of fossil fuels raises substantial concerns regarding climate justice, particularly for the vulnerable Pacific Island nations. This chapter critically examines the inconsistencies in Australia's climate policies and fossil fuel export strategies, assessing their implications through the frameworks of climate and energy justice. It highlights the detrimental impacts on Pacific Island countries, while also considering the perspective that Australia's fossil fuel exports contribute to energy access in the Asia–Pacific region. The discussion highlights the intricate interplay between economic interests, energy justice, and climate considerations, providing insights into the multi-faceted challenges of pursuing sustainable development.

Keywords

Climate justice · Energy justice · Australia · Pacific Island nations

14.1 Introduction

The Intergovernmental Panel on Climate Change (IPCC) has emphasised the imperative of achieving net-zero greenhouse gas emissions by 2050 to mitigate the average global temperature increase (IPCC 2019). Accordingly, signatories to the Paris Agreement are aiming to achieve this objective, a commitment echoed by numerous nations globally. Australia, having ratified the Paris Agreement, acknowledges the need to restrict global temperature rise to 2 °C, recognising that achieving net-zero emissions by 2050 is an essential precondition for this objective (Wood et al. 2021).

Australia is a country with an abundance of renewable energy sources with a substantial potential to both instigate and benefit from a 'clean energy revolution' (McKenzie 2021). Historically, Australian governments have prioritised the interests of fossil fuel industries and have been reluctant to implement effective emission reduction mechanisms (Guglyuvatyy

E. Guglyuvatyy (✉)
Department of Business Law and Taxation, School of Business, Monash University, Kuala Lumpur, Malaysia
e-mail: evgeny.guglyuvatyy@monash.edu

A. Trakic et al. (eds.), *Energy Justice*, Sustainable Development Goals Series,
https://doi.org/10.1007/978-981-97-6059-6_14

and Stoianoff 2020). As a result, the effectiveness of Australia's climate change regime and its subsequent ability to achieve greenhouse gas emission reduction targets has been significantly affected. Despite this historical disposition, the Australian government has instituted commendable policies incentivising renewable energy, resulting in approximately 35% of the country's electricity being derived from renewable sources (Clean Energy Council 2023). This trajectory fosters the transition to clean energy, concurrently promoting energy justice for Australians.

The recent policy initiatives of the Federal government also exhibit and support ambitious climate change goals. Paradoxically, however, Australia intends to persist in extensive fossil fuel exports, including gas, oil, and an augmented export of coal (Crowe 2021). This dichotomy in Australia's stance on climate change, coupled with its historical reluctance to implement effective greenhouse gas mitigation mechanisms, contravenes the interests of neighbouring Pacific Islands, thereby raising questions relating to climate justice. Climate change poses a direct threat to the livelihoods and sovereignty of Pacific Islanders, potentially resulting in population displacement. Consequently, Australia's substantial fossil fuel exports are expected to exacerbate the challenges posed by climate change, particularly impacting its immediate neighbours, the Pacific Islands.

Conversely, acknowledging the significant reliance of the Asia–Pacific region on coal and other fossil fuels, one might contend that Australia's fossil fuel exports contribute to energy justice by providing vital energy access, particularly for marginalised and economically disadvantaged segments of the population. This article scrutinises the inconsistencies within Australia's climate change policies and its traditional approach to fossil fuel exports. An analysis of the merits and drawbacks of such policies is conducted in the context of climate and energy justice, considering the Pacific Islands countries affected by Australia's position. The article ends with some arguable conclusions, shedding light on the intricate and multifaceted nature of the interplay between energy justice, the economy, climate considerations, and other pertinent factors.

14.2 Energy Justice and Fossil Fuels

The concept of energy justice has become a prominent cross-disciplinary research focus (Bickerstaff et al. 2013), aiming to apply principles of justice to various facets of the energy landscape. This includes but is not limited to energy policy (McCauley et al. 2013), production and systems (Heffron and McCauley 2014), security (Sovacool et al. 2013),and the political economy of energy (Jenkins et al. 2016a, b). Moreover, the concept of energy justice is connected to environmental justice and climate justice, encompassing a wide range of relevant factors (Heffron and McCauley 2017) and giving rise to complex implications associated with the term (Evans Osei Opoku and Acheampong 2023).

The literature commonly identifies three fundamental concepts within the realm of energy justice: distributional justice, procedural justice, and recognition justice (McCauley et al. 2013). Distributional justice centres on achieving an equitable allocation of benefits and burdens arising from energy activities among stakeholders. Procedural justice emphasises the enhancement and development of decision-making processes related to energy, with an emphasis on ensuring impartial and inclusive participation by all relevant stakeholders. Recognition justice is oriented towards acknowledging and comprehending the distinct needs of minorities, older individuals, and other marginalised groups of populations, particularly in the context of energy services (Evans Osei Opoku and Acheampong 2023). Consequently, whether energy is derived from fossil fuels or renewable sources, it interconnects with the encompassing framework of energy justice.

Fossil fuels are commonly linked to issues of injustice, primarily owing to their inherently carbon-intensive nature (McCauley 2018). This association prompts deep considerations regarding the detrimental impacts, necessity, safety, and sustainability of oil, gas, and coal production. However, this assessment represents only a partial perspective (McCauley 2018). The accessibility constraints and cost prohibitions associated with renewable energy persist in many Asia–Pacific and other countries (EU-ASEAN 2023). Consequently, fossil fuels remain an indispensable energy source, particularly in the short to medium term. The regressive nature of energy costs further complicates the situation, as any increase disproportionately affects households with lower incomes. For Asia–Pacific nations, dealing with such increases poses a formidable challenge, considering their economic development priorities and demanding cautious and multifaceted resolutions (Baum et al. 2023). In the light of these considerations, it is essential to examine how Australia, characterised by its abundant natural resources and significant role as a fossil fuel exporter, confronts the challenges presented by climate change and attempts to navigate the transition towards sustainable, green energy. The subsequent section delves into an analysis of Australian climate change policy.

14.3 Australian Climate Change Policy

Australia has implemented several federal policies aimed at reducing greenhouse gas (GHG) emissions and incentivising investments in green energy and carbon reduction projects. Particularly, one of the most successful policies is the Renewable Energy Target (RET). Introduced by the Australian Labour Party's government in 2009, the RET scheme builds upon the earlier mandatory renewable energy target regulation (Guglyuvatyy and Stoianoff 2016). The primary objective of the RET is to fulfil the government's commitment to ensuring that 20% of Australia's electricity is sourced from renewable sources by the year 2020.

Under the Renewable Energy Target Scheme, liable entities such as electricity retailers and large industrial users are required to procure a specific percentage of their electricity annually from renewable sources (St John 2014). In January 2011, the scheme was split into two: the Large Scale Renewable Energy Target (LRET) for large renewable energy developments like wind and solar farms, and the Small Renewable Energy Scheme (SRES) for smaller installations, including rooftop solar panels and solar water heaters (St John 2014). Liable entities discharge their obligations by acquiring renewable energy certificates generated by both large and small renewable energy producers. The RET generates financial incentives for the establishment or expansion of renewable energy facilities, such as wind and solar farms.

According to a recent report from the Clean Energy Regulator, renewable energy sources constituted approximately 35% of total generation in the national electricity market (Clean Energy Council 2023). While the Renewable Energy Target (RET) remains operational, the Australian Federal government has displayed hesitancy in establishing a new target for renewable energy. The LRET's 33,000 GWh target was met in 2019, but the RET scheme continues to require electricity retailers to meet their obligations under the policy until 2030 (St John 2014). The Small Renewable Energy Scheme (SRES) is also scheduled to run until 2030, albeit with a diminishing level of subsidy for the purchase of renewable energy each year until the programme's conclusion. Although there exists no imposed limit on the quantity of renewable energy that can be generated under the SRES, the reduced government subsidy diminishes the incentive for the installation of new solar panels.

On 15 September 2022, the newly elected Australian Federal government implemented significant climate law reforms through the Climate Change Act 2022 and the Climate Change (Consequential Amendments) Act 2022. These Acts incorporate emissions reduction targets of 43% below 2005 levels by 2030, as well as confirm the plan to achieve net-zero emissions by 2050 (Climate Change Act 2022). This represents a substantial increase from the previous government's GHG reduction target of 26–28% by 2030. The implementation of these new laws is a significant improvement in Australia's climate change policy. The government proposed to reform a range of existing policies, such as the Carbon Farming Initiative and the Safeguard Mechanism (Safeguard Mechanism Reforms 2023). An expert panel reviewed the Australian Carbon Credit Units (ACCUs) under the Emissions Reduction Fund to ensure their integrity (Chubb et al. 2022).

Generally, effective policies to reduce greenhouse gas emissions require a limit or cap on total emissions and mechanisms to prevent sources of emissions from exceeding the limits. The Australian climate regime did not include emission caps or instruments to ensure that polluters limit their GHG emissions. To address this issue, a Safeguard Mechanism was announced in 2016 (Safeguard Mechanism). This mechanism applies to businesses that are required to report under National Greenhouse Gas and Energy Reporting regulations and emit more than 100,000 tonnes of CO_2-equivalent emissions in a fiscal year (National Greenhouse and Energy Rule). This applies to businesses across a wide range of industrial sectors, including power generation, mining, oil and gas production, manufacturing, transportation, and waste. The Safeguard Mechanism requires the largest polluters in Australia to maintain their emissions within baseline levels, and it covers around 215 entities (Clean Energy Regulator).

The Safeguard Mechanism reforms proposed to provide a clear direction towards achieving net-zero emissions in 2050 while giving businesses the flexibility to invest in abatement measures at a manageable pace (Safeguard Mechanism Reforms 2023). The proposed changes also intend to send a strong signal to encourage investment. The Safeguard facilities will contribute a proportionate share of the national target of reducing net emissions by 2030, with emissions covered by the Safeguard falling from an estimated 143 million tonnes in 2022–23 to no more than 100 million tonnes by 2030 and capped at 1,233 million tonnes between 2021 and 2030 (Safeguard Mechanism Reforms 2023). The reformed Safeguard Mechanism is expected to achieve an estimated 205 million tonnes of abatement by the end of the decade.

In the 2022–23 Budget, the Rewiring the Nation initiative was established, which will provide AUD$20 billion in low-cost finance to modernise and expand Australia's electricity grids at the lowest possible cost. The first transmission investments under the Rewiring the Nation programme were announced, which will provide concessional financing for the Marinus Link between Tasmania and Victoria, offshore wind projects and Renewable Energy Zones in Victoria, as well as the Victoria-NSW Interconnector (VNI West) KerangLink (Safeguard Mechanism Reforms 2023). To reduce bills, cut emissions, and relieve pressure on the grid, AUD $224.3 million over four years has been allocated in the 2022–23 Budget to deploy 400 Community Batteries across Australia. The Community Solar Banks initiative has been allocated $102.2 million over four years in the 2022–23 Budget (Safeguard Mechanism Reforms 2023).

Australia's recent shift in its GHG mitigation approach is characterised by a connection with existing climate change policies that have demonstrated limited effectiveness. Notably, the primary policy instrument, the Safeguard Mechanism, focuses exclusively on the largest polluters, imposing emissions restrictions on major industrial facilities only, representing roughly 28% of the nation's total emissions (Safeguard Mechanism Reforms 2023). Although there is evident progress in Australia's climate policy, it remains a topic of contention, particularly due to the Australian government's reluctance to implement effective carbon mitigation measures that would encompass a broader spectrum of emitting industries.

The fragmented nature of Australia's climate policy not only poses risks to its international standing and reputation but also raises concerns about exacerbating global temperature increases. However, despite the perceived inefficiency in greenhouse gas mitigation policies, Australia is actively undergoing a transition towards renewable energy sources. This apposition highlights the complexity and multifaceted nature of Australia's environmental policies, with implications for both global energy justice and climate considerations.

14.4 Australian Fossil Fuel Exports

Despite positive developments in the transition towards green energy, the Australian government persists in supporting traditional energy exports. Australia's contribution to global fossil fuel mining, as measured by CO_2 potential, stands at around 4% (Swann 2019). The country extracts 57 tonnes of fossil fuel-related CO_2 per year per Australian, a figure ten times higher than the global average (Swann 2019). Nevertheless, the ongoing exportation of coal and gas from Australia remains a crucial pillar supporting employment and regional communities (Satherley 2022).

As the leading global exporter of coal, the second largest exporter of gas, and the highest per capita emitter of greenhouse gases within the Organisation for Economic Cooperation and Development (OECD), Australia plays an important role in the international fossil fuel landscape (Edney-Browne 2021). The cumulative impact of its domestic emissions, coupled with the potential emissions from fossil fuel exports, positions the country as the fifth largest emitter worldwide (Swann 2019). Despite escalating social and political scrutiny, Australia has sustained and expanded its dominance in the fossil fuel sector (Wright et al. 2021).The nation's historical role as a primary exporter of fossil fuels has been a recurring focal point in discussions of economic contribution, with industry communications emphasising the fiscal gains derived from mining royalties and tax revenue accruing to the government (Satherley 2022).

The Australian fossil fuel sector, along with associated political discourse, has employed broader moral arguments to justify its significance. Particularly, this includes emphasising the sector's claimed leadership role in developing technical solutions and global benefits for the universal challenge of climate change (Warren et al. 2016). Critics argue that the Australian Federal government's investments in 'false solutions', such as carbon capture and storage, serve to extend the reliance on fossil fuels (Warren et al. 2016). This stance contradicts the post-COP26 decision, explicitly advocating for countries to 'phase down coal and fossil fuel subsidies' (Morton 2021). In contrast to this, the Australian government directs its investments towards technological developments that enable the fossil fuel industry to persist in emitting greenhouse gases (Satherley 2022). Simultaneously, the government subsidises processes aimed at carbon reduction to offset these emissions.

During the 2021 UN Climate Change Conference in Glasgow, the Conference of the Parties (COP26), several nations, including the European Union and the United States, endorsed the Global Methane Pledge, which aims to reduce methane emissions by at least 30% by 2030 (Rucci 2021). Initially resistant, Australia, under the recent Labour government, has now become a signatory to this commitment (Morton 2021). Additionally, numerous countries pledged substantial financial commitments for global forest protection, intending to bolster forest management and restoration in the ensuing years (Rucci 2021). Although Australia supported the COP26 ban on deforestation, it remains one of the global leaders in land clearing (Pickering and Gugulyvatyy 2019). Remarkably, Australia's existing and proposed climate change strategy lacks a specific goal or policy to mitigate land clearing.

The Australian government plans to subsidise the construction of new gas-fired power plants (Morton 2021). While several countries have pledged to cease funding fossil fuels in the near future, Australia is currently considering about 100 new fossil fuel projects that would result in an annual increase in greenhouse gas pollution of around 1.7 gigatonnes of CO_2 equivalent (Morton 2021). Moreover, the government intends to permit the development of new gas basins, exemplified by the Beetaloo Basin in the Northern Territory, which has received subsidies amounting to up to $226 million (Verschuer et al. 2021). This initiative has the potential to generate an additional 100 Mt of greenhouse gas emissions, comprising 40 Mt from Australia (equivalent to 7.3% of total annual emissions) and another 60 Mt from importing countries (Ogge 2021). Additionally, the approval of Woodside's

Scarborough gas field in Western Australia is anticipated to result in an additional 1.6 billion tonnes of emissions over its lifetime (The Australia Institute 2021). The collective emissions from proposed gas projects in Australia are estimated at 329 million tonnes of CO_2 per year, constituting two-thirds of Australia's current annual emissions, equivalent to the emissions of 46 coal-fired power plants (Swann 2020).

One of the most striking strategies of the government is the 'business as usual' approach to energy exports, meaning Australia will continue its traditional fossil fuel exports. Despite international calls for transitioning away from fossil fuels, the Australian government consistently expresses support for the sector and reaffirms its commitment to exporting coal, gas, and oil in the coming decades (Crowe 2021). A concrete manifestation of this commitment occurred in March 2022 when the Australian Prime Minister announced a surge in coal exports to Europe (Mazengarb 2022). The logic behind this policy is that if Australia exits the market, other countries will fill the gap. Therefore, if a country is going to benefit from energy exports, let it be Australia, especially given its competitive advantage (Kouchakji 2023). Some commentators have argued that Australian coal is generally 'cleaner' than coal from other countries (Hunt 2016). Besides, if other energy-exporting countries have lower climate change ambitions than Australia, it would be better for Australia to benefit from the energy market, thereby providing more revenue that can be reinvested in renewables (Hunt 2016). This pragmatic stance, which greatly benefits Australia, underpins the government's efforts to maintain the country's prominence on the global fossil fuel stage despite growing environmental concerns and international calls for emissions reductions.

14.5 Australia's Climate Policy, Fossil Fuel Exports, and Their Implications

It could be argued that the Australian government's position is at odds with global efforts to mitigate climate change. The recently enacted Climate Change Act and reform of existing mechanisms is a positive development in Australia's climate change policy, but Australia's continued export of fossil fuel export contributes significantly to GHG emissions. The government declares it is making unprecedented investments in climate change adaptation and resilience and is working internationally to reduce emissions. For example, the Australian government's $1.9 billion investment (Australian Government 2020) in protecting the Great Barrier Reef is positive, but such adaptation measures can only slow the damage and not prevent widespread coral bleaching, especially given Australian fossil fuel exports, which increase greenhouse gas emissions and further exacerbates coral bleaching (Boström-Einarsson 2020).

One of the top national interests of the Australian government is maintaining a stable Indo-Pacific region (Australian Government 2016), which is among its highest priorities. A white paper on foreign policy stresses the necessity to collaborate with Pacific Islands, acknowledging their security and stability as fundamental to Australian strategic interests (Australian Government 2016). The integration of Pacific islands into the Australian economy is deemed crucial for regional stability and economic prospects, but it is unclear whether the islands will join this push for integration, especially where they feel their sovereignty is under threat (Wallis and Powles 2018). Despite the Australian government's prioritisation of various issues related to the Pacific Islands, the primary concern for the islands themselves is climate change (Pacific Islands Forum). Given their status as developing nations, the islands depend on support from more economically advanced countries to formulate effective mitigation and adaptation strategies for addressing climate change (The Paris Agreement). Consequently, a conflict of interests emerges between Australia and the Pacific Islands regarding what is deemed in the best interest of the Pacific nations.

The Pacific Islands are very different in terms of political status, population, development, and migration prospects, encompassing dependent territories, states in free association with the

United States or New Zealand, and fully independent countries (Firth 2018). While Australia has historically played a significant role in supporting Pacific Island nations on the global stage, recent years have witnessed increased self-determination and autonomy among these nations, particularly in advancing their own climate change priorities (Hayward-Jones 2019). Despite Australia being a key donor in the region, it is no longer the exclusive partner of choice, as China has expressed a willingness to support the island nations. The prevailing 'global order based on agreed rules', rooted in Western principles (Wallis and Powles 2018), is contrasted with China's preferences, raising concerns for the Australian government amidst the rising influence of China (Dobell 2019). Australia's emphasis on order and stability may not necessarily align with the priorities of the island nations. Illustratively, the recently established security agreement between China and the Solomon Islands has generated apprehension in Australia, with fears that this pact could amplify China's influence in the South Pacific region (News.com.au 2022).

The Australian government's concurrent pursuit of easing island nations' climate change concerns and implementing stringent border protection policies at their territorial borders suggests a precedence of Australia's interests over those of the islands (Edney-Browne 2021; Hayward-Jones 2019). This prioritisation occurs despite the island's heavy reliance on agriculture, fishing, and tourism, which are vital to their economies and susceptible to the impacts of climate change (Marles 2017). Tuvalu's foreign minister emphasised the urgency of climate change at the UN COP26 summit by delivering a speech while standing knee-deep in seawater, symbolising the vulnerability of low-lying Pacific Island nations in the face of climate change (ABC 2021).

Climate change poses a dual threat to Pacific Islanders, jeopardising both livelihoods and national sovereignty, potentially leading to population displacement. Australia's excessive fossil fuel exports and reluctance to mitigate GHG are anticipated to exacerbate health and economic challenges for its Pacific Island neighbours. The consequential mass migration within the Pacific region is likely to impact regional security, stability, and Australia's national interests (Burkett 2011). Australia's position on energy exports and its controversial climate policy contradicts both the principle of climate justice and the priority interests of island nations, especially as their sovereignty faces a tangible threat.

Australia's foreign policy implies a global order founded on agreed rules, yet there are numerous international agreements and climate change initiatives available for Australia to align with in support of its neighbours. A widely accepted principle of international law holds that states bear an obligation to conduct themselves without detriment to the interests of other states. Therefore, although sovereign states like Australia have the autonomy to determine their actions in the international arena, they also have a simultaneous obligation to refrain from actions that infringe the rights or interests of other states.

Examining the positive aspects, the Australian government has conveyed its intention to improve strained relationships with Pacific nations, marking a constructive development. During Foreign Minister Penny Wong's visit to Samoa, emphasis was placed on the Labour party's commitment to addressing climate change and pursuing more ambitious emission reduction targets (Malcolm 2023). This signals a proactive approach to repairing relations with Pacific nations and countering Chinese influence in the region. Additionally, the Australian Prime Minister is actively seeking support from Pacific nations for Australia's bid to host the world's largest climate summit in 2026 (Packham 2023). This initiative is accompanied by a commitment to reinstate financial support for the United Nations' controversial Green Climate Fund (Kaminski 2023). These developments precede the upcoming Pacific Islands Forum, where the Australian Prime Minister is expected to face renewed appeals from regional counterparts, urging Australia to impose restrictions on new fossil fuel projects (Kaminski 2023).

14.6 Australian Policies and Energy Justice

The Australian government's persistent commitment to ongoing energy exports has prompted enquiry from various commentators. Nevertheless, proponents argue that this stance is justified by the ongoing global demand for Australian coal and gas, especially in developing countries. For instance, the Asia–Pacific region's substantial dependence on coal provides an appropriate context for investigating pathways towards a just transition (Ritchie and Roser 2017). Historically, contemporarily, and prospectively, this region has demonstrated a commitment to utilising coal as a primary fuel to foster economic development (UNESCAP 2021), with Asia–Pacific countries contributing to over 75% of the global coal generation capacity (IEA 2021).

Despite substantial commitments to curtail coal dependency (Pressburger et al. 2022), there remains scepticism regarding the sincerity of some governments' intentions to reduce reliance on coal (Han 2022). Moreover, the transition away from coal poses significant challenges, particularly concerning issues such as energy access and the potential for job losses. These challenges underscore the necessity for a scrupulous examination of relevant aspects in the pursuit of a just transition to renewable energy (International Labour Organisation 2022).

Beyond coal, the energy transition encompasses various dimensions, especially the increasing adoption of renewable sources in the Asia–Pacific region. Despite rapid growth in modern renewable energy, such as solar and wind, their contribution to total electricity generation consistently remains below 10% (Malik and Bertram 2022). Further inquiry is needed to understand the implications of the region's ongoing reliance on fossil fuels and the potential of renewable alternatives (McCauley and Pettigrew 2023). Examining these dynamics is crucial as a just transition necessitates not only the adoption of clean technologies but also the cultivation of fairness and equity (World Bank 2021). Several studies emphasise the pivotal role of fairness and equity in either hindering (Sovacool 2021) or facilitating (Djanibekov and Gaur 2018) the energy transition in the Asia–Pacific (Siemens 2022). The literature emphasises the criticality of fossil fuel dependence, pinpointing specific contexts where this reliance is most pronounced (Kleovoulou et al. 2021; Sabyrbekov and Ukueva 2019; Al-Saidi 2022). The transition towards cleaner, renewable energy introduces new dimensions of injustice for the Asia–Pacific region, demanding comprehensive scholarly examination (McCauley and Pettigrew 2023).

The ongoing disparities affecting individuals and communities in both developing and developed nations are expected to undergo substantial transformation associated with the departure from fossil fuels (McCauley and Heffron 2018). Effectively addressing the social dimensions of such transition presents a difficult challenge. The extraction of fossil fuels generates benefits for certain segments of the population, such as extraction workers, while simultaneously inflicting harm upon others, particularly communities affected by pollution. A successful transition requires the equitable protection and promotion of the rights of both beneficiary and adversely impacted groups, recognising that climate impacts disproportionately burden the world's impoverished populations (Muttitt and Kartha 2020).

Despite the apparent contradiction between Australian fossil fuel exports and climate policies, there exists a potential avenue for fostering energy justice. Each facet contributes to energy justice in a distinct manner; specifically, Australian climate policies encourage a just transition towards renewable energy within the nation, while fossil fuel exports potentially enhance access to energy for nations in the Asia–Pacific region. This paradoxical finding underscores not only the multifaceted nature of energy justice but also presents a philosophical dilemma that is notably contentious and frequently overlooked by prominent experts in the field.

The substantial contribution of Australian energy exports to greenhouse gas (GHG) emissions directly contributes to the rise of sea levels and other consequences of climate change. This impact is particularly pronounced in Pacific Island nations, resulting in increased tensions in international relations and exacerbating challenges in achieving climate justice for these nations. Australia's climate regime and fossil fuel exports play a dual role, promoting energy justice within the country and across Asia–Pacific nations, especially benefiting economically disadvantaged people by providing access to energy. Conversely, particularly for small Pacific Island nations, the climate justice dimension is adversely affected by Australia's climate policies and export strategies.

This inherent paradox necessitates careful consideration of the perspectives of all involved parties, demanding a judicious approach to comprehensively evaluate potential measures to mitigate the adverse impacts of Australian policies and exports. While Australia has already emerged as a significant financial donor, aiming to bolster resilience in Pacific nations against the consequences of climate change, it is unequivocally evident that more comprehensive efforts are imperative to facilitate adaptation. However, this paper refrains from proposing specific solutions to address the resulting contradictory dichotomy concerning energy and climate justice. Instead, it emphasises the intricate nature of justice issues, whether pertaining to energy or climate. Moreover, it establishes a foundation for further research in this expansive realm, emphasising the imperative for continued exploration and examination of these complex and interconnected issues.

14.7 Conclusion

Australia's transition to renewable energy marks a significant step towards a cleaner energy future and helps to address domestic energy justice concerns. However, a paradox emerges in Australia's continued export of fossil fuels, particularly coal and gas, despite ambitious climate goals. This paradox raises questions about Australia's commitment to climate justice, particularly concerning its Pacific Island neighbours. The potential adverse effects of Australia's fossil fuel exports on Pacific Islands, such as rising sea levels and climate-induced challenges, underscore the intricate relationship between Australia's climate policies and its role as a major fossil fuel exporter. This paradox highlights the often-overlooked complexity and philosophical dilemmas in discussions of energy justice. The concept of energy justice provides a valuable framework for analysing Australia's policy landscape. Examining fossil fuels within the context of energy justice reveals the challenges associated with transitioning away from carbon-intensive sources, especially in economically constrained regions.

Australia's climate policy and fossil fuel exports present a multifaceted scenery, characterised by commendable initiatives, paradoxical approaches, and potential implications for global justice. The complex interplay between environmental, economic, and social factors underscores the necessity for continued research and examination of these issues. Despite Australia's commendable contributions to climate resilience in Pacific nations, the inherent dichotomy surrounding energy and climate justice persists. This study refrains from proposing specific solutions, emphasising the intricate nature of justice issues in both energy and climate realms. The paper underscores the need for a comprehensive examination of the social dimensions of the energy transition, considering both the benefits and harms associated with fossil fuel extraction and the shift towards renewable energy. The complex nature of justice issues, whether related to energy or climate, demands continued exploration to inform comprehensive and equitable policy decisions in the face of global environmental challenges.

References

ABC (2021) Tuvalu's foreign minister Simon Kofe gives COP26 speech knee-deep in the sea to show nation on frontline of climate crisis. Available from https://www.abc.net.au/news/2021-11-10/tuvalu-minister-makes-cop26-speech-from-sea/100608344

Al-Saidi M (2022) Energy transition in Saudi Arabia: giant leap or necessary adjustment for a large carbon economy? Energy Rep 8:312–318

Australian Government (2016) Defence white paper. Australian Government. 2017 Foreign policy white paper. Commonwealth of Australia
Australian Government (2020) Working together to protect the Reef in 2020–21. Media Release. Available from https://minister.awe.gov.au/ley/media-releases/working-together-protect-reef-2020-21
Baum F, McGreevy MP, MacDougall CM, Henley M (2023) Energy as a social and commercial determinant of health: a qualitative study of Australian policy. Int J Health Policy Manage 12:7193
Bickerstaff K, Walker G, Bulkeley H (2013) Introduction: making sense of energy justice. In: Bickerstaff K, Walker G, Bulkeley H (eds) Energy justice in a changing climate. Zed Books, London
Boström-Einarsson L et al (2020) Coral restoration—a systematic review of current methods, successes, failures and future directions. PLoS ONE 15(1): e0226631
Burkett M (2011) The nation ex-situ: on climate change, deterritorialised nationhood and the post-climate era. Climate Law 2:345–374
Chubb I, Bennett A, Gorring A, Hatfield-Dodds S (2022) Independent review of ACCUs. Department of Climate Change, Energy, the Environment and Water, Canberra
Clean Energy Council (2023) Clean Energy Australia. Report 2023
Clean Energy Regulator. The safeguard mechanism
Climate Change (Consequential Amendments) Act 2022 (Cth)
Climate Change Act 2022 (Cth)
Crowe D (2021) Scott Morrison backs coal wealth for decades to come. The Sydney Morning Herald
Djanibekov U, Gaur V (2018) Nexus of energy use, agricultural production, employment and incomes among rural households in Uttar Pradesh, India. Energy Policy 113:439–453
Dobell G (2019) Australia's pacific pivot: destiny, duty, denial and desire. Report, Australian Strategic Policy Institute
Edney-Browne A (2021) Pacific bully and international outcast: how Australia's climate policies isolate it from the pacific and the world. Greenpeace Australia Pacific
EU-ASEAN Business Council (2023) Energy transition in ASEAN
Evans Osei Opoku E, Acheampong AO (2023) Energy justice and economic growth: does democracy matter? J Policy Modell 45(1):160–186
Firth S (2018) Instability in the pacific islands: a status report. Lowy Institute
Guglyuvatyy E, Stoianoff N (2016) Carbon policy in Australia—a political history. In: Stoianoff N, Kreiser L, Butcher B, Milne JE, Ashiabor H (eds) Green fiscal reform for a sustainable future—reform, innovation and renewable energy. Edward Elgar Publishing
Guglyuvatyy E, Stoianoff N (2020) Australian carbon policy: two steps forward, one step backwards? In: Zachariadis T, Kreiser L, Milne J E, Ashiabor H (eds) Green fiscal reform for a sustainable future—reform, innovation and renewable energy. Edward Elgar Publishing
Han H (2022) Climate hypocrite? A case of Korea's involvement in coal capacity expansion in southeast Asia. Korean J Int Stud 20(2):275–299
Hayward-Jones J (2019) Cross purposes: why is Australia's pacific influence waning? Austr Foreign Affairs 6:29–50
Heffron R, McCauley D (2017) The concept of energy justice across the disciplines. Energy Policy 105:658–667
Heffron R, McCauley D (2014) Achieving sustainable supply chains through energy justice. Appl Energy 123:435–437
Hunt G (2016) Fact check: does Australia export cleaner coal than many other countries? ABC News. Available from https://www.abc.net.au/news/2015-11-27/fact-check-is-australias-export-coal-cleaner/6952190
IEA (2021) World energy statistics. International Energy Agency
Intergovernmental Panel on Climate Change (2019) Global warming of 1.5 °C. IPCC special report
International Labour Organisation (2022) A just energy transition in Southeast Asia: the impacts of coal phaseout on jobs. Regional Office for Asia and the Pacific Report
Jenkins K, Heffron R, McCauley D (2016) The political economy of energy justice in Canada, the UK, and Australia: a nuclear energy perspective. In: Van de Graaf T, Sovacool B, Ghosh A, Kern F, Klare MT (eds) Palgrave handbook of the international political economy of energy. Palgrave Macmillan
Jenkins K, McCauley D, Heffron R, Stephan H, Rehner R (2016b) Energy justice: a conceptual review. Energy Res Soc Sci 11:174–182
Kaminski I (2023) UN's green climate fund too scared of risk, finds official review. Climate Home News
Kleovoulou EG, Konstantinou C, Constantinou A, Kuijpers E, Loh M, Galea KS, Stierum R, Pronk A, Makris KC (2021) Stakeholders' perceptions of environmental and public health risks associated with hydrocarbon activities in and around the Vasilikos energy center, Cyprus. Int J Environ Res Public Health 18(24):13133
Kouchakji K (2023) Is the tide turning for Australia's fossil fuel exports? Energy Monitor
Malcolm NY (2023) Wong reiterates climate change commitment. The Australian
Malik A, Bertram C (2022) Solar energy as an early just transition opportunity for coal-bearing states in India. Environ Res Lett 17(3):034011
Marles R (2017) Richard Marles on Australia and the Pacific. Speech, Lowy Institute
Mazengarb M (2022) Morrison rejoices in high fossil fuel prices, but Ukraine coal gesture will last just days. Renew Econ
McCauley D (2018) Fossil fuels and energy justice: coal, oil and gas. In: Energy justice palgrave

McCauley D, Heffron D (2018) Just transition: integrating climate, energy and environmental justice. Energy Policy 119:1–7
McCauley D, Pettigrew K (2023) Building a just transition in asia-pacific: four strategies for reducing fossil fuel dependence and investing in clean energy. Energy Policy 183:113808
McCauley D, Heffron R, Stephan H, Jenkins K (2013) Advancing energy justice: the triumvirate of tenets. Int Energy Law Rev 32(3):107–110
McKenzie A (2021) Towards Glasgow: why Australia's climate policy is risking our future. Austr Foreign Affairs 12:69–88
Morton A (2021) Australian government refuses to join 40 nations phasing out coal, saying it won't 'wipe out industries.' The Guardian
Muttitt G, Kartha S (2020) Equity, climate justice and fossil fuel extraction: principles for a managed phase out. Climate Policy 20(8):1024–1042
News.com.au (2022) China inks security deal with Solomon Islands. Available at https://www.news.com.au/technology/innovation/military/china-inks-security-deal-with-solomon-islands/news-story/dd0e2432cb21e60090831f8313f74d4b
Ogge M (2021) Wrong way, go back: why the gas-fired recovery plan will fail to reduce energy prices or create jobs but will increase emissions. The Australia Institute
Packham B (2023) Green climate fund cash pledge to secure pacific support for climate summit bid. The Australian
Pickering C, Guglyuvatyy E (2019) Negative impact of land clearing and deforestation on the great barrier reef: assessing the effectiveness of Queensland's vegetation management act 1999 (Qld). Carbon Clim Law Rev 13(3):195–207
Pressburger L, Evans M, Yu S, Cui Y, Somani A, Iyer G (2022) A comprehensive economic coal transition in South Asia. PNNL 32383:1910045
Ritchie H, Roser M (2017) Fossil fuels. Our World in Data. https://ourworldindata.org/fossil-fuels
Rucci T (2021) The contribution of the European Union to the COP26, Eyes on Europe. 29 Nov 2021
Sabyrbekov R, Ukueva N (2019) Transitions from dirty to clean energy in low-income countries: insights from Kyrgyzstan. Central Asian Survey 38(2):255–274
Safeguard Mechanism Reforms (2023) Position paper. Australian Government Department of Climate Change, Energy, the Environment and Water
Safeguard Mechanism. Available from https://www.dcceew.gov.au/climate-change/emissions-reporting/national-greenhouse-energy-reporting-scheme/safeguard-mechanism#toc_0
Satherley T (2022) Coal, gas and decarbonisation—challenges and policy choices. Parliament of Australia
Siemens (2022) Asia pacific energy transition readiness index: geared up for change. Siemens Energy Plc., pp 1–22
Sovacool B (2021) Clean, low-carbon but corrupt? Examining corruption risks and solutions for the renewable energy sector in Mexico, Malaysia, Kenya and South Africa. Energy Strat Rev 38:100723
Sovacool B, Sidortsov R, Jones B (2013) Energy security, equality and justice. Routledge, London
St John A (2014) The renewable energy target: a quick guide. Parliament of Australia
Swann T (2019) High carbon from a land down under quantifying CO_2 from Australia's fossil fuel mining and exports. The Australia Institute
Swann T (2020) Weapons of gas destruction: lifting the lid on greenhouse gas emissions from Australian fossil gas projects and resources. The Australia Institute
The Australia Institute (2021) Woodside's Scarborough gas field equivalent to 15 new coal power plants, risks Murujuga rock art. Media Release.
UNESCAP (2021) Coal phase out and energy transition pathways for Asia and the pacific. United Nations Economic and Social Commission for Asia and the Pacific, pp 1–82
Verschuer R, Ogge M, Campbell R (2021) Subsidising fracking in the Beetaloo basin. The Australia Institute
Wallis J, Powles A (2018) Australia and New Zealand in the pacific islands: ambiguous allies. Report No. 43, Strategic & Defence Studies Centre: ANU College of Asia & the Pacific
Warren B, Christoff P, Green D (2016) Australia's sustainable energy transition: the disjointed politics of decarbonisation. Environ Innov Societal Trans 21:1–12
Wood T, Reeve A, Ha J (2021) Towards net zero: practical policies to reduce transport emissions. Grattan Institute
World Bank (2021) Just transition for all. Just transition for all: the World Bank supporting countries transitioning away from coal
Wright C, Nyberg D, Bowden V (2021) Beyond the discourse of denial: the reproduction of fossil fuel hegemony in Australia. Energy Res Soc Sci 77:102094

Evgeny Gugluvatyy is an Associate Professor at the Department of Business Law and Taxation, School of Business, Monash University, Malaysia. He specialises in climate change law and policy, taxation, forest regulations and other thematic areas. Evgeny's teaching and research experience spans over a number of years at various universities, including the University of New South Wales, the University of Technology Sydney, the University of Western Australia, Monash University and others. He has extensively contributed to the academic discourse through the publication of numerous peer-reviewed articles and chapters across a wide range of academic journals and publishers.

15 The Energy Injustice of the Orang Asli in Peninsular Malaysia

Audrey K. L. Siah, Lilian M. L. Chan, and Grace H. Y. Lee

Abstract

This chapter explores the energy injustices faced by the Orang Asli, the indigenous peoples of Peninsular Malaysia. Despite their rich cultural heritage and historical significance, the Orang Asli endure severe socio-economic marginalisation, exacerbated by inadequate access to energy. This energy poverty starkly contrasts with Malaysia's status as a major energy producer. The chapter highlights how energy justice, which advocates for equitable access to energy resources and technologies, is crucial for addressing the disparities faced by the Orang Asli. Key issues include the impact of land ownership disputes, palm oil plantations, and hydropower projects on their livelihoods and rights. The transition to renewable energy presents both opportunities and challenges, as it must be managed to avoid further marginalising the Orang Asli. The chapter calls for inclusive energy policies that reduce the energy burden on marginalised communities, ensuring their active participation in energy justice discussions. By integrating the perspectives and traditional knowledge of the Orang Asli, Malaysia can develop sustainable and equitable energy solutions that respect their human rights and cultural integrity. Achieving energy justice for the Orang Asli involves not only providing reliable electricity but also recognising their role in a sustainable future and ensuring their rights are upheld in the transition to green energy.

Keywords

Energy justice · Orang Asli · Indigenous rights · Renewable energy · Socio-economic marginalisation

15.1 Introduction

In the diverse tapestry of Peninsular Malaysia's population, the Orang Asli hold a unique yet often overlooked position. This chapter delves into the intricate world of the Orang Asli of Peninsular Malaysia, the indigenous peoples whose existence is as rich in culture and tradition as it is complex in socio-economic challenges. The Orang Asli of Peninsular Malaysia are not a monolithic people but a collection of 18 subgroups, each with its own unique language, culture, and history. According to the Department of Orang Asli Development (JAKOA) of Malaysia, the Orang Asli number is

A. K. L. Siah · L. M. L. Chan · G. H. Y. Lee (✉)
Monash University Malaysia, Kuala Lumpur, Malaysia
e-mail: grace.lee@monash.edu

A. K. L. Siah
e-mail: audrey.siah@monash.edu

L. M. L. Chan
e-mail: lilian.chan@monash.edu

A. Trakic et al. (eds.), *Energy Justice*, Sustainable Development Goals Series,
https://doi.org/10.1007/978-981-97-6059-6_15

209,575 or only 0.6% of the total population. From coastal fishing villages to forest-dwelling tribes practising subsistence agriculture, this variety covers the Negrito, the Senoi, and the Aboriginal-Malay groups, showcasing a plurality of lifestyles. They have played a crucial role in the area's history for over five thousand years. It is believed that, for many, their ancestry began in China and Tibet, and they settled across Southeast Asia, including Indonesia and Peninsular Malaysia (Masron et al. 2013). Despite having been there for a long time, the Orang Asli of Peninsular Malaysia are still severely marginalised culturally and economically (Mat Dong et al. 2022); problems with land ownership, education, and employment exacerbate their plight.

Central to this examination is the concept of energy justice, a pressing concern that is gaining public attention worldwide. Driven by global inequality and poverty, climate change and environmental degradation, and the need for successful economic transitions in the energy sector, energy justice highlights the importance of equitable access to energy resources and technologies. For the Orang Asli, this manifests as an acute challenge, contrasting sharply with Malaysia's status as a significant oil and natural gas producer and its advancements in electrification. The Orang Asli communities in Peninsular Malaysia endure disproportionate energy poverty and unfair energy burdens, illustrating the stark disparities in energy access and distribution. This injustice could intensify as Malaysia transitions towards renewable energy. Important challenges influencing Orang Asli's rights and way of life include land ownership, the effects of palm oil plantations, and the building of hydropower dams. This chapter aims to provide a picture of the Orang Asli's existing condition in Peninsular Malaysia, shedding light on their challenges and resilience by comprehending these diverse concerns.

15.2 Background on the Orang Asli in Peninsular Malaysia

The Orang Asli of Peninsular Malaysia comprise three main ethnicities, which consist of 18 subgroups among them: the Negrito (or Semang), the Senoi, and the Aboriginal-Malay (or Melayu-Proto) groups. As of 2022, the total population of Malaysia stands at 33.94 million, with only 209,575 being Orang Asli. Table 15.1 shows that the Senoi are the largest ethnic group, followed by the Melayu-Proto, with the Negrito being the smallest. Among the nine states of Peninsular Malaysia, Pahang has the highest proportion of Orang Asli, followed by Perak.

The lifestyles and methods of subsistence among the Orang Asli of Peninsular Malaysia are diverse and differ greatly from one community to another. Coastal communities, as you might expect, mainly rely on fishing as their primary occupation, while other communities engage in permanent agriculture like rubber, oil palm, or

Table 15.1 Number of indigenous people by ethnic group and state in 2022

State	Senoi	Melayu-Proto	Negrito	Total
Johor	49	15,773	3	15,825
Kedah	30	10	296	336
Kelantan	15,382	73	2,032	17,487
Melaka	23	1,809	1	1,833
Negeri Sembilan	110	12,107	4	12,221
Pahang	36,002	41,485	1,128	78,615
Perak	57,747	667	2,811	61,225
Selangor	5,425	15,532	4	20,961
Terengganu	944	85	43	1,072
Total	115,712	87,541	6,322	209,575

Source The Department of Orang Asli Development (JAKOA)

cocoa (Masron et al. 2013). Indigenous communities that reside predominantly in or near forested areas tend to practice hill rice cultivation and hunting and gathering (Omar et al. 2021). They also generate cash income by trading produce such as petai, durian, rattan, and resins. Only a minority maintain a semi-nomadic lifestyle, relying on the seasonal resources provided by the forest (Masron et al. 2013).

15.3 The Benefits of Better Access to Reliable Energy

In today's world, access to modern energy, namely electricity, is essential for achieving a decent standard of living. The provision of basic needs like healthcare, sanitation, clean water, and vital modern services like transport and telecommunications all depend upon the availability of electricity (IEA 2020). United Nations Conference on Trade and Development (UNCTAD) reported that access to reliable and quality energy supply 'drives industrialisation, boosts productivity and economic growth, spurs human development, and is crucial to achieve almost all of the United Nations Sustainable Development Goals (SDGs)' (UNCTAD 2023). For example, the lack of access to electricity can significantly limit a person's educational opportunities and, consequently, their employment and income levels. These, in turn, would logically weaken the individual's nutrition and health, and this would further reduce their productivity and employability—creating a vicious cycle that could adversely impact generations, if unchecked.

Figure 15.1 shows an adaptation of a basic economic model that follows a circular flow. If you recall the old adage that 'What goes around, comes around', you should easily be able to appreciate what this model tries to explain.

If we start from the most humble of circumstances, access to a reliable electricity supply would allow an individual to study and learn at their own time to attain a level of training or education that would make them more employable. Employment would translate their work into earnings that would enable that person to afford better consumption of goods and services (and so, improve their nutrition, health, etc.) and, possibly, accumulate some savings. The accumulated savings can be invested in assets with an expectation of higher consumption in future. Besides the ability to invest in physical assets, a person may also invest in further education, which would feed into this circular flow all over again, with the added benefit of potentially making wiser consumption and investment choices, further raising their standard of living.

Better decision-making could benefit not only the individual's socio-economic circumstances but also reduce environmental degradation.

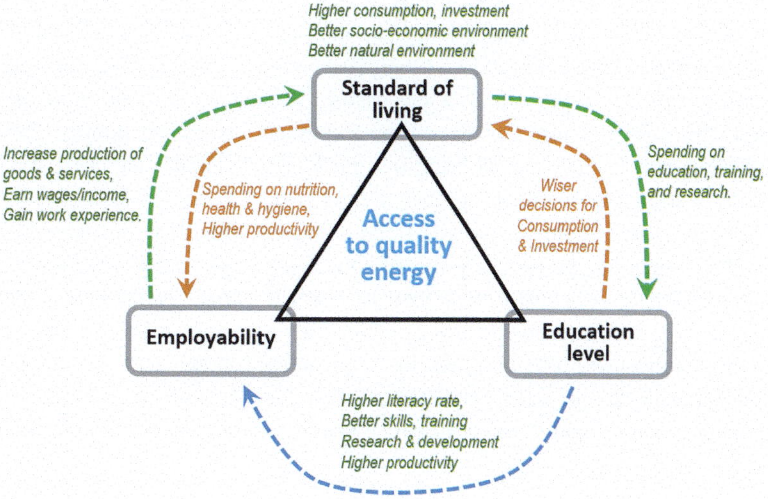

Fig. 15.1 A circular flow model for some benefits of energy access

Table 15.2 Access to electricity in Malaysia

	Percentage (%)
Access to electricity (% of population)	100
Access to electricity, urban (% of urban population)	100
Access to electricity, rural (% of rural population)	100

Source World Bank Development Indicator

Economists call these beneficial 'side effects' a *positive externality*. Measuring the benefits of energy access is somewhat complicated because there are numerous ways in which the positive externality might manifest and in amounts that are difficult to quantify. For example, the positive externality may be passed down to the next generations because higher-educated parents tend to rear higher-educated offspring.

We may also expect positive macroeconomic effects from having a higher-skilled labour force by attracting more foreign direct investment (FDI) and creating more jobs in the economy. A higher employment rate can increase the government's income tax revenues, lower crime rates, and reduce public expenditure on crime-prevention and unemployment benefits—freeing up public resources that may be better utilised elsewhere. The ripple effect of positive externality arising from better access to quality energy can be far-reaching.

According to the World Bank's data presented in Table 15.2, Malaysia's rural and urban populations are officially 100% connected to an electricity supply. However, these statistics are inaccurate or misleading because as recently as November 2023, a leading Malaysian national daily reported that at least 137 Orang Asli villages across Malaysia still lack electricity (The Star 2023). The Ministry of Rural and Regional Development is currently working out the feasibility of adding these villages to the national grid. However, the remote locations of the villages could hinder that prospect. Some of those villages used to have access to solar energy, but that ended when the equipment broke down due to poor maintenance. The villagers in those areas use kerosene lamps, palm seeds, or candles for their lighting needs. They typically cook their meals using bamboo or wood fire.

Because most of the Orang Asli inhabit the marginal and remote areas of Peninsular Malaysia, they struggle daily with energy poverty. Although the government was able to connect a few villages to the national grid and provided diesel-powered generators to some larger villages, the smaller villages were left out. Many Orang Asli cannot afford the amount of energy required to meet their essential needs (Pye et al. 2015), and so their health, education, and overall standard of living tend to fall below the nation's average. Many indigenous communities in Malaysia are classified as hard-core poor and live in remote rural or mountainous regions without basic infrastructure and electricity (Bala and Tan 2021).

15.4 Disproportionate Energy Burden Leading to Energy Poverty

The Malaysian government regulates the country's electricity tariff through the Energy Commission (EC) using Incentive-Based Regulation (IBR) to ensure that electricity prices are fair. The tariff is divided into two parts: the base tariff, which is fixed for three years, and the imbalance cost pass-through (ICPT), which is reviewed every six months. The ICPT is adjusted based on global fuel price fluctuations of coal and gas. Table 15.3 displays the current monthly tariff rates for household electricity usage. Different electricity tariff rates apply for other purposes (e.g. commercial, industrial, street lighting, etc).

As you can see, the electricity fees are the same for all domestic usage in Malaysia, regardless of household income level. This tariff standardisation implies that low-income households bear a greater energy burden because they

Table 15.3 Domestic tariff rates for 2023

Tariff block	Current rate (RM/kWh)
For the first 200 kWh (1–200 kWh) per month	0.218
For the next 100 kWh (201–300 kWh) per month	0.334
For the next 300 kWh (301–600 kWh) per month	0.516
For the next 300 kWh (601–900 kWh) per month	0.546
For the next kWh (901 kWh onwards) per month	0.571

Source Tenaga Nasional Berhad (https://www.tnb.com.my/)
The minimum monthly charge is RM3.00

spend a larger proportion of their income on their electricity bills compared with high-income households. This is particularly true for the poorest and the marginalised indigenous groups.

Based on the Malaysia Millennium Development Goals Report by the Economic Planning Unit in 2015, the indigenous communities have the lowest income ranking. Out of all the indigenous peoples, 50.92% of them live below the poverty line and 34.34% fall below the hard-core poverty line. Recent research conducted by the Centre for Orang Asli Concerns (COAC) in 2019 found that 99.2% of the indigenous population in Peninsular Malaysia are categorised under the B40 income bracket, with earnings below RM4000 per month. Some indigenous communities in Malaysia have a much lower income than the national poverty line of RM2208. For example, the Semaq Beri indigenous communities in Terengganu earn less than RM880 per month (Ramli 2020). Similarly, the indigenous communities of Lake Temenggor in Perak rely on subsistence fishing and hunting, and they mostly have no cash income.[1]

Low-income households, particularly the indigenous communities, bear an unfairly high burden of energy costs compared to the national average. Due to their extremely low income, they cannot afford the high cost of electricity supply. As a result, these energy-insecure residents are forced to prioritise food, clothing, and other basic necessities over utility bills. This leaves them with no choice but to live without access to electricity. The relatively high energy burden pushes these vulnerable Orang Asli into energy poverty.

15.5 Encroachment on Orang Asli Land

As of December 2023, there is no comprehensive, accurate, official map of the boundaries of all Orang Asli customary land. Modernisation and land-encroaching projects often threaten their customary dependence on their land for their livelihood (Mat Dong et al. 2022). Examples of such projects include land conversion for commercial crops or building infrastructure like highways or dams. Leading newspapers in Malaysia recently reported that a group representing 1000 Orang Asli lodged a police report against the logging and farming activities carried out on Orang Asli land in Kelantan by a commercial developer firm (Bernama 2023).

15.6 The Transition Towards Renewable Energy

The Ministry of Economy of Malaysia recently announced a new renewable energy target that aims to achieve 40% renewable energy in the power mix by 2035 and 70% by 2050, according to the Enerdata report.[2] As of 2021, Malaysia has generated about 17% of its electricity from renewable sources (16% from hydro, 1% from

[1] https://panasonic.net/sustainability/en/lantern/2015/01/malaysia.html.

[2] https://www.enerdata.net/publications/daily-energy-news/malaysia-sets-new-target-reach-70-renewables-power-mix-2050.html.

solar, and 1% from biomass). The country has also set a target of attaining carbon neutrality by 2050. The shift from fossil fuel-based energy to cleaner, low-carbon sources will have winners and losers, meaning that not everyone will benefit equally or even benefit at all.

In Peninsular Malaysia, about 17% of Orang Asli villagers (that is, more than 35,000 people) still live without electricity (Nadia 2023). Unfortunately, the move towards a greener future has resulted in the infringement of Indigenous rights and the dislocation of Indigenous people from their ancestral lands.

Although Malaysia's common law recognises the customary communal land rights of the Orang Asli under the Aboriginal Peoples Act 1954 (Act 153), Orang Asli communities often face issues of land tenure as their ancestral lands are not recognised by state governments, despite being inherited through generations. It is possible for the government to revoke their claims and provide compensation upon eviction. In addition, there are currently insufficient or unclear guidelines for the computation of said compensation, nor are the affected Orang Asli consulted before the process (Abu Bakar et al. 2022).

The Orang Asli often find their lands redesignated as forest reserve or state land. Instead of using the Aboriginal Peoples Act 1954, the state government typically designates the land as a State reserve under Section 62 of the National Land Code (Daud et al. 2020). This means that the Director-General of the Department of Orang Asli Development (Malay: Jabatan Kemajuan Orang Asli, JAKOA) has no power over the land. The 1954 Aboriginal Peoples Act fails to adequately protect Indigenous peoples, especially regarding land usage and ownership. Only a small percentage of Indigenous people have ownership of land, and < 20% of Indigenous villages in Peninsular Malaysia are gazetted and retained as Indigenous reserve land.[3] As such, the Indigenous peoples are just occupying the land at the discretion of the state government.

The expansion of oil palm plantations in Malaysia for palm oil as a renewable energy source has significantly impacted the habitats of Orang Asli living in remote areas.[4] The displacement of Indigenous peoples is caused by the lack of legal land rights, preventing them from securing ownership. Taking Kelantan as an example, JAKOA states that there are no Orang Asli Reserves in the state. Instead, the state has only designated Orang Asli Areas. This implies that the state has the authority to lease the land to external parties. Since 2015, about 200 indigenous families from two villages in Kelantan lost their livelihoods due to alleged encroachment on their land for oil palm plantation, logging, mining, and durian plantation.

A renewable energy project that posed a threat to the livelihood of Indigenous communities was the construction of hydroelectric dams. Hydroelectric dams often symbolise modernity and progress that contribute towards green growth strategies. However, many studies have been done all over the world on such projects that reveal the induced displacement and resettlement of communities that are already marginalised, like Indigenous Peoples, adversely affecting them (Schapper and Urban 2021). The Ulu Geruntum hydroelectric dams in Gopeng, Perak, scheduled to start operating in 2023, are prime examples. These dams were built to meet the Kyoto Protocol's sustainability requirements, but in 2018, the local Indigenous people protested against them. They blamed the dam's construction for causing devastation to their local ecosystems.[5] The Nenggiri Hydroelectric Dam, expected to commence operations in 2027, is estimated to generate 600 GWh of clean energy annually in Gua Musang, Kelantan. However, this dam poses a threat to the livelihoods, rights, and culture of approximately 5000 Indigenous

[3] https://minorityrights.org/minorities/orang-asli/#: ~ :text=Very%20few%20Orang%20Asli%20have,tenants%20on%20their%20traditional%20lands.

[4] https://www.aljazeera.com/news/2018/8/17/palm-oil-threatens-indigenous-life-in-malaysia.

[5] https://cleanmalaysia.com/2018/02/27/orang-asli-villagers-cry-foul-new-hydroelectric-dam-perak/.

people living in southern Kelantan.[6] The Indigenous communities have complained that they were not fairly compensated for the construction of hydroelectric dams.

15.7 Conclusion and Policy Implication

The situation of the Orang Asli in Peninsular Malaysia, as discussed in this chapter, highlights a fundamental nexus of cultural integrity, socioeconomic struggle, and energy justice. The Orang Asli tale is one of resilience in the face of energy shortages and social marginalisation, as well as the cultural richness and historical depth. This scenario reflects a greater global dilemma in which policies aimed at internalising the environmental costs of energy production risk aggravating the energy burden on poor people. As a result, it is critical to identify and assist marginalised populations such as the Orang Asli in programmes that offer sustainable energy solutions.

The Malaysian government should prioritise inclusive energy policies that address the particular issues of the Orang Asli. These measures must attempt to lower the energy burden, particularly given the possibility of increasing energy costs due to environmental policy changes. Furthermore, Orang Asli peoples should be included in global conversations about energy justice. Their involvement is crucial for meaningful discourse and action in keeping with Sustainable Development Goal 7's goal of universal energy access. The Orang Asli must be included in current and future energy collaborations, guaranteeing their active involvement and decision-making influence.

The transition to renewable energy must be carried out with the utmost regard for the human rights of the Orang Asli. State and business sector actions should be aligned with human rights standards, with the Orang Asli recognised as vital players in a sustainable future. In the transition to green energy, it is critical to include the opinions of indigenous peoples such as the Orang Asli. Development efforts to mitigate climate change must engage these communities to avoid the repetition of power disparities observed in conventional energy sectors. Their perspectives and traditional knowledge are crucial in developing long-term and fair energy solutions.

Attaining energy justice for the Orang Asli involves more than simply providing inexpensive electricity; it is also about recognising their importance as a community, respecting their rights, and recognising their role in a sustainable future. As Malaysia and the rest of the globe move towards greener energy options, the participation, respect, and empowerment of the Orang Asli should be prioritised. To satisfy the energy demands of everyone, a comprehensive, inclusive, and respectful strategy is required, as is ensuring a fair and equitable transition for vulnerable groups and creating a precedent for the treatment of indigenous populations in the global energy and environmental sustainability debate.

[6] https://www.thevibes.com/articles/news/62704/temiar-history-threatened-orang-asli-protest-nenggiri-dam-project-on-gua-musang-rainforest.

References

Abu Bakar N, Hussain F, Sufian A, Hashim HN (2022) Malaysian experience of Orang Asli involuntary resettlement compensation package. Asian J Environ Behav Stud 7(21):53–65. https://doi.org/10.21834/ajebs.v7i21.407

Bala P, Tan C (2021) Digital inclusion of the Orang Asli of Peninsular Malaysia: remote virtual mechanism for usability of telecentres amongst indigenous peoples. Electron J Inform Syst Develop Countr 87(4):e12171. https://doi.org/10.1002/isd2.12171

Bernama (2023) Orang Asli group lodges police report over land encroachment activities. New Straits Times

Daud W, Mohsin AB, Rahman MKA (2020) Land ownership for orang asli in Malaysia: current situation. Int J Sci Technol Res 9(3):4998–5002

IEA (International Energy Agency) (2020) Defining energy access: 2020 methodology. IEA. https://www.iea.org/articles/defining-energy-access-2020-methodology. Accessed 16 Dec 2023

Masron T, Masami F, Ismail N (2013) Orang Asli in Peninsular Malaysia: population, spatial distribution and socio-economic condition. J Ritsumeikan Soc Sci Human 6:75–115

Mat Dong M, Midmore P, Plotnikova M (2022) Understanding the experiences of Indigenous minorities through the lens of spatial justice: the case of Orang

Asli in Peninsular Malaysia. Reg Sci Policy Pract 14 (5):1223–1240. https://doi.org/10.1111/rsp3.12512

Nadia A (2023) Will solar power shine a light on Orang Asli villages left behind? Malaysiakini

Omar K, Kamaruddin A, Md Razali M, Ab Manaf N, Halim M, Sa'at N (2021) Natural resources dependency: Are indigenous communities ready to switch to new career? Stud Appl Econ. https://doi.org/10.25115/eea.v39i4.4589

Pye S, Dobbins A, Baffert C, Brajkovic J, Miglio R, Deane P (2015) Energy poverty and vulnerable consumers in the energy sector across the EU: analysis of policies and measures. Policy 2:64–89

Ramli MW (2020) Keadaan ekonomi komuniti orang asli Semoq Beri di Kampung Sungai Berua Terengganu. e-BANGI J 17(6):68–79

Schapper A, Urban F (2021) Large dams, norms and indigenous peoples. Dev Policy Rev 39:O61–O80. https://doi.org/10.1111/dpr.12467

The Star (2023) More than 130 Orang Asli villages still have no electricity. https://www.thestar.com.my/news/nation/2023/11/02/more-than-130-orang-asli-villages-still-have-no-electricity. Accessed 16 Dec 2023

UNCTAD (United Nations Conference on Trade and Development) (2023) Improving energy access key to meeting development goals in Africa. UNCTAD. https://unctad.org/news/improving-energy-access-key-meeting-development-goals-africa. Accessed 16 Dec 2023

Audrey K. L. Siah is a Senior Lecturer and Department of Economics Coordinator at Monash University Malaysia. Her research focuses on applied economics, with a particular emphasis on important areas such as globalisation, economic growth and development, and social economics. Audrey is committed to interdisciplinary research and has presented her research findings at international conferences. Her research has also been published in renowned international journals such as Management Science, Common Law World Review, Personality and Individual Differences, Journal of the Asia Pacific Economy, and Singapore Economic Review. Before starting her academic career, she was an entrepreneur who regularly interacted with a diverse group of stakeholders.

Lilian M. L. Chan is an economist with more than 25 years of experience teaching economics at undergraduate and postgraduate levels in Malaysia and abroad. She is currently a Lecturer of economics in the School of Business at Monash University Malaysia. Lilian received her bachelor's and master's degrees in Economics from excellent universities in England and earned her Ph.D. from Universiti Putra Malaysia for pioneering a Contingent Valuation study for the conservation of a living heritage in George Town, Penang. Lilian's current research interests include but are not limited to environmental and natural resource economics, tourism economics, and learning and teaching. In her spare time, Lilian enjoys reading, music, and "a good cuppa" but not necessarily in that order.

Grace H. Y. Lee is a Professor of Economics and Head of the Department of Economics with Monash University Malaysia. In addition, she is the principal investigator (PI) of the World Values Survey (WVS) for Malaysia (Wave 6, 7 and 8), and is one of the elected executive members of the WVSA (2022–2027). Her principal areas of interest are applied economics, experimental and behavioural economics. She has published numerous articles in top-tier academic journals, including Management Science, Journal of Economic Behavior and Organization, and Public Choice. As a supervisor of many honours and PhD students, she has been recognised for her outstanding contributions to teaching, earning awards such as Teaching Prizes, PVC's Awards for Excellence in Teaching and Research, the Vice-Chancellor's Award for Honours Supervision, and the Dean's Award for Equity, Diversity, and Social Inclusion Research Excellence.

Part IV

Findings and Recommendations

Building a Just Energy Future: Findings and Recommendations

16

Adnan Trakic, Ridoan Karim, and Pervaiz K. Ahmed

Abstract

This chapter presents an ethical framework for achieving energy justice, summarising key points from earlier chapters. It explores the link between energy access and human rights, energy and environmental justice, and social and economic factors influencing energy access. The framework emphasises integrating technological innovations, financial mechanisms and regulatory frameworks to ensure fair access to clean and affordable energy. A comprehensive approach is advocated, considering socio-economic conditions, technological advancements and political influences, supported by robust regulations, financial incentives and societal values. This approach aims to ensure energy security, affordability and climate justice, promoting a fair and sustainable energy future. The chapter provides guidance for policymakers, researchers and practitioners to navigate the complex landscape of energy justice and contribute to a just energy transition.

Keywords

Energy justice · Environmental justice · Ethical framework · Regulatory framework

16.1 Introduction

In this chapter, we seek to provide an overall summary of the main points discussed in each substantive chapter, accompanied by the proposed recommendations. While we have endeavoured to convey the contents of the chapters as accurately as possible, we strongly encourage readers to refer to the actual chapters for a more precise account of the discussions presented by their respective authors. Additionally, the summary and findings represent the outcome of our review, and it is important to acknowledge that they may be influenced by our individual perspectives.

Before we proceed with the summary and recommendations section, we wish to briefly revisit the aim we set in chapter one of this book. As promised, the book undertakes an analysis of the concept of 'energy justice', drawing on insights from various pertinent disciplines. This analysis aims to provide society with an ethical framework that can effectively guide decision-making in the energy sector.

The concept of energy justice can only be fully understood and appreciated when examined holistically from the perspectives of various

A. Trakic (✉) · R. Karim · P. K. Ahmed
Monash University Malaysia, Kuala Lumpur, Malaysia
e-mail: adnan.trakic@monash.edu

R. Karim
e-mail: ridoan.karim@monash.edu

P. K. Ahmed
e-mail: pervaiz@sunway.edu.my

A. Trakic et al. (eds.), *Energy Justice*, Sustainable Development Goals Series,
https://doi.org/10.1007/978-981-97-6059-6_16

disciplines. The book began with the conceptual analysis of energy justice and reiterated the need for it to be elevated to a level of human rights. This required a legal examination of the relevant national and international legal instruments that broadly deal with human rights. An interesting perspective was offered in the argument that the 'right to life', which has been embedded in almost all national constitutions, must be interpreted liberally to include 'quality of life'. The courts in many countries have noted that quality of life is a broad category and, as such, it could accommodate the right to access clean and affordable energy.

When discussing the pursuit of affordable, reliable and modern energy for all, as embedded in SDG 7, there is an aspect that is most effectively scrutinised through the lens of economists. This is where we have substantive contributions from economists who have examined patterns of access to energy, specifically electricity, in many countries in the context of energy justice and its distributional, procedural and recognition justice principles. While the transition to clean energy is necessary, it must be done in a way that distributes energy burdens fairly. There are also major concerns that the people most affected by the transition to clean energy are often not adequately compensated or consulted when these clean energy projects are being developed. The book has identified that pursuing energy justice objectives has the potential to not only foster equity and fairness in the production, distribution and consumption of energy among low-income and high-income households but also to empower marginalised individuals, including women and children, in many communities to realise their full potential.

The book also features chapters contributed by technologists, engineers, and financial and tax professionals, providing, among other things, intriguing perspectives on the utilisation of the latest technologies in the renewable energy sector. These insights explore how technology can enhance the production and distribution of green energy in a more equitable manner.

With all this being said, we now proceed to the summaries and recommendations from the substantive chapters as we see it. Again, these are our own impressions and understanding of what was discussed in these chapters. We also have not hesitated to offer our own commentary where we deemed it appropriate.

16.2 Discussion of Findings

16.2.1 Access to Energy as a Basic Human Right

In Chap. 2, entitled, 'Energy Justice and Human Rights', Muhamad Hassan Ahmad, Ashgar Ali Ali Mohamed, Mohamed Hanipa Maidin and Muhammad Naqib Eishan Jan examine whether access to energy could be considered a basic human right protected under international law as well as under the municipal laws of selected jurisdictions. They first clarify that no binding international instrument or local constitution contains the exact phrase 'energy justice' or 'access to energy' under protected rights and liberties. As such, one may be under the impression that access to energy cannot be regarded as a basic human right. That, however, is not the case.

The authors argue that access to energy is encompassed under the broader category of 'right to life', which has been explicitly included in most countries' legislations. The right to life, as they explained, must not be understood restrictively, only to mean the physical existence in that law protects life, and anyone who takes life without due process of law will be accountable before the law. According to the authors, life must be interpreted much more liberally to include the quality of life. Therefore, rights such as the right to livelihood, the right to live in a reasonably healthy and pollution-free environment, the right to reasonable accommodation and the right to dignity are also protected categories of rights under the 'right to life'.

The authors made reference to articles 21 and 5(1) of the Federal Constitutions of India and Malaysia, respectively, which guarantee the right to life. The courts in both countries have adopted liberal interpretations of 'right to life' provisions to include all other facets that are integral parts of life itself and that contribute to the quality of life. In India, the authors cited the case of *Virendra Gaur v State of Haryana*, where the Supreme Court held that under article 21, the right to life includes the petitioner's right to be treated in a public hospital. In this case, the petitioner was refused treatment in public hospitals due to the lack of facilities and was forced to seek help in a private hospital. The courts in Malaysia have also adopted a similar approach, as can be seen from the Federal Court decision in *Tan Tek Seng v Suruhanjaya Perkhidmatan Pendidikan* in which the court interpreted the 'right to life' under article 5(1) to include the right to livelihood, quality of life and reputation. Interestingly, in a recent decision in the case of *Maria Chin Abdullah v Ketua Pengarah Imigrasen*, the Federal Court held that the right to life now includes 'the right to travel abroad', and the list is not exhaustive.

Following the courts' decisions, the authors suggest that 'access to energy' should also be added as a new category under 'right to life'. They argued that access to energy is an enabler of other human rights, such as rights to food, shelter, clothing, water, sanitation, medical care, schooling and access to information. As such, multiple UN declarations and resolutions have urged the national governments to guarantee these basic human rights to all its people. The authors also explored the latest developments in the EU and the US and practical steps they have undertaken in recent years to help their people deal with energy poverty. They note that the experiences in those jurisdictions are worth exploring and emulating.

16.2.2 Energy Justice Vis-à-Vis Environmental Justice

In Chap. 3, entitled 'Does Energy Justice Trump Environmental Justice', Maizatun Mustafa and Farahdilah Ghazali try to answer a long-standing question about the compatibility between energy justice and environmental justice paradigms. In the first instance, as the authors mention, it may appear that the two doctrines have different objectives. It may be argued that environmental or climate justice places the preservation and protection of the environment in the first place. On the other hand, energy justice puts people's well-being and their right to access affordable and reliable energy above everything else. So it seems as if the public interest in having access to reliable and affordable energy is irreconcilable with the protection of the environment as most of the energy produced today for electrification, transportation, and, quite frankly, everything else comes from fossil fuels like oil, natural gas and coal which are harmful to the environment. As such, climate justice advocates for the wide-spread use of renewable energy. Currently, however, the availability of renewable energy falls short of meeting the demand for affordable access on a large scale. Achieving climate justice requires not only affordability but also an adequate supply to ensure that everyone can access renewable energy at reasonable prices and in significant quantities.

That being said, the authors' central argument in this chapter is that energy justice and environmental justice should not be approached in an exclusionary way, as if the two concepts cannot be reconciled. They strongly believe that the protection of the environment and energy security and affordability are not only reconcilable but the two objectives can only be achieved if they are considered together. In other words, they are dependent on each other. Preserving the environment is for the benefit of people and future generations. Thus, environmental justice is about the well-being of people as much as it is about the well-being of the environment. Disregarding apparent inequalities regarding access to energy and its affordability and instead only prioritising renewables is not a sustainable or humane solution. What is required is a balanced approach where both objectives are going to be considered to provide the most optimal solutions.

In order to tackle energy poverty and security, on the one hand, and air and water pollution, greenhouse gas emissions and habitat destruction, on the other hand, the authors proposed a mixed use of energy sources as a solution. Governments should continue to incentivise renewable energy production while ensuring that there is sufficient energy so that everyone can have access to it. They also did not reject the use of nuclear energy if the circumstances demand its usage, but they warned about the dangers that come with nuclear toxic waste. The authors also recognise that renewable energy sources come with certain drawbacks. They cite examples of wind farms, which have not only led to an increase in bird mortality rates in the surrounding areas but also have adverse visual effects on the landscape.

16.2.3 Access to Electricity and Clean Cooking Fuels

In Chap. 4, entitled 'Patterns of Global Disparities in Access to Electricity and Clean Cooking Fuels', Rajesh Ramachandran examines the association between socio-economic status, rural residence, education and gender of household heads with access to electricity and clean cooking fuels. The selection of electricity and clean cooking fuels for this study is not accidental. Rajesh chose these energy sources due to their importance in our lives. Electricity plays a crucial role in enabling economic activity and growth, as well as supporting essential services like health care and education. For those of us who have always had access to electricity, it is easy to take for granted the impact it has on our daily lives. However, in parts of the world where electricity is not universally accessible, it is clear that reliable power is essential for progress and development. However, Rajesh points to some rather disturbing statistics regarding access to electricity in Sub-Saharan Africa (SSA) countries. The data taken from the Demographic and Health Surveys covering 66 countries and over 1.67 million households showed that less than 10% of the households in Chad, Burundi, and Malawi have access to electricity. In Ethiopia, the percentage of households having access to electricity is 35%, while in Nigeria, Africa's most populous country, the share is 59%. In other SSA countries, the percentages are slightly higher but still far below the universal coverage available in most countries around the globe.

The decision to investigate access to clean cooking fuels, such as liquefied petroleum gas, biogas, electricity and natural gas, was a deliberate one. This is because the type of cooking fuel used has a direct impact on public health and the environment. Cooking with non-clean fuels, such as wood and charcoal, can cause severe respiratory diseases and even premature deaths due to indoor pollution. Children and women are most susceptible to the effects of indoor pollution. Additionally, harmful emissions from burning these fuels can damage the environment. The data again showed the bleakest picture in the SSA countries when it comes to access to clean cooking fuels. With the exception of South Africa and Gabon, with shares of 81% and 89%, respectively, no other SSA country has more than 50% of households having access to clean cooking fuels. In 21 countries, Rajesh notes, the share is less than 10%, which is very disheartening.

After having examined the data in detail for all the 66 countries surveyed, Rajesh concludes that there is a significant correlation between households' socio-economic status and urban location with access to two energy sources. He also found that households led by more educated individuals and women exhibit a higher rate of using modern energy sources. This finding, Rajesh argues, highlights the significance of education and gender in attaining greater distributional justice when it comes to access to electricity and clean cooking fuels.

16.2.4 Access to Electricity and Educational Outcomes

In Chap. 5, entitled 'Exploring the Dynamics of Access to Electricity and Educational Outcomes in Asia', Pauline Khoo examines electricity's transformational influence on Asia's socio-economic landscape. She refers to studies showing that increased access to electricity directly correlates with improvements in people's health care, standards of living and education. It is, thus, no surprise that developed countries, which consume more energy per person than their developing counterparts, are doing much better in all three dimensions. In other words, there seems to be a direct correlation between the economic prosperity and the increased energy consumption.

Be that as it may, Pauline decided to focus this study on exploring if there is a direct correlation between an increase in access to electricity and improvements in educational outcomes, such as improved school attendance, completion and attainment of better educational opportunities. One can understand the motivation behind this study. Many of us have witnessed that children from impoverished remote villages tend not to do as well in terms of schooling and academic performance as their counterparts from urban areas. There could be multiple reasons for that, as children in villages often have less time to dedicate to their studies as they need to help their parents cultivate land and do other household chores. If those households have no access to electricity, children's time for study will be substantially reduced.

Pauline limits her study to three Asian countries: India, Nepal and Pakistan. All three countries have made remarkable progress in the electrification of their communities in the past two decades. Currently, the electrification rates stand at around 80% in India, 90% in Nepal and 80% in Pakistan. The question, however, is whether the improved access to electricity translated into improved educational outcomes in those countries. According to Pauline, the statistics show that the educational outcomes in all three countries have substantially improved, along with the improved access to electricity. The improvement was particularly notable in the lower-income and middle-income groups. This finding is significant, especially when considered in the context of distributional justice principles.

Distributional justice advocates the equitable distribution of energy benefits so that all communities can fully benefit from it. At the moment, as Pauline correctly highlights, there is a huge imbalance in energy consumption globally. The top 10% of the world's wealthiest people consume 20 times more energy than the bottom 10%. As statistics in India, Nepal and Pakistan have shown, the more equitable spread of energy consumption could potentially correct the imbalance and produce more just outcomes in health care, standard of living and educational outcomes.

At the end of her chapter, Pauline notes that her study is limited to exploring the correlation between improved electricity access and educational outcomes. The extent of her analysis was not, as she clarifies, sufficient to prove the causation, which would require a higher standard of analysis and more detailed exploration.

16.2.5 Hydrogen Production and Export

In Chap. 6, Diane Kraal and Richard Taylor provide a very insightful commentary on Australia's potential to be one of the leading countries in hydrogen production and export and how those aspirations could, if not appropriately managed, adversely affect the environment and the people, particularly Australia's First Nations people. Hydrogen fuel is often seen as a viable alternative to fossil fuels. While burning fossil fuels releases harmful gases like carbon dioxide, carbon monoxide and methane into the atmosphere, burning hydrogen, on the other hand, only produces water and heat, both of which are environmentally friendly. This fact on its own suggests that hydrogen fuel is undoubtedly a preferred energy source.

Hydrogen as a chemical element is not readily available in its original pure form on Earth. It only exists as a building block with other elements, like in water, where it is coupled with oxygen. Splitting hydrogen from oxygen in the water is a very energy-intensive process that can only be done by using electricity in a process known as electrolysis. As a side note, it is worth mentioning that scientists in Germany have been performing small-scale experiments in which only sunlight is used in the process, thereby sidelining the need for the production of electricity to produce hydrogen (Devlin 2017). If the experiment proves to be a viable alternative for a large-scale production of hydrogen, this could be a game changer for the entire industry. However, at the moment, the hydrogen production process remains electricity dependent, and only if that electricity is produced from renewable energy sources like solar panels and wind turbines could the hydrogen fuel be labelled 'green hydrogen'. In order for countries to reach their targets of net-zero carbon emissions, they would need to ramp up their efforts in producing green hydrogen fuel.

One major problem with all this, as Kraal and Taylor also point out, is that most of the hydrogen fuel produced today is not green. In fact, most of it is produced using gas or other fossil fuels like coal. Thus, green hydrogen remains a very scarce commodity and is hardly able to meet local demand, let alone have enough of it for export. That said, the authors explained that Australia is making notable progress in developing a strategy for large-scale green hydrogen production and export. In 2019, The Australian Government's Energy Council set up a Hydrogen Working Group chaired by renowned scientist Dr Alan Finkel to explore the viability of large-scale hydrogen fuel production. Some Australian state governments have also taken their initiatives in this regard. For instance, the government of South Australia introduced a Hydrogen and Renewable Energy Bill 2023 to regulate the issuing of licenses to companies owned by the state government over land and water for the purposes of green hydrogen production.

We wish to add that the prospect of producing and storing green hydrogen fuel in the context of what is often termed 'sector coupling' is very attractive. Sector coupling is a term used to denote a sustainable cycle of renewable energy production and environmentally friendly emissions. For example, the green hydrogen fuel produced in the electrolysis process from renewable energy sources like solar panels and wind turbines could have multiple uses. When solar panels and wind turbines are not able to meet the electricity consumption demands on days when there is no sunlight or wind, hydrogen fuel can be used to make electricity to make up for the deficit. Hydrogen may also be used for other purposes like transportation, heating and even for heavy industry usage, such as in the factories producing cement and steel.

Besides green hydrogen's obvious environmental benefits, Kraal and Taylor have also pointed to some additional benefits of using hydrogen, like increased economic opportunities in developing the associated infrastructure necessary for large-scale hydrogen production, transport and storage. Countries like Australia might particularly reap huge economic benefits due to their large land mass, abundant resources, high solar potential and good infrastructure. The Australian government has recognised this huge economic potential in hydrogen production and export and has made significant investments in it in the last few years. For instance, since 2021, hydrogen-related investments in Australia, Kraal and Taylor noted, have almost doubled from A $133–185 billion in 2021 to A$230–303 billion in 2022.

The authors also point out the disadvantages of large-scale hydrogen production. Producing and storing green hydrogen is an expensive procedure. There are major issues associated with infrastructure in terms of storage and transportation of hydrogen. According to Kraal and Taylor, the costs associated with the production and export of green hydrogen are different from country to country. Several countries have been identified as potential exporters of hydrogen due to their favourable geographical position. These are countries like Egypt, Algeria, Morocco and Australia. Australia, in particular, has a large land space, a steady source of sunlight throughout the year and abundant water resources. These favourable conditions are not present in the countries situated in the northern hemisphere. This is why countries in the northern hemisphere like Germany, the world leader in hydrogen consumption, have expressed their wish to import green hydrogen from countries like Australia. In fact, this point admittedly triggered Kraal and Taylor to write this chapter.

It is interesting that the authors considered this development in the context of colonisers asking their former colonies to produce hydrogen and export the same to the colonisers' land. This may have a resemblance to the past, where the colonisers exploited the colonised territories by ruining their natural habitat and taking their resources. The production of green hydrogen at a scale sufficient for exports to the countries that import energy, like Germany, Korea, Japan, China and Singapore, cannot be done without significant effects on the land, natural resources and people. The authors argue that the land to build the solar panels would normally be in rural areas outside the cities. As such, the people who would be the most affected are the First Nations people. The companies involved will usurp their precious land, and often, the economic benefits will not reach displaced individuals. Water sources, already scarce in many of those places, will also be affected. This is where the authors make a very strong point that any strategy on green hydrogen fuel production and export developed by the government must consider the principles of energy justice: recognition justice, distribution justice, restorative justice and cosmopolitan justice.

Recognition justice demands that the government, its energy-related agencies and companies that have been given licenses to operate energy facilities recognise that the land they use has been occupied and preserved by the First Nations people for the past 65,000 years. Their approach to land, as described by Bruce Pasco in his book Dark Emu, which the authors admittedly used as an inspiration for this chapter, was the light footprint approach (Pascoe 2014). First Nations people cultivated land, and yet, at the same time, they preserved it for future generations.

Distribution justice stresses that the distribution of benefits and costs of the negatives from the energy sector must be done fairly. As such, the government must ensure that the people affected the most, particularly First Nations people, can also share the monetary benefits of green hydrogen production and export. One way of doing so, the authors suggested, is through the collection of revenues through secondary taxation, which will be distributed fairly. Hydrogen fuel is not subject to taxation, unlike the resource tax currently in place for mineral resources and transformed fuels.

The authors also note that the secondary taxation of hydrogen production could solve the problems associated with restorative justice. This concept calls on those who harm the environment or people to provide adequate remedies. For instance, in the process of desalination to produce hydrogen, one of the hazardous elements that comes out of that process is toxic brine, which may harm the environment. There is also a danger of hydrogen leakage into the atmosphere, which is, as the authors stated, 40 times more damaging than carbon dioxide. Tax funds from hydrogen could be used to restore the damage to the extent to which it is possible.

There is also a question of cosmopolitan justice, which concerns the well-being of people. Kraal and Taylor questioned the need for countries like Australia to make significant sacrifices so that the problems faced by other wealthy nations are solved. Every country and community needs to share the burden of ensuring that the net-zero emissions target is achieved by 2050. The authors concluded by urging the Australian government to take a more cautionary approach and not be quick in exporting the hydrogen. They argued that if hydrogen fuel is to be produced to help the country reach its net-zero emission targets, its production should only be limited to local consumption and not for export purposes.

16.2.6 Nuclear Energy Justice

In Chap. 7, entitled 'It is Time to Address Nuclear Energy Injustices', Ridoan Karim argues that a new paradigm for the nuclear energy industry is necessary. He contends that nuclear energy can only be seen as a viable energy source when viewed through energy justice principles. Three principles are of particular relevance here, namely distribution justice, procedural justice and recognition justice. The author challenges the notion that nuclear energy should not be considered a clean energy source and argues that it can be a viable option under sustainable development goals despite major safety concerns. This belief has led some countries to abandon plans for nuclear power plants and shift focus to other forms of clean energy, as seen in the case of Malaysia.

Malaysia had plans to build nuclear power plants and made concrete steps to ensure that it had the capacity to do so efficiently and safely (Chatzis 2016). However, when the new government came into power in 2018, the plans were halted due to safety concerns (Mohamed Radhi 2020). Now, the current government is considering reviving its nuclear energy plans again (Ikram and Aziz 2023). This chapter is a timely reminder of some of the principles that need to be considered if the plan is to succeed.

The government's plans to reintroduce nuclear energy have not been received well by everyone. Sahabat Alam Malaysia and Greenpeace Malaysia (2023), two prominent NGOs, issued a joint statement rebuking the plans by saying that nuclear energy should not be an option not only due to perceptions by the public about its possible safety issues but, as they say, due to actual evidence that countries which already have nuclear power plants have not yet found ways in which nuclear waste can be disposed of in a safe and environmentally friendly manner. They also refuted an argument that nuclear energy is a cheaper source of energy in comparison with fossil fuels and even less so in comparison with renewables like water, wind and solar.

Be that as it may, Ridoan has taken a position that is more in line with the current Malaysian government, which is one of not ruling out nuclear energy provided it can be produced in a safe, sustainable and just manner. This begs the question of what nuclear energy is and whether it can be considered clean energy. Nuclear energy is stored in the nucleus of atoms, which are the units that form all matter. There are two kinds of atomic reactions that create nuclear energy. One is called nuclear fusion, where atoms are joined together. It is our understanding that nuclear power plants currently do not have the capability to produce nuclear energy safely and reliably using this method. This brings us to the second method, known as nuclear fission. This is a method currently used by nuclear power plants. Nuclear fission is a reaction in which atoms split apart, releasing huge amounts of heat energy. This heat is used to warm the cooling agent, usually water, thereby creating steam. The steam is channelled through the pipelines to turn turbines, initiating the electricity production process. As can be seen, the outcome is clean and pollution-free energy with no greenhouse emissions. However, there is one big problem, and that is nuclear waste.

In order to have atoms split, nuclear power plants use pallets of uranium as fuel to produce nuclear fission. The uranium pallets are highly toxic and can stay active for thousands of years. If not stored properly, they can easily contaminate the environment and water, causing severe long-term and generational health-related hazards. While disposal of nuclear waste is a major concern, the daily operation of the nuclear power plant is not immune from the so-called normal accidents. These are considered minor radioactive emissions caused normally by the ageing of the reactors, cooling system leaks and accidental fires. All these are legitimate concerns that regulators must address.

Ridoan's proposal is to inject energy justice principles to solve some of these concerns. For example, the distribution and recognition justice principles are necessary to ensure that communities most affected by the construction and operation of a nuclear power plant and disposal of nuclear waste are properly rewarded and acknowledged. Ridoan highlights the example of the Yucca Mountain Nuclear Waste Repository in the state of Nevada, USA, which was selected as a safe place for nuclear waste disposal. The people of Nevada, particularly those living in the vicinity of the repository, strongly protested, questioning why the waste produced elsewhere in other states should put their lives in danger when they themselves have no benefit from it. Ridoan argues that there needs to be a fair distribution of benefits and possible detriments. Those who are disenfranchised and those who live in remote areas are likely to have to pay a higher price relative to benefits compared to the people living in the urban centres who end up benefiting the most, with little or no setbacks.

Ridoan stresses the importance of the procedural justice principle, which emphasises the need to consult with all stakeholders who may be affected by a nuclear power plant. Without this consultation and understanding of safety concerns, people are likely to object to nuclear energy, even in situations where its benefits far outweigh its disadvantages. This objection is understandable, given major nuclear power plant disasters like Chernobyl and Fukushima. However, Ridoan argues that with the latest technological advancements, such as the use of AI to prevent nuclear meltdowns and improved next-generation nuclear power plants with fusion and molten salt reactors, nuclear energy can be a safe, clean and reliable source of energy for all.

16.2.7 Financial Sector and Energy Justice

In Chap. 8, titled 'The Role of the Financial Sector in Energy Justice', Mirzet Seho and Sabina Hodzic discuss how financial institutions can assist individuals in addressing issues related to affordable and clean energy access, as well as supporting countries in their transition to more sustainable and environmentally friendly energy sources. The financial institutions in question include, but are not limited to, commercial, national and multilateral development banks, investment firms and impact investors. Mirzet and Sabina emphasise that these institutions have traditionally played a crucial role in financing large-scale energy projects. In the past, those energy projects were mostly connected to fossil fuels like coal mines, oil and natural gas fields, coal-fired power plants and oil refineries. Gradually, as the world became more climate-conscious, financial institutions began to integrate environmental criteria into their lending and investment decisions. The authors cite several examples of financial institutions that have successfully provided environmental and social investments, like Germany's Community Bank for Landing and Giving, the Indonesian Tropical Landscape Finance Facility and the European Investment Bank. At the core of their business lies providing access to funds for green energy-related projects.

One of the most interesting parts of the chapter discusses the 'democratisation' or 'decentralisation' of the energy sector through financial innovations. The authors identified three financial innovations: blockchain technology, mobile payment platforms and pay-as-you-go models, and crowdfunding. According to the authors, blockchain technology can create secure and transparent peer-to-peer trading platforms. For example, these platforms have already been used by PowerLedger in Australia and Brooklyn Microgrid in New York to allow consumers to buy and sell electricity directly among themselves. SunContract is another example, which tokenised solar panels and rooftops, enabling individuals and businesses to invest in solar farms and earn passive income. The authors are particularly impressed with SunContract's project in the Slovenian village of Zavrate, which used to rely on the traditional grid but now has become 100% self-sufficient through solar energy. According to reports, the residents of Zavrate now produce their own electricity and trade surplus within the village. It is worth exploring this model further to determine whether it can be implemented on a larger scale.

The authors also refer to the example of M-Pesa, a mobile payment platform in Kenya. This platform enabled consumers to pay for energy services via mobile devices without involving the traditional banking industry. It is said that M-KOPA Solar, a company that utilises the M-Pesa platform, served over 3 million customers across Kenya, Uganda, Nigeria and Ghana, impacting over 16 million lives. When considering the impact of these platforms, coupled with various crowdfunding initiatives, particularly in Sub-Saharan Africa, where conventional sources of funds for energy investments are lacking, one cannot help but be amazed at the profound effect they have had on the lives of ordinary people.

The authors conclude by emphasising that these financial innovations have successfully democratised energy financing and fostered renewable energy production. They also underscore the active role financial institutions need to play in promoting distributive, procedural and recognition justice within the energy justice paradigm.

16.2.8 Energy Technology Innovation as Enabler of Justice

In Chap. 9, entitled 'The Role of Energy Technology Innovation in Solving Energy Injustice', Manjeevan Seera brought to light the pleas of the segments of society that tend to be unjustly and disproportionately affected in countries' energy transition endeavours. As part of the green energy transition plans, countries worldwide have been making significant investments in energy technology innovations (ETIs) to optimise the efficiency of clean energy production and distribution. Manjeevan highlights that the sudden shift to clean energy brings along a host of challenges, most notable of which are the disproportionate access to clean energy sources, inequitable energy costs and disproportionate environmental burdens associated with energy production.

In other words, if not properly managed, the clean energy transition could lead to severe energy poverty among the segments of society that are already struggling with poverty and neglect. This is why, Manjeevan notes, energy poverty is explicitly mentioned as an area requiring special attention in the EU-mandated National Energy and Climate Plans. If not adequately addressed, energy poverty, Manjeevan argues, could hinder access to essential services like health care and education to vast segments of society and, at the same time, result in the denigration of the environment as those unable to afford energy prices would have no choice but to resort to non-sustainable alternatives; for example, cut trees for wood to survive cold winters or to feed their families.

Manjeevan's central argument in this chapter is that ETIs can eliminate energy poverty and help countries meet green energy transition targets. This is particularly true for ETIs in renewable energy generation (photovoltaic technology, wind turbines and bioenergy), energy storage, energy efficiency improvements and smart grid technologies. These can enable countries to achieve more sustainable and equitable energy systems. ETIs associated with renewable energy generation have been singled out by Manjeevan as an area with the greatest potential to achieve energy justice ideals. For instance, rural electrification using solar panels installed in remote communities, often forgotten by the urban centres, remedies some of the injustices. Manjeevan cites successful examples in several developing countries, including China, India and Pakistan. China, in particular, has an ambitious plan to make the best use of the emerging ETIs in the production and distribution of renewable energy. ETIs in wind and solar energy generation were identified as the most promising ones.

Manjeevan also discusses the emergence and impact of machine learning (ML) algorithms in renewable energy. In particular, he singles out technologies associated with the 'neural networks' as the ones with the greatest potential. Neural networks, as subsets of ML algorithms, are computational models designed to recognise patterns and make predictions based on data. They have been said to have great success in optimising energy output and lowering maintenance costs. Manjeevan adds that neural networks have been shown to be very accurate not only in predicting patterns in relation to weather (solar and wind), thereby increasing efficiency in energy production, but also consumption patterns to improve efficiency in energy distribution.

One of the challenges associated with ETIs like neural networks is dealing with large amounts of data in a safe and secure manner. Manjeevan highlights cybersecurity as being at the top of the list of challenges. Another area where improvements are necessary is in the regulatory space to enable targeted subsidies, which would encourage investments in ETIs. In fact, in 2023, the Malaysian Ministry of Economy, in its National Energy Transition Map (2023: 64), remarked that the 'slow gradual uptake of sustainable practices within domestic industries impedes the swift transition to cleaner energy sources'. Perhaps the subsidies could incentivise some of the industry players to embrace ETIs and take a proactive role in the green energy transition.

16.2.9 Taxation and Green Energy Transition

In Chap. 10, 'How to Tax Our Way to Energy Justice by Francesco Cannas', Francesco Cannas begins by acknowledging the conundrum between the need for affordable and reliable energy on the one hand and the urgency of meeting net-zero emissions for climate protection on the other. He identifies two major challenges in this regard. First is the commitment by countries to comply with SDG 7, which advocates access to clean and affordable energy for all. While clean energy is preferred over energy derived from fossil fuels, clean energy is nowhere close to being sufficient to satisfy the population's growing needs worldwide. We must still rely on energy derived from non-renewable sources like fossil fuels and nuclear power to meet our daily needs.

There are valid concerns around the renewability of renewable energy. Despite being called 'clean' energy, the materials used to build solar panels, turbines and batteries must still be mined and manufactured, which has environmental impacts. Additionally, the production of renewable energy requires a significant amount of land, often in rural areas, which can negatively impact the lifestyles of indigenous people. As noted in Kraal and Taylor's chapter, when these renewable energy sources wear out, they must be disposed of properly. However, the remains of these sources, made of toxic materials, are often buried and can remain there indefinitely. Despite these concerns, it is crucial that countries continue to pursue green energy transition efforts if they are serious about meeting their net-zero emissions targets by 2050. However, this transition must be done in a just and sustainable manner to ensure that people have uninterrupted access to an affordable and reliable source of energy.

Access to energy brings us to the second major challenge Francesco identified, and that is the Russia-Ukraine war. Francesco, being a legal scholar residing in Italy, is in a position to see the conundrum that the Ukrainian crisis has caused in Italy and neighbouring countries when it comes to access to gas. Before the war, 46% of Italy's gas was imported from Russia. Half of the gas (49% precisely) in Germany was imported from Russia. The sudden breakdown of the relations between the EU countries and Russia due to war meant that those countries, particularly Italy and Germany, needed to quickly find alternative sources of energy, which surely could not have been sourced in the amount required from clean energy sources. The alternative source of gas and oil imports also affected other countries that competed for the same, leading to a sharp rise in prices.

However, one positive outcome of this crisis is a sharp increase in the demand for renewable energy in Europe. It is reported that renewable capacity additions have gone up to 40% after the war, as many countries witnessed a drastic increase in solar and wind uptake to reduce their reliance on Russian gas (IEA 2023).

After considering the backdrop of these challenges, Francesco takes the position that the transition to green energy is necessary, but it must be done in a manner that will not affect people's livelihoods. In fact, we submit that the right to energy can be equated to our constitutional right to life, which also includes the right to quality of life. There is no quality of life if we have to turn off our lights and heaters in order to reduce carbon emissions. There has to be a solution where peoples' access to affordable and reliable energy is guaranteed while, at the same time, ensuring that significant progress is made in the sustainable energy sector. We should strive for a future where everyone can access 'energy-light' (hopefully powered using clean energy sources) and not a future with less light. A darker world is reminiscent of the past and not the future.

So, the central question that Francesco deals with in this chapter is the role of the taxation system in the transition to green energy. He explains that tax systems in the world currently could be divided into two types: those with 'revenue-generating taxes', which are meant to finance the country's budgets, and the 'regulatory taxes', which are geared towards corrective measures by, for instance, correcting people's behaviour in the hope that they will refrain from polluting the environment. It is these regulatory taxes that are often pointed out by many as a tool that governments could effectively use to promote and encourage the green energy transition. For example, tax incentives to subjects involved in the production and distribution of clean energy have been suggested by some as the way to go.

Francesco seems to adopt a more cautionary approach to using regulatory taxation as a means of creating effective government policy. While he agrees that taxation can play a constructive role in promoting a sustainable energy sector, its effects, however, are very limited. The primary driver of the green energy transition, Francesco argues, has to be the government and its agencies responsible for creating and implementing policies in that regard. The taxation system cannot replace that vital role of the government. Instead, it can only supplement it. He reminds readers that, after all, the main purpose of taxation is the collection of revenue for the financing of the government and public sector and not the creation of a strategy such as the one that deals with the green energy transition.

Francesco provides two notable examples that seem to demonstrate some limitations when it comes to the use of regulatory taxation in promoting the clean energy sector. The first challenge is making sure that the government funds that have been advanced by way of tax incentives for research in the green energy sector actually produce a positive impact on the environment. It is understandably difficult for the government sector to provide this assurance. There is also an issue of market distortion whereby giving tax incentives only to subjects dealing with green energy transition makes it selective. Another issue is that tax incentives are supposed to act to 'incentivise' the polluters to change their behaviour. In other words, they are supposed to serve as a preventive measure. When pollution occurs despite the tax incentives, the incentives lose their purpose as their original purpose of preventing pollution failed to be realised.

Francesco also talks about the need to consider taxation in the context of 'environmental equity' and 'environmental neutrality' concepts. He explains in detail what he meant by these terms and notes that countries need a more robust and inclusive approach when drafting policies on the transition to green energy. Currently, the decisions seem to be made exclusively from an economic point of view. That needs to change. The pool of expertise should extend to others, including legal scholars and practitioners. If counties are taking their net-zero emissions commitments seriously, perhaps they should consider, Francesco suggests, enacting their commitment to green energy transition in their written constitutions. As such, any future law, regulation or policy would have constitutional backing and naturally would have greater recognition.

16.2.9.1 Energy Justice in Islam

In Chap. 11, entitled, 'Islamic Approach to Energy Justice', Jasmin Omercic explores whether energy justice principles can find support in Islamic thought and jurisprudence. He begins by stating that the perceived energy-related injustices are mostly a result of human factors. As such, instead of placing the entire emphasis on the externalities and technological advancements, a more robust analysis of normative, motivational factors represents a noble approach to the problem. This study is limited to the analysis of Islam, a religion of over 1.8 billion Muslims worldwide. But its unique approach to energy and climate justice problems could be a model for similar studies of other world religions and their approaches to these important issues of today, which, for the most part, are caused by human imperfections and poor choices.

Jasmin's central thesis in this chapter is that energy justice, especially its sub-forms like distributive, procedural and recognition justice, are consistent with Islamic teachings. He invoked several basic Islamic principles in support of this argument. First is the principle of stewardship of vicegerency (*khalifah*), which is premised on the belief that God ordained man to be the vicegerent on the earth. As such, he is ultimately responsible for its care and preservation. In other words, man must demonstrate utmost care in the utilisation of the earth's resources so that they are preserved for the welfare of current and future generations. This is entirely in line with the UN sustainability agenda. Another principle is justice (*adl*), which advocates the equitable distribution of resources, including energy ones. Public interest (*maslahah*) is another Islamic tenant denoting social responsibility. It gives public interest precedence over the interests of individuals. Therefore, energy and its resources should be rendered to the public, not individuals. This principle also rejects all forms of monopoly, oligopoly and even market competition if it will enrich individuals at the expense and impoverishment of others.

Among other Islamic principles, Jasmin notes the relevance of moderation (*wasatiyah*) and consultation (*shura*). Jasmin cited the Qur'anic verse 143, Chap. 2, in which the word '*wasatiyah'* was translated as 'justly balanced'. This is particularly relevant in the context of wastefulness. Qur'an calls upon people to be mindful of the production and consumption of resources so that they are produced and consumed in moderation, i.e. sustainably. We witness the incredible wastefulness of resources, especially by the societies in developed countries. At the same time, many parts of the world, like Sub-Saharan Africa, are left without electricity and are deprived of decent living conditions. Another principle that is very much related to procedural justice is consultation (*shura*). Jasmin argues that a justly balanced society is expected to consult before and during energy-related projects so that all relevant stakeholders can be involved in the decision-making process.

Islam provides a solid ethical basis for the production and distribution of energy and, as such, is consistent with the SDGs and ESGs. Islam can play a major role in the energy transition, especially in Muslim-majority countries, just like it did in the transition to Islamic banking and finance, as well as insurance (*takaful*). In fact, some of the Muslim countries have already taken major practical steps in this direction. The best example is the issuance of Islamic bonds (*sukuk*), estimated to reach USD$2.4 trillion in 2024, to fund sustainable and green projects. Jasmin called on the Muslim nation-states to continue their efforts in transitioning to renewable energy. He commends them for their current efforts, especially countries like Saudi Arabia and Qatar, which have made substantial investments in renewable energy projects.

16.2.9.2 Tackling Energy Poverty in ASEAN

In Chap. 12, entitled 'Challenges in Addressing Energy Injustice in ASEAN', Mohamed Hanipa Maidin discusses the challenges faced by the countries in ASEAN, most of which have the status of 'developing nations' except Singapore, which is considered a developed economy. The author is in a position to offer a unique perspective, as he was a former Deputy Minister of Law in Malaysia and a former Member of Parliament. Throughout this chapter, he emphasises the need to reconcile two seemingly irreconcilable objectives. On the one hand, ASEAN countries have committed to reducing their greenhouse gas emissions and accelerating their transition to green energy. On the other hand, how can we talk about renewables and green energy when many communities in ASEAN have no access to any form of energy and live in substantial poverty?

While there are many challenges that Mohamed Hanipa identifies, energy poverty stands out as the most pressing one. Electrification is particularly an issue in ASEAN as many communities live in remote places and islands. Providing electricity to them is a very costly undertaking. As a result, many of these communities still live in darkness without access to electricity. Mohamed Hanipa highlights some staggering statistics in Indonesia, where it is estimated that some 2500 villages still do not have access to electricity. The author also points out some discouraging figures in his home country, Malaysia, where about 137 Orang Asli villages live without electricity. The relevant government agencies have often helped those communities by providing them with diesel generators, which are admittedly not a sustainable and environmentally friendly source of electricity, but with no other options left, they become a necessity.

Electrification of those remote villages is not the only problem. Mohamed Hanipa remarked that most electricity powering ASEAN communities and economies comes from coal. In fact, he reiterated that coal is the second-largest energy source in ASEAN, after gas. The sharp increase in demand for electricity also means an increase in the construction of coal-powered power plants and coal mines. So, most of the ASEAN economies have become so dependent on coal that any attempt to reduce that dependency will have a serious economic consequence, which politicians and leaders are hesitant to bear. Lower reliance on coal means the closure of coal mines and coal-powered plants, leading to job insecurity and energy poverty. The projections in Indonesia are that around 30,000 workers from the coal sector could be laid off in the next 20 years. The author noted that the effects will be felt not only by the coal mine and coal power plant workers and their families but also by the entire ecosystem of workers and the industries along the coal value chain.

The solution, according to Mohamed Hanipa, must be one that will see the reduction of energy poverty in ASEAN and then also the reduction in greenhouse gas emissions. To achieve this aim, countries in ASEAN must be given more assistance in the form of renewable technology transfers and flexibility to sustain their green energy transition plans. This transition will take time and must not lead to energy poverty. The countries in ASEAN, too, he noted, must be serious in tackling the challenges and must not only talk about the plans without actions.

16.2.9.3 Litigating Against Energy Polluters in the US

In Chap. 13, entitled 'Efficacy of Citizen Suits Against Energy Polluters in the US', Julio C Colon talks about the developments in the US when it comes to remedying the harmful effects of energy production on human health and the environment. As Julio points out, the remedies ought to be observed through the lens of distributional justice, one of the most important tenets of energy justice. It is only fitting that those who live near the energy polluters and who are likely to be affected by the pollution are adequately compensated for the harm caused by the polluters. That said, Julio is not putting the blame on the polluters. In fact, he acknowledges that we need energy. Without electricity, our lives would be unimaginable. It is simply unthinkable for a modern society to live and prosper without access to electricity. So, in that context, Julio argued, one cannot reasonably oppose energy production. The argument, instead, must be focussed on strict compliance with the standards adopted by the regulators for that particular industry so that the pollution is mitigated and adequately remedied when it causes harm.

Julio limits his discussion primarily to coal and coal-fired power plants, as coal is one of the predominant sources of energy in the US. The coal-burning process produces carbon and coal ash, both of which are harmful to the health and environment. There have been multiple suits initiated by citizens against polluters in the US. Julio discusses these cases with reference to the court judgments, and it is best that the readers refer to the original chapter to find out the details. Suffice to say here that the legal battles have been based predominantly on three federal legislations: the Clean Air Act, the Resource Conservation and Prevention Act, and the Clean Water Act. Julio clarified that all three legislations contain provisions on standards, enforcement and permitting.

Julio contends that the provisions on standards tend to be too scientific to be understood by average citizens. For instance, for a citizen to prove that the chemicals emanating from the facility are in contravention of acceptable standards requires sophisticated measurement and testing, which most average citizens are not in a position to do. This is perhaps why many cases brought by the citizens are based on other statutes and common law torts like nuisance, negligence and trespass. These common law torts are much easier to prove and understand as the party knows that to succeed, they need to prove that they have been deprived of the use and enjoyment of private property. The provisions on the enforcement authorise the US Environmental Protection Agency (EPA) to take the necessary actions in cases of violation of any of the provisions of the acts. These actions could include civil and criminal penalties and sometimes even confiscating the permit to operate. All three legislations also contain the so-called permitting provisions allowing citizens to initiate private actions in their states under other statutes and common law.

16.2.9.4 Australia's Paradox: Fossil Fuel Exports and Climate Justice

In Chap. 14, titled, 'Examining the Paradoxical Nexus: Energy Justice amidst Australia's Growing Fossil Fuel Exports and Its Implications for Climate Justice', Evgeny Guglyuvatyy brings to light the difficulties Australia faces when it comes to maintaining its place as one of the worlds' most dominant fossil fuel exporters while concurrently trying to cut down the greenhouse gas emissions in line with climate justice objectives. He called it a paradox in that, on the one hand, Australia has made significant investments in renewables to meet its net-zero target emissions by 2050 while, at the same time, it has shown no intention to cut down on its fossil fuel exports. Quite to the contrary, as Evgeny notes, Australia is expanding its fossil fuel exports to new markets and considering funding about 100 new fossil fuel projects. It is also paradoxical, Evgeny argues that Australia supported COP26's ban on deforestation while it remains the global leader in land clearing.

That being said, Evgeny praises the Australian government's recent climate justice efforts, which are evident in several policy changes and legislative enactments. He noted that the Climate Change Act 2022 and Climate Change (Consequential Amendments) Act 2022, in particular, have included very ambitious greenhouse gas emission reduction targets of 43% below 2005 levels by 2030, as well as the renewed commitment to achieving net-zero emissions by 2050. The emission reduction targets are indeed very ambitious, given that previous governments' targets were 26–28% by 2030. Besides, Evgeny talks about many other specific government policies to intensify renewable energy production. It is worth noting that 35% of Australia's electricity is derived from renewable sources.

Australia's strong stance in the green energy transition, however, becomes somewhat perplexing when one considers the statistics on its fossil fuel exports. According to Evgeny, Australia's contribution to fossil fuel mining stands at 4% (57 tonnes of fossil fuel), which is ten times higher than the global average. Besides, Australia is the leading global exporter of coal and the second-largest exporter of gas. When the current Australian government announced plans to increase exports of coal to Europe due to heightened demand, many critics argued that Australia was not serious enough in its commitments to meet net-zero target emissions. However, Australia's Minister for Climate Change and Energy's response was rather pragmatic 'Europe is energy hungry and capital rich, Australia's energy rich and capital hungry, and that means that there's a lot that we can do together' (Mathiesen 2023). Evgeny also points to the arguments that some have raised in the defence of coal export, of which two are most dominant. First, Australia's coal is 'cleaner' than the coal from other supplying countries. Second, Australia will reinvest the funds to spur the growth of renewables in a transparent and efficient manner.

Australia's fossil fuel exports, Evgeny explains, are also directly impacting its relationship with neighbouring countries, in particular the Pacific Islands. The Pacific nations are facing an existential crisis due to climate change. It has been reported that many nations could be completely wiped out if the international community does not take drastic actions to cut greenhouse gas emissions, which are contributing the most to climate change (Salem 2020; Guardian 2021). One can see that, as Evgeny notes, Australia, being the fifth largest emitter worldwide, would need more than simple talk to convince its neighbours that their existential national interest is also a priority for Australia. This is why Australia has traditionally tried to exert a strong influence in the region by being the key donor and partner. But, its influence, Evgeny contends, has drastically diminished to the extent that it is no longer seen as the exclusive partner of choice. China, in particular, made significant inroads to project its influence, such as the recent security agreement between the Solomon Islands and China.

Evgeny concludes by saying that he aimed to demonstrate the tension between the need to provide an affordable and reliable source of energy in many impoverished communities like those in the Asia–Pacific region and the urgency to cut emissions so that Pacific nations and future generations could live in a better world. This tension, Evgeny concludes, is complex and not easy to solve. Most of Australia's fossil fuel exports, Evgeny notes, go to countries in Asia–Pacific, where many communities rely on the exported fossil fuel as the main energy source. So, balancing access to energy with climate justice is an issue that will remain to be debated and will require continued research and examination to come to an optimal solution. While we may not know what the optimal solution is at this stage, we are confident, however, that both energy security and climate justice concerns must be addressed adequately by that solution.

16.2.9.5 The Energy Injustice of Orang Asli in Malaysia

In Chap. 15, entitled 'The Energy Injustice of Orang Asli in Peninsular Malaysia', Audrey KL Siah, Lillian ML Chan and Grace HY Lee bring the plight of the Orang Asli people to light. Orang Asli are Malaysia's indigenous communities, which have lived and occupied the land for over 5,000 years. They comprise three main ethnicities, which consist of 18 subgroups, each with a unique language, culture and history. Among the three ethnic groups, the Senoi are the largest, followed by the Aboriginal-Malay (or Malayu-Proto). The Negrito (or Semang) are the smallest of the three. According to the latest statistics, 209,575 Orang Asli live in Peninsular Malaysia, which accounts for only 0.6% of Malaysia's overall population.

The authors explain that Orang Asli have been marginalised in Malaysia. They cite some discouraging statistics. They said that 99% of the Orang Asli people are under the so-called B40 income bracket (with earnings below RM4000 per month). Furthermore, 34% fall below the hard-core poverty line. Energy poverty is particularly an issue. The authors argue that any other socio-economic issues would automatically be solved in solving energy poverty. The authors demonstrated this through a circular flow model where at the heart of the problem and solutions lies access to quality energy. They highlight that access to energy would enable the Orang Asli to have better access to education. They would be able to acquire skills and know-how, making them more competitive in the employment market. Getting a well-paying job would consequently allow them to have better earnings, leading to improved living standards. The surplus of earnings could be invested in the education of their children and other members of their community. The circle then repeats repeatedly, leading to the more equitable spread of resources and improved living standards. But, as the authors note, at the heart of this model lies access to quality energy.

While the World Bank's data indicates that access to electricity rates in Malaysia for both urban and rural areas is 100%, the authors note that this is not the case. The authors argue that the World Bank's data is misleading, citing a more recent report published by a local newspaper mentioning 137 Orang Asli villages, which still do not have access to electricity. The main reason is said to be the remoteness from the available electricity grids. Some of those villages used to have solar panels, but they have been non-functional due to lack of maintenance. For lighting, those communities rely on kerosene lamps, palm seeds and candles, while for cooking, they use bamboo and woodfire. All this shows that energy poverty remains a major problem for the Orang Asli communities.

The authors note that to help these communities, the government should introduce targeted subsidies. Currently, electricity tariffs are the same for low-income and high-income households, which means that the energy burdens for low-income families are much greater. They will need to spend a larger proportion of their income on electricity bills compared to high-income households. The authors note that encroachment on Orang Asli lands has also been a major problem for them. Most of the land and forest they have occupied for the past 5000 years has been designated as state reserve land under Section 62 of the National Land Code. This allows the states to give licenses to logging and mining companies as well as to use land for the construction of hydroelectric dams. Only less than 20% of the land Orang Asli occupy has been gazetted as indigenous reserve land under the Aboriginal Peoples Act 1954 and, as such, is under the jurisdiction of the government agency called the Department of Orang Asli Development (JAKOA).

16.3 The Energy Justice Framework

Following the discussion of the book chapter findings, and based on these, we move on to synthesise our thoughts into an ethical framework for energy justice. In Chap. 1, we highlight three key principles of justice: distributive, procedural and recognition justice. The three are principal or primary modalities that govern the nurturance of energy justice. Encapsulated within the three modalities are secondary modalities, as component parts supportive of the primary modalities. For instance, restorative justice and cosmopolitan justice, as introduced by Diane Krall and Richard Taylor, are secondary or supportive modalities.

To build a just energy future, we must deliver distributive, procedural and recognition justice to both humans and the environment. This is the Person-Environment bind, in that any consideration of justice for humans must provide justice for the environment, and vice versa. In other words, we cannot optimise the primary modalities without simultaneous consideration of both humans and the environment. We propose an ethical framework that brings together key drivers and associated facilitating conditions to enable energy justice through the joint consideration and optimisation of the principal modalities (see Fig. 16.1).

16.3.1 Drivers of Energy Justice

The driver component in the framework is made up of three factors:

Socio-economic Factors. Socio-economic factors play a major role in influencing energy justice outcomes. Economic and social disparities have been found to play an important role in determining the level of access to energy, and inequalities in these can jeopardise the attainment of justice across any single or combination of the modalities. For instance, individuals with low income located in poor geographies, such as developing countries and rural areas, suffer from a range of disadvantages. Low-income individuals and vulnerable communities tend to have poor health and education outcomes compared to those living in prosperous communities.

In this book, a discussion of these factors takes place in Chaps. 4, 5, 12 and 15.

Science, Technology and Innovation Factors. Technology is becoming an increasingly influential factor in energy justice. Innovations in energy sourcing, manufacture and delivery are changing production envelopes. For instance, higher efficiency in producing energy makes it possible for energy prices to be lowered to an affordable level, thereby allowing greater access to wider segments of the global population. Moreover, alternative energy sources, such as solar, wind, hydrogen and nuclear, open up environmentally friendly pathways. Adoption of advances in mining and circularity principles within conventional fossil fuel extraction can also allow for more environment-friendly interim solutions in the long transition to renewable energy solutions.

Discussion of these factors takes place in Chaps. 6, 7 and 9.

Socio-political and Geo-political Factors. With emerging evidence that many planetary boundaries have been crossed and quite a number are close to being crossed, the world is facing a crisis whose magnitude threatens the existence of humans and the planet itself. In response, the global community has come together to define societal objectives to bring climate change under control. The goals encapsulated in the net-zero

Fig. 16.1 Ethical framework for energy justice

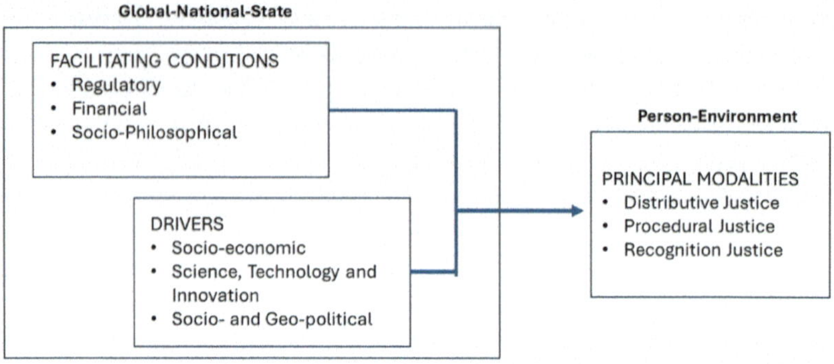

by 2050 aspiration are driving the transition to planet-friendly energy alternatives. The global pact has rippled down to national and local initiatives and requires the backing of political masters to support the change. Unfortunately, the shift to renewable energy sources is neither simple nor without drawbacks. For instance, renewable technologies such as electric vehicles and solar energy themselves are reliant on rare earth materials, increasing the level of extractive mining. The scarcity and difficulty in extraction coupled with high demand for these rare earth elements have sky-rocketed prices and led many nations blessed with these natural resources scrambling to profit and take advantage of their resource basins. Unfortunately, this has been at the expense of the environment and the health and well-being of communities that live in or near these locations. The battle for rare earth elements has heated up and escalated into geo-political contestation to dominate the sector. This form of global, regional, national and state competition has major implications for communities, society and the environment.

Discussion of these factors takes place in Chaps. 3 and 14.

16.3.2 Facilitating Conditions for Energy Justice

Drivers do not deliver energy justice by themselves but are governed by the environmental context in which they are situated. The environmental context can either facilitate or constrain the drivers' ability to deliver energy justice in accordance with the distributive, procedural and recognition modalities. Facilitating conditions are important in the well-functioning of the drivers. The framework highlights three facilitative factors:

Regulatory Factors. Global, regional, national and state-level regulations can impact the ability and nature of energy justice delivery. The regulatory environment sets the standards and rules of operation in the transition to energy justice. Regulations and standards can either be punitive to rectify transgressions that compromise any of the three primary modalities and secondary modalities, or they can incentivise the shift to secure these modalities in the process of driving energy transitions.

Discussion of these factors takes place in Chap. 13 and parts of 14.

Financial Factors. Energy transitions and initiatives cannot be realised without funding. Lack of funding can stultify even the best of innovations, technologies and transition plans. Devising funding arrangements that incentivise renewable energy transitions and energy efficiency innovations is crucial in the move to energy justice. Funding can be provided either to drive the shifts to renewable energy or environmentally friendly conventional energy by focussing on the supply side of energy production or, on the demand side, by improving market access by ameliorating the condition of vulnerable segments of society through subsidy or other novel mechanisms facilitating affordability.

Discussion of these factors takes place in Chaps. 8 and 10.

Societal and Philosophical Factors. At the broad level, the values that society subscribes to have a pervasive influence on the nature of justice that prevails in that society. Social norms and values are embedded in the behaviours and actions of people and institutions. These value orientations are a consequence of complex interactions and events over history. Among the many, one key influence in defining values is religion. Other important influences are the nature of governing institutions, such as democratic values, and socially embedded interactions that sediment into cultural values. For instance, indigenous communities, such as Orang Asli in Malaysia, subscribe to environmental protection as part of their spiritual obligations. Religions, such as Christianity and Islam, emphasise the need to protect the rights of the vulnerable. Islam accords special rights to nature and the environment, while Christianity adopts a perspective of dominion over the earth. The values set influence the nature of justice and to whom it is accorded.

Discussion of these factors takes place in Chaps. 2, 3 and 11.

The facilitating conditions and driver factors are interconnected, working together to influence outcomes. Their interaction is not only simultaneous but also hierarchical, encompassing various levels from global to national, state and even more localised contexts. This layered interaction plays a crucial role in shaping the magnitude and direction of their influence and impact.

The framework illuminates several critical insights. First and foremost, it shows that achieving energy justice requires a multifaceted approach. Simply focussing on one aspect or a group of factors would not suffice. For true progress, we need to design facilitators that amplify the positive influences of driving factors. Secondly, both driver and facilitating factors must consistently incorporate and optimise the three key modalities in each and every energy delivery decision as we navigate the future. Without a comprehensive approach that considers these modalities, the quest for energy justice will likely falter, both immediately and in the long term. Ignoring these modalities in our actions and decisions will perpetuate a world marked by growing disparities and divisions, where tensions and conflicts are rampant, and a small, wealthy elite continues to thrive at the expense of the increasingly marginalised poor, who lose not only their voice but also any chance for improvement.

16.4 Conclusion

In conclusion, we want to emphasise a crucial point. There may be a misconception that energy justice, in advocating for a more equitable and fair production, distribution and consumption of energy sources, harbours a hidden agenda to impede the transition to renewable energy sources. It might be perceived as if those of us who assert that access to energy and electricity is a fundamental human right and that we want to see 100% access to electricity in every part of the world are proposing to slow down the transition to renewables and escalate the production and export of fossil fuels. We unequivocally state that this assumption is erroneous.

Addressing climate change and transitioning to clean energy sources remains, in our view, one of the most urgent global priorities. However, we contend that solving climate change by advancing the transition to green energy without considering the well-being of individuals and their ability to cope with rising energy prices, and in some cases, even accessing basic energy sources like electricity, is not ethically sound. Preserving the environment on the one hand and ensuring energy security and affordability on the other are compatible objectives. In fact, we argue that not only are they compatible, but one cannot be optimally addressed without considering the other.

Therefore, we hope that the book has done justice to both energy justice and climate justice principles, both of which represent some of the most pressing problems in the world that needs resolution.

References

Chatzis I, IAEA Department of Nuclear Energy (2016) IAEA starts review of Malaysia's Nuclear Power Infrastructure Development. https://www.iaea.org/newscenter/news/iaea-starts-review-of-malaysias-nuclear-power-infrastructure-development. Accessed 10 Oct 2016

Devlin H (2017) Let there be light: Germans switch on 'largest artificial sun. The Guardian. https://www.theguardian.com/science/2017/mar/23/worlds-largest-artificial-sun-german-scientists-activate-synlight. Accessed 2 Jan 2024

International Energy Agency (2023) Renewable power on course to shatter more records as countries around the world speed up deployment. https://www.iea.org/news/renewable-power-on-course-to-shatter-more-records-as-countries-around-the-world-speed-up-deployment. Accessed 4 Jan 2024

Ikram I, Aziz A (2023) Nuclear still on the cards in Malaysia's energy transition. https://theedgemalaysia.com/node/680568. Accessed 29 Aug 2023

Mathiesen K (2023) In from the coal: Australia sheds climate pariah status to make up with Europe. Politico. https://www.politico.eu/article/coal-australia-climate-pariah-europe-eu-brussels-energy/. Accessed 1 Feb 2023

Ministry of Economy (2023) National energy transition roadmap: energising the nation, powering our future. Ministry of Economy

Mohamed Radhi NA (2020) Malaysia won't use nuclear power, says PM. https://www.nst.com.my/news/government-public-policy/2020/02/564295/malaysia-wont-use-nuclear-power-says-pm. Accessed 10 Feb 2020

Pascoe B (2014) Dark Emu: black seeds: Agriculture or accident? Magabala Books, Brunswick

Sahabat Alam Malaysia and Greenpeace Malaysia (2023) A dangerous position to keep: nuclear power as an option in Malaysia. https://www.greenpeace.org/malaysia/press/51399/a-dangerous-position-to-keep-nuclear-power-as-an-option-in-malaysia/. Accessed 23 Oct 2023

Salem S (2020) Climate change and the Sinking Island States in the Pacific. E-International Relations. https://www.e-ir.info/2020/01/09/climate-change-and-the-sinking-island-states-in-the-pacific/. Accessed 9 January 2020

The Guardian (2021) Tuvalu minister to address Cop26 knee deep in water to highlight climate crisis and sea level rise. The Guardian. https://www.theguardian.com/environment/2021/nov/08/tuvalu-minister-to-address-cop26-knee-deep-in-seawater-to-highlight-climate-crisis. Accessed 8 Nov 2021

Adnan Trakic is an Associate Professor and Head of Business Law and Taxation Department at the School of Business, Monash University Malaysia. He graduated with an LLB (Hons), a Master of Comparative Laws (MCL), a Doctor of Philosophy (Ph.D.) from the International Islamic University Malaysia (IIUM) and a Graduate Certificate in Higher Education (GCHE) from Monash University. He has taught and researched in the fields of conventional and Islamic finance law, business law, and dispute resolution. A co-author of several books, his published work extends to many articles, book chapters, and other contributions.

Ridoan Karim is a Lecturer at the School of Business, Monash University Malaysia, where he teaches and conducts research in the fields of business and cyber law. He is a passionate academic and researcher who has published widely in peer-reviewed journals and presented papers at numerous national and international conferences. His areas of interest include "energy justice", "science, technology and law", "privacy and data protection law", and "Asian and comparative law", among others. Dr. Karim has served as a Consultant and Fellow on projects funded by the Ministry of Education Malaysia, University of Malaya and the Monash Data Futures Institute (MDFI). He earned his Doctor of Philosophy (PhD) from the University of Malaya, Malaysia, as well as a Master of Business Administration (MBA) from the University of Chichester, UK, a Master of Comparative Laws (MCL) from International Islamic University Malaysia (IIUM), and a Bachelor of Laws (LLB) from BRAC University, Bangladesh.

Pervaiz K. Ahmed is a Distinguished Professor and Director of the Sunway Institute for Global Strategy and Competitiveness, Sunway University. Pervaiz is a Management scholar exploring innovation and strategic dynamics of entrepreneurial change. His outlook on management is colored by moral responsibility, particularly ethics of care and sustainability. Pervaiz has held several senior positions in the UK before coming to Malaysia. He has published extensively in international journals and is a regular keynote speaker with a number of academic awards for his research. In addition, he has served as editor and sits on the editorial boards of several international journals. He has extensive experience working with and advising blue-chip companies and public sector organizations, such as Unilever, Ford, AT&T, NCR, British Telecommunications, and the NHS in Europe. He has also been involved with corporate clients in Asia, as well as government agencies, such as the Singapore National Productivity Council. In recent times Pervaiz has been involved with the Government of Dubai's Public Sector Innovation and Improvement Initiative, and also served on projects commissioned by Economic Planning Unit (EPU), Ministry of International Trade (MITI), and APEC Secretariat.

Correction to: Does Energy Justice Trump Environmental Justice?

Maizatun Mustafa and Farahdilah Ghazali

Correction to:
Chapter 3 in: A. Trakic et al. (eds.), *Energy Justice*, Sustainable Development Goals Series,
https://doi.org/10.1007/978-981-97-6059-6_3

In the original version of the book, the following belated corrections have been incorporated: In chapter "Does Energy Justice Trump Environmental Justice?", the affiliation "International Islamic University Malaysia, Kuala Lumpur, Malaysia" of the author "Farahdilah Ghazali" has been changed to "School of Business and Economics, Universiti Putra Malaysia, Serdang, Selangor, Malaysia". The book has been updated with the changes.

The updated version of this chapter can be found at
https://doi.org/10.1007/978-981-97-6059-6_3

A. Trakic et al. (eds.), *Energy Justice*, Sustainable Development Goals Series,
https://doi.org/10.1007/978-981-97-6059-6_17

The manufacturer's authorised representative in the EU is Springer Nature Customer Service Centre GmbH, Europaplatz 3, 69115 Heidelberg, Germany. If you have any concerns regarding our products, please contact ProductSafety@springernature.com

Printed and bound by CPI Group (UK) Ltd, Croydon, CR0 4YY
01/10/2026
02225962-0020